W0050372

Wolfgang Seeger

Standard Variants
of the Skull and Brain

Atlas for Neurosurgeons and Neuroradiologists

Springer-Verlag Wien GmbH

Professor Dr. med. WOLFGANG SEEGER
Prof. em. für Neurochirurgie der Universität Freiburg i. Br.
Au, Federal Republic of Germany

This work is subject to copyright.
All rights are reserved, whether the whole or part of the material is concerned, specifically those of translation, reprinting, re-use of illustrations, broadcasting, reproduction by photocopying machines or similar means, and storage in data banks.

Product Liability: The publisher can give no guarantee for all the information contained in this book. This does also refer to information about drug dosage and application thereof. In every individual case the respective user must check its accuracy by consulting other pharmaceutical literature. The use of registered names, trademarks, etc. in this publication does not imply, even in the absence of a specific statement, that such names are exempt from the relevant protective laws and regulations and therefore free for general use.

© 2003 Springer-Verlag Wien
Originally published by Springer-Verlag Wien New York in 2003
Softcover reprint of the hardcover 1st edition 2003

Printed on acid-free and chlorine-free bleached paper

SPIN: 10923016

With 177 partly coloured Figures

CIP data applied for

ISBN 978-3-7091-7224-7   ISBN 978-3-7091-6071-8 (eBook)
DOI 10.1007/978-3-7091-6071-8

Preface

This book tries to take into consideration the most important anatomical variants with reference to their inherent advantages and disadvantages during surgery for the clinician, especially for neurosurgeons and neuroradiologists. Many variants, rare and frequent ones, are usually omitted because they appear clinically irrelevant. Phylogenetic and ontogenetic aspects are mentioned for better understanding of some variants. The precise listing of the frequency of occurrence of variants in the literature does not fulfill the requirements of clinicians. During frequent surgeries, e.g. the pterional approach, even rare variants are likely to be observed more often. In the case of rare surgeries, e.g. the transcondylar approach to the premedullary area, even frequent variants in this region, such as atypical courses of the N. hypoglossus, will not be known well in the clinical situation. They may lead to errors which will be long forgotten at the time of the next surgery. Therefore, the variants will be distributed in three groups according to their frequency:

- rare variants   < 10 %
- common variants   10-50 %
- normal findings   > 50 %

The authors successor in Freiburg, Professor Dr. J. Zentner, made available rooms and materials for the anatomical dissections and demonstrations, as he has been doing for more than 5 years of his chairmanship in Freiburg.

The director of the neuroradiological section at the Department of Neurosurgery in Freiburg, Professor Dr. M. Schumacher and his coworkers, provided the MRI (MRT) images shown in this book. The author found a good translator in Dr. A.Weyerbrock, which helped to improve the neurosurgical aspects of presentation of this book. I am grateful to Mrs. E. Rotermund, Professor Zentner's secretary for typing and for better presentation of the manuscript. I would especially like to thank the Springer-Verlag WienNew York for the continuous good cooperation, help and excellent preduction of the book.

Freiburg, August 2003   *Wolfgang Seeger*

Contents

Chapter 1
Variants of the frontobasal – frontodorsal – areas
(Figs. 1 to 47)

Fig. 1. Frontodorsal and frontobasal microsurgical approaches

Schematic presentation

FIG. 1

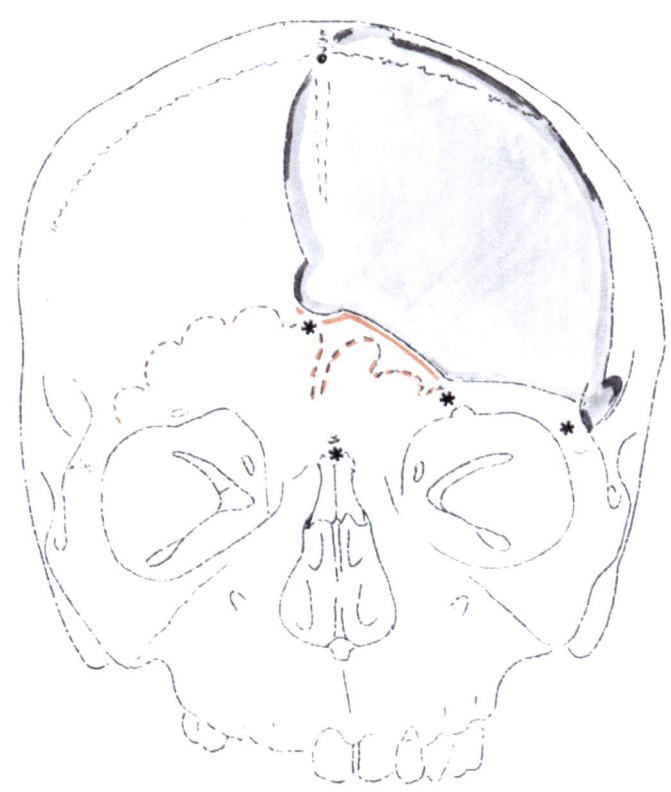

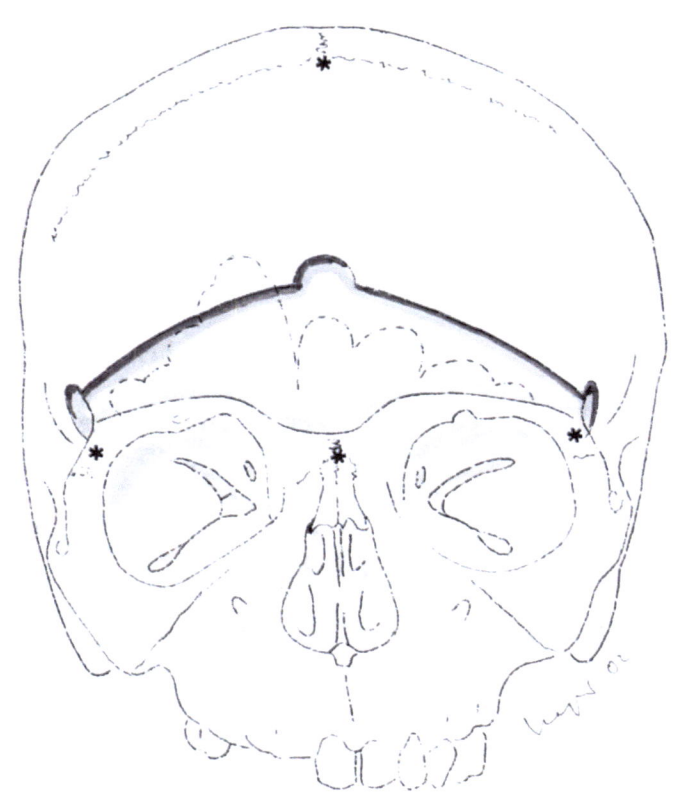

Figs. 2–22. **Variants of the anterior skull base** and adjacent structures

Fig. 2. **Foramen supraorbitale – Incisura supraorbitalis**

Anatomical dissections

These common variants often exist in the same individual

In mammals, multiple variants may exist simultaneously

Clinical aspects:
- **Incisura less problematic for preparation of vessels and nerves**
- **Foramen increases risk for bleeding and damage of nerves with consecutive neuromas**

FIG. 2

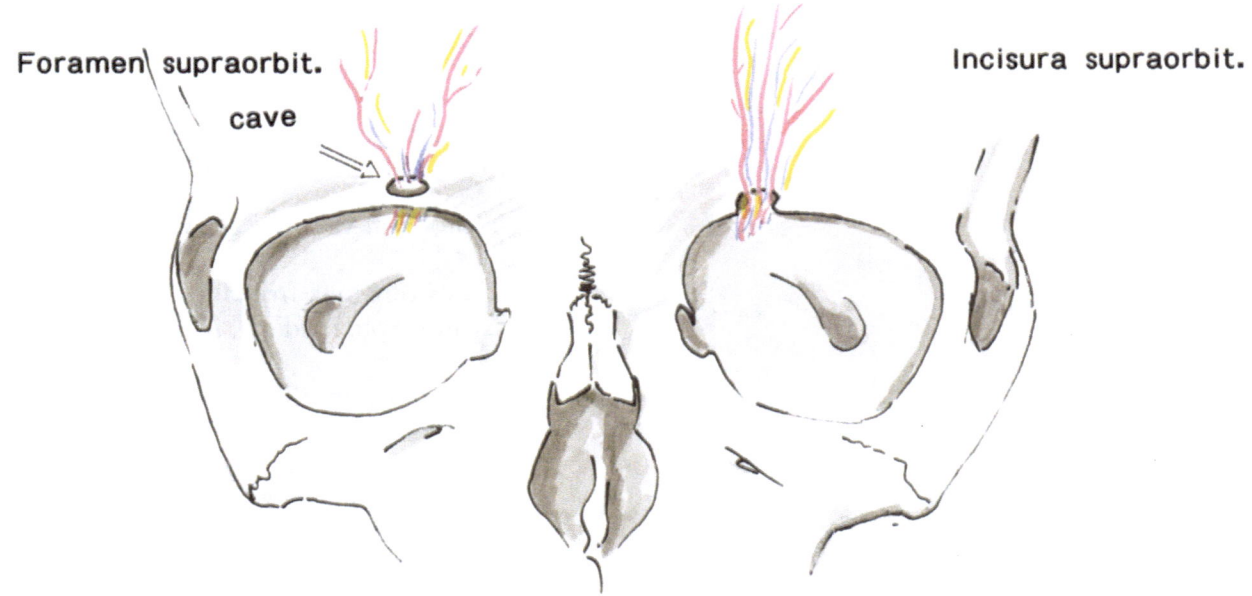

Foramen supraorbit. cave

Incisura supraorbit.

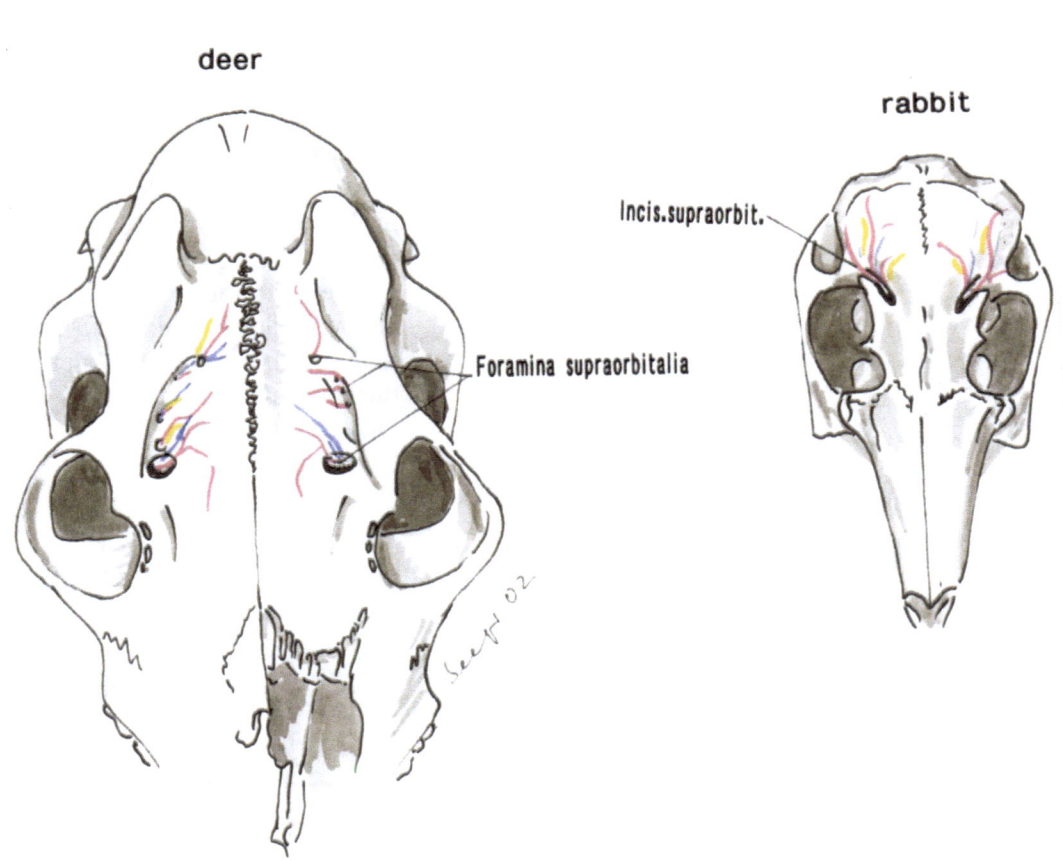

deer

rabbit

Incis.supraorbit.

Foramina supraorbitalia

Fig. 3. Frontobasal variants, Survey

A Variants of Sinus frontalis, Cellulae ethmoidales, Sinus sphenoidalis, and adjacent structures

    1 Extension of Sinus front. into the orbital roof

    2 Extension of Cellulae ethmoidales into the orbital roof

    3 Pneumosinus of Proc. clinoideus ant. by widening of Cellulae ethmoidales

    4 Ethmoid canals

    5 Pneumosinus of Proc. clinoideus ant. by widening of Sin. sphenoidalis

    6 Small and asymmetric pneumosinuses

    7 Connection of For. coecum with Cavum nasi

    8 Pneumosinus of Crista galli

    9 Congenital defect of basal dura close to Lamina cribrosa

    10 General widening of Sinus sphenoidalis

    11 Close relationship of N.opticus and A.(Foramen) ethmoidalis (e) post.

B Sinus sphenoidalis does not exist (normal finding in children < 4 years)

    12 Corpus sphenoidale filled with Spongiosa

C **13** Basal position of Diaphragma sellae

D **14** Senile atrophic pituitary. Typical widening of Cisterna sellae

E **15** Diaphragma sellae complete; opening for the pituitary stalk is narrow

F **16** Diaphragma sellae incomplete; opening for the pituitary stalk is wide

G Numerous variants of A. ophthalmica, in this sketch:

    17 Origin of A. ophthalmica from A. meningea media

Clinical aspects:
- **CSF leak** (see A and F)
- **Orientation problems (see B)**
- **danger for cranial nerves II and V/3,** see A (5, 11, 4, 10, and E, and G (embolization)

Fig. 4. **Large widening of pneumosinuses**

Anatomical drawings
A   Transparent presentation
B   Variant of Canalis rotundus (Seeger 2000)

Clinical aspects:
Danger for CSF leak after trauma or surgery

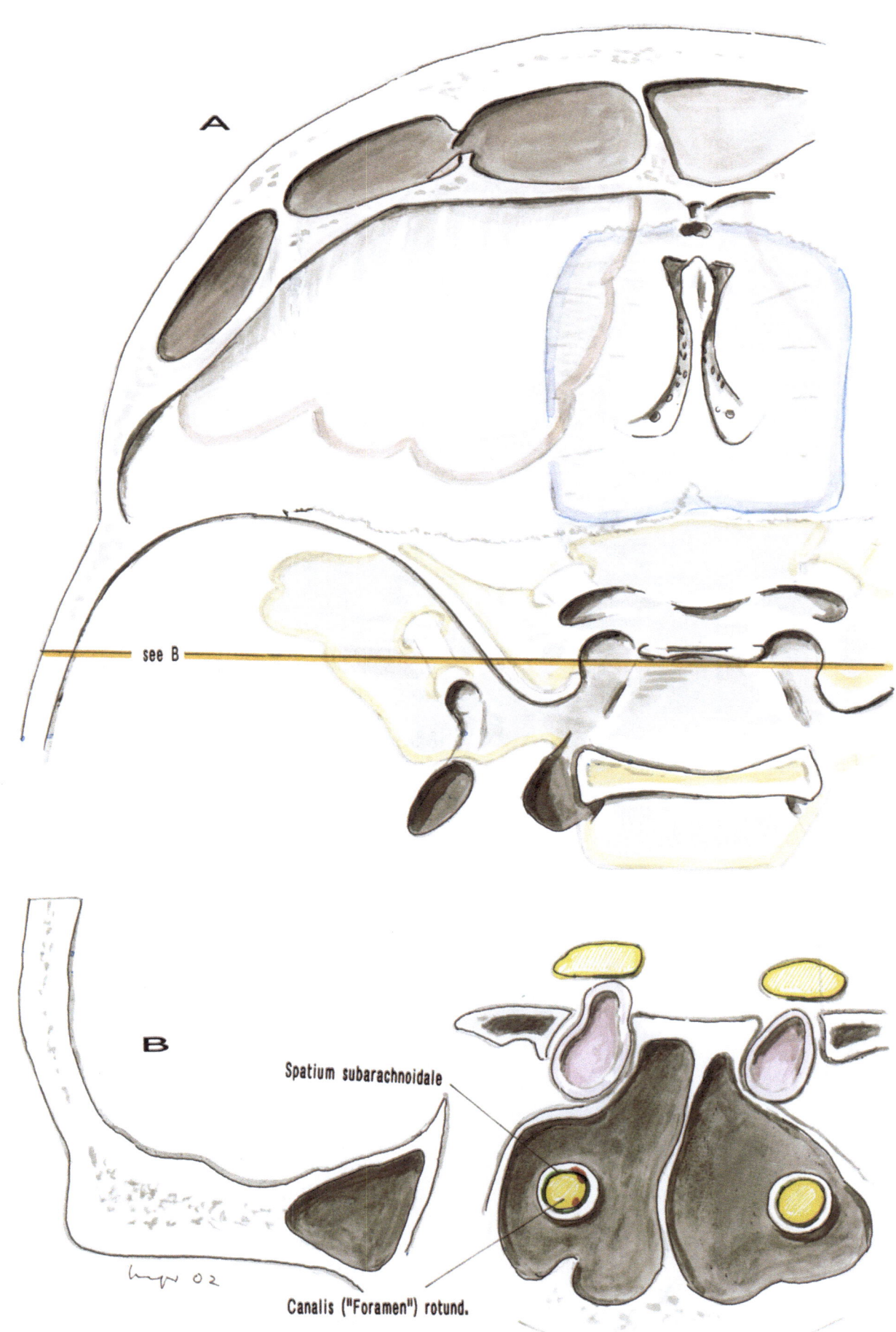

A

see B

B

Spatium subarachnoidale

Canalis ("Foramen") rotund.

+ N. V2

Fig. 5. **Widening of Sinus sphenoidalis** (pneumatized Crista galli added) Anatomical dissection

Microsurgical aspects: Danger for bleeding, danger for lesioning of N.II and for N.V/2, and for CSF leak after trauma

A   Skull dissection
B   As A, sectional enlargement

Abbreviations
a   open For. caecum
a'   Lamina cribrosa, as it is well known
bb'   Can. ethmoidalis ant. (projection)
cc'   Can. ethmoidalis post. (projection)
d   thin wall of Can. opticus
e   arrow: Pneumosinus of Processus clinoideus ant.
e'   as e
f   thin wall of Sinus sphenoidalis overlying A. carotis int.
g   thin wall overlying N. maxillaris and its leptomeningeal sheet with CSF
g'   as g, segment of Canalis rotundus
g"   as g, segment of Fossa pterygopalatina

Arrows: Routes of CSF with and without CSF leak

FIG. 5

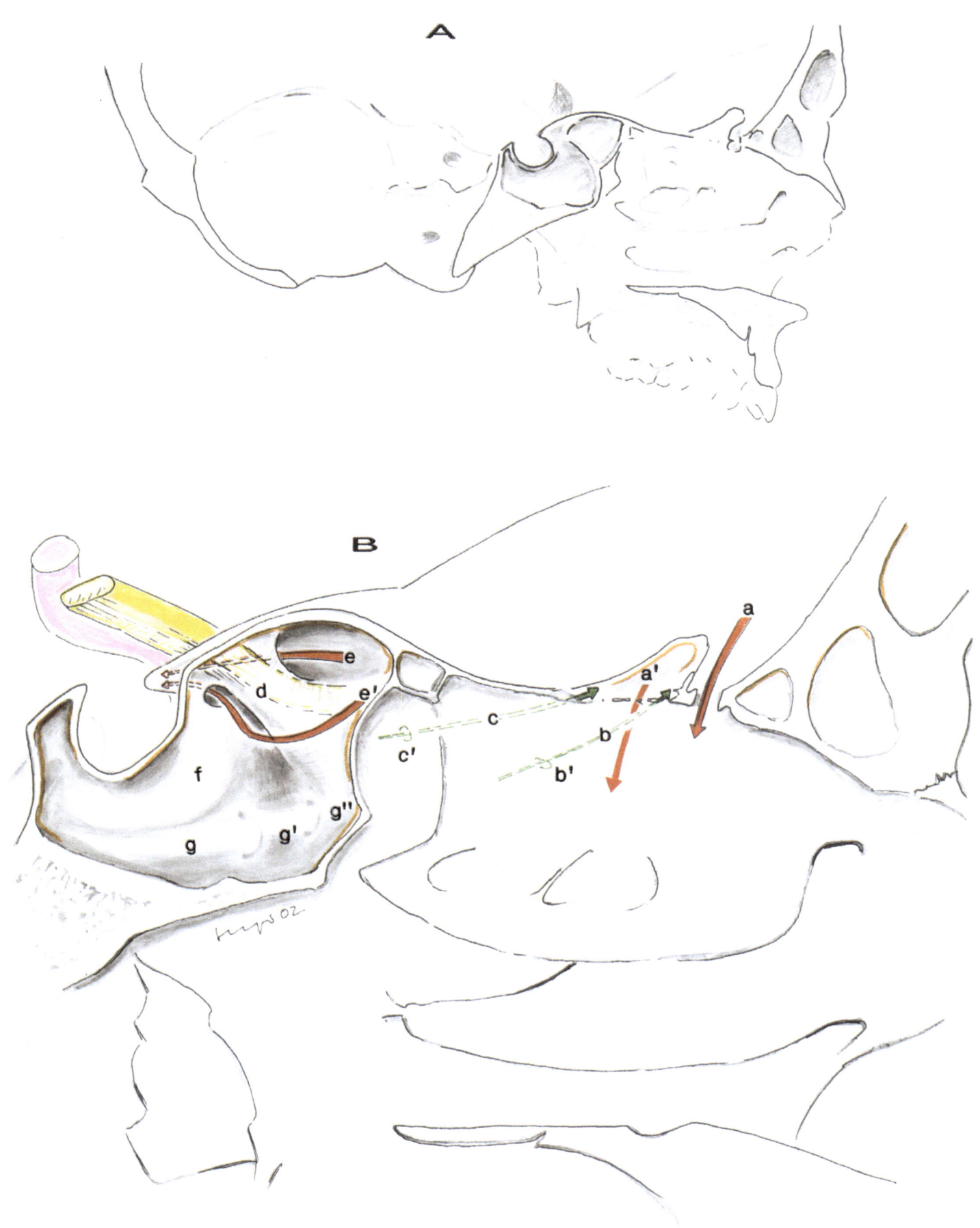

A

B

Fig. 6. **Pneumatization of the orbital roof**

A   By widening of Sinus frontalis (common variant)

B   By widening of Cellulae ethmoidales (rare)

Arrows: Connections with Cavum nasi

Clinical aspects: CSF leak following trauma or surgery
- **well known after lesion of A**
- **nearly unknown after lesion of B**

FIG. 6

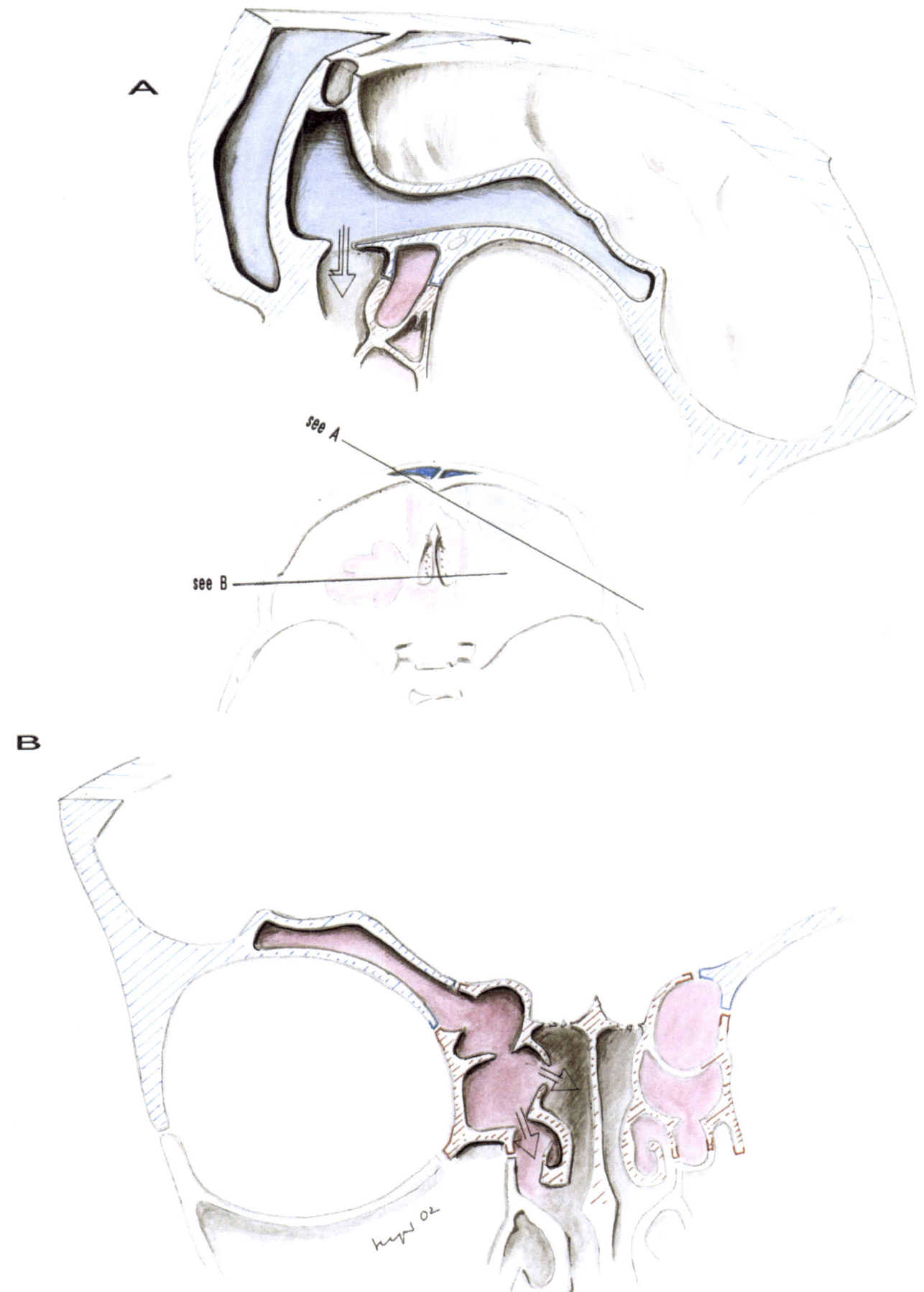

A

see A

see B

B

Fig. 7. **Pneumatization of the orbital roof and of Processus clinoideus ant.**

A Widening of Sinus frontalis and Cellulae ethmoidales are combined – anatomical sketch –

B Pneumatization of Processus clinoideus ant. by widening of Cellulae ethmoidales.
Arrow: Connection
For pneumatization of Processus clin. ant. from a widened Sinus sphenoidalis see D in Fig. 8
* Floor of the left Canalis opticus

B' Anatomical sketch for understanding B
a Sinus frontalis
b Fissura orbitalis superior

Clinical aspects
- **for A see** arrows in Fig. 8, A and C
- **for B and B': Danger for CSF leak after microsurgery of ophthalmic aneurysms, decompression of Canalis opticus (especially Pneumosinus dilatans – see Hoydt -) perhaps posttraumatic.**

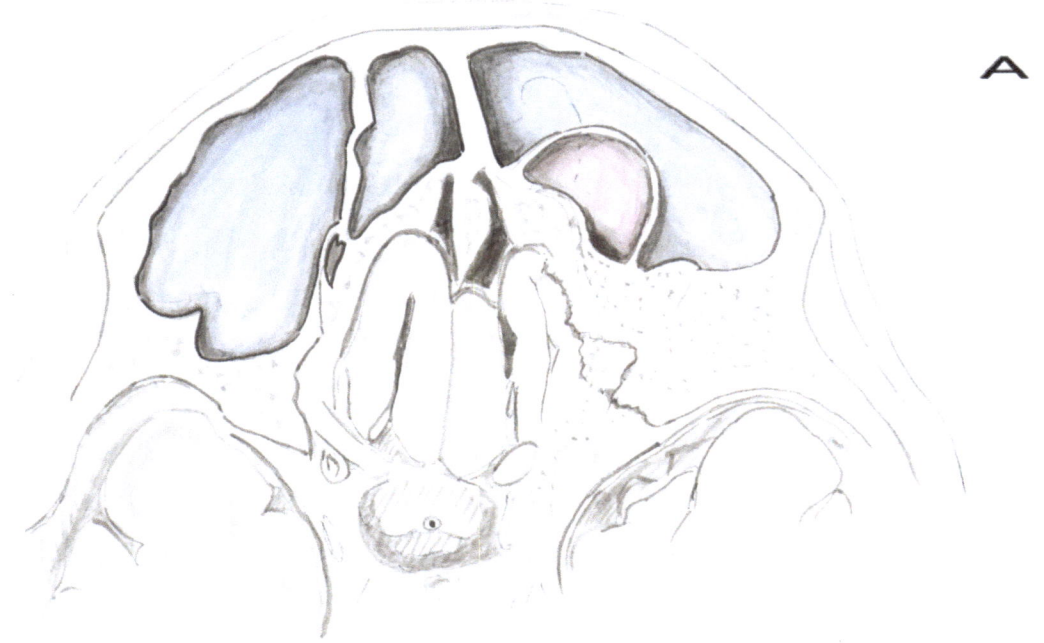

A

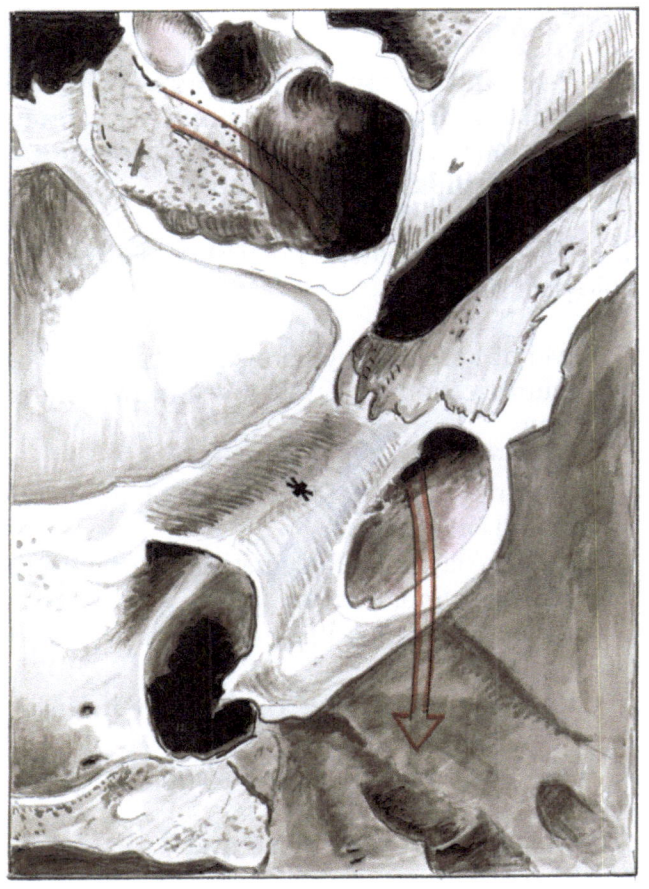

B

A + B according to Lang (1981)

p 457 and p 81

modif.

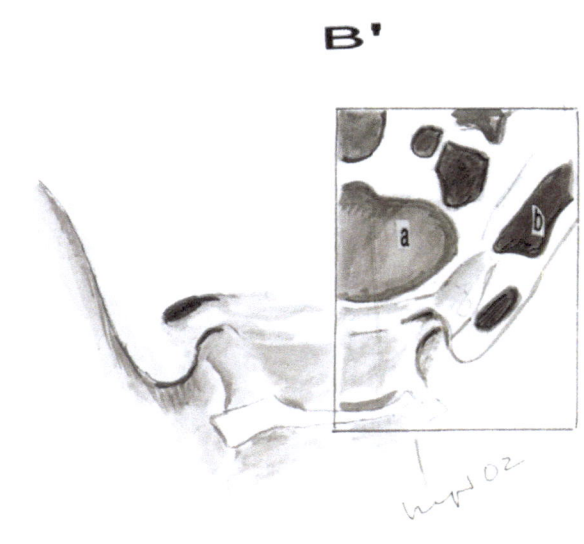

B'

Fig. 8. **Pneumatization of the orbital roof by a widened Sinus frontalis, combined with pneumatization of the orbital roof by a widening of Cellulae ethmoidales.** CT scan

S., K., 7/17/1977, female, CT 12/2001). History: No pathological findings

A   Level of Foramen jugulare and Foramen ovale. Pneumatization of orbit by Sinus frontalis (arrow)

B   Basal level of the temporal lobe

C   Orbito-meatal level. Pneumatization of orbit by ethmoid cells (arrow)

D   Level of Ala minor. Pneumatization of the right Processus clinoideus ant.
     In other cases, this pneumatization is present bilaterally
     (Hoydt; Unsöld, 1982, Unsöld, 1983, Unsöld and Seeger, 1989, p. 67)

E   Wide temporobasal pneumatization

For clinical aspects see Figs. 6 and 7

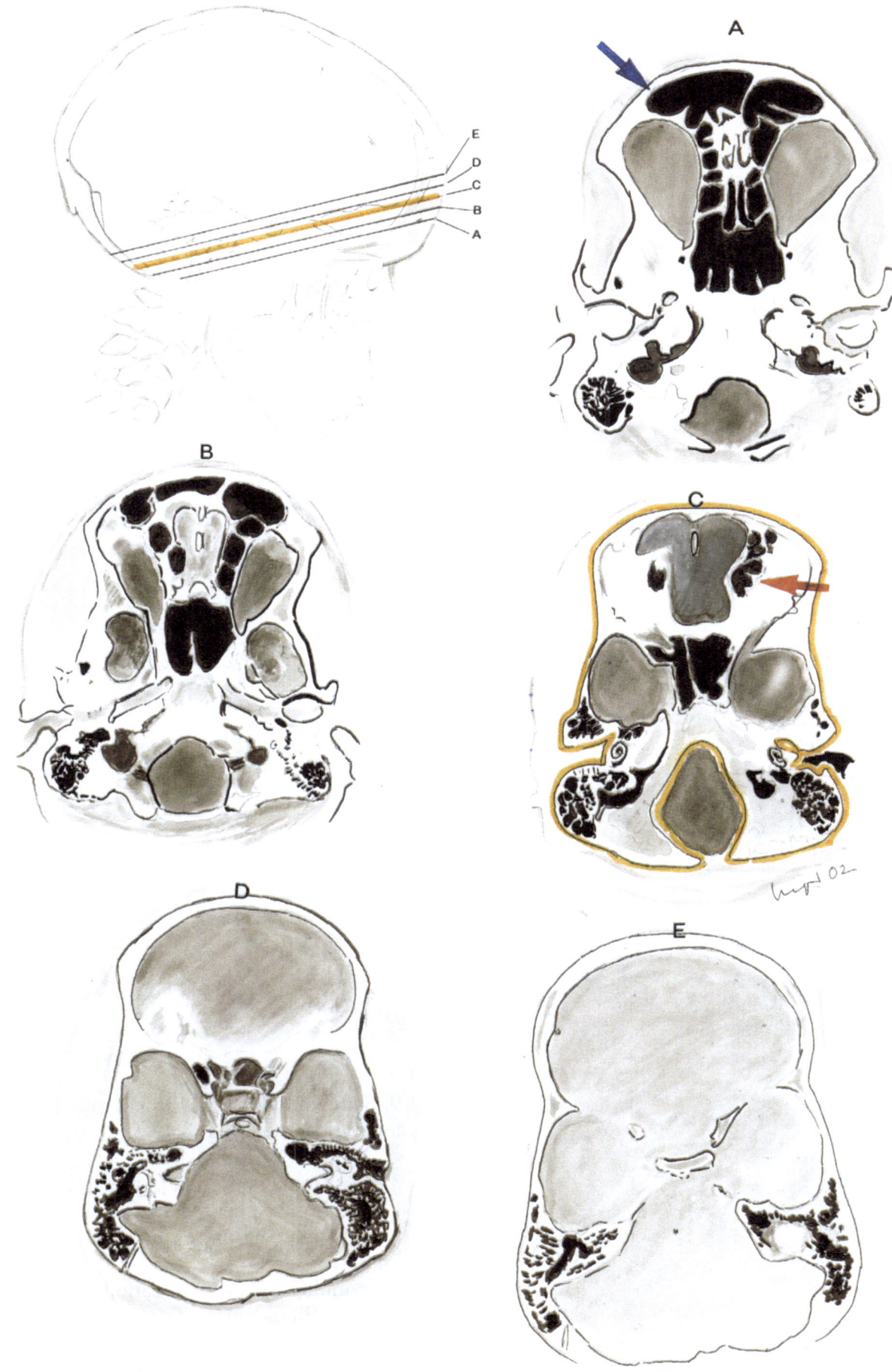

Fig. 9. Same individual in 1985 (8 years)
For comparison with Fig. 8

F   Sinus frontalis present in this level only
G   Wide Sinus sphenoidalis in F to H (beginning
    of pneumatization normally at age 4 of child-
    hood)
H   Beginning of pneumatization of the right Pro-
    cessus clinoideus ant.
I   Pneumatization of the upper temporal area

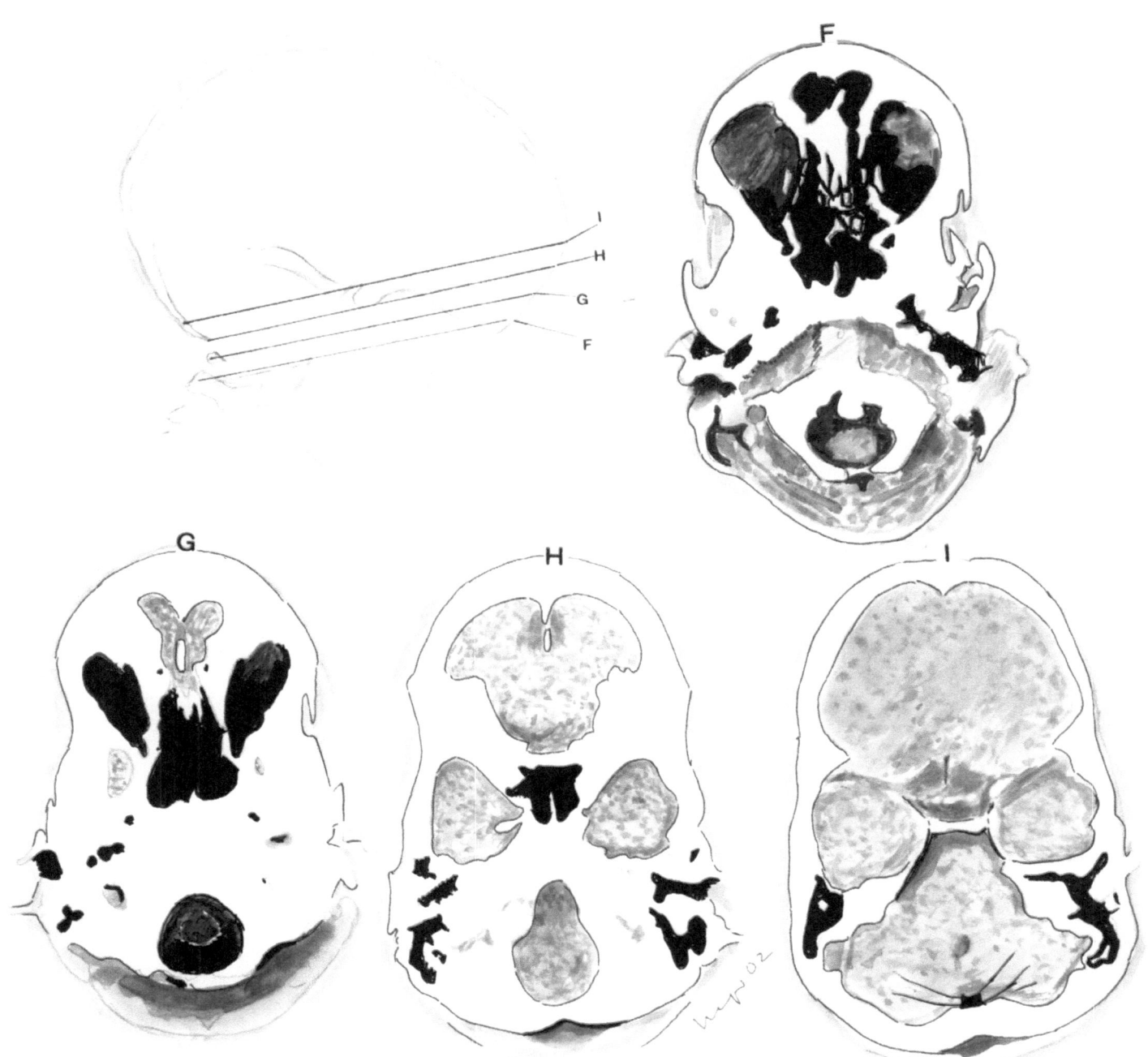

Fig. 10. **Atypical route of CSF leakage from the orbital roof through the ethmoid cells**

This may occur, if ethmoid cells are extended into the orbital roof.

A   Red colored: Ethmoid cells
       Blue colored: Sinus frontalis
B   Arrow: Route of rhinoliquorrhoea bypassing Sinus frontalis into the ethmoid cells and Cavum nasi

Clinical aspects:
A surgically performed obstruction of Ductus nasofrontalis is not able to prevent a CSF leak

FIG. 10

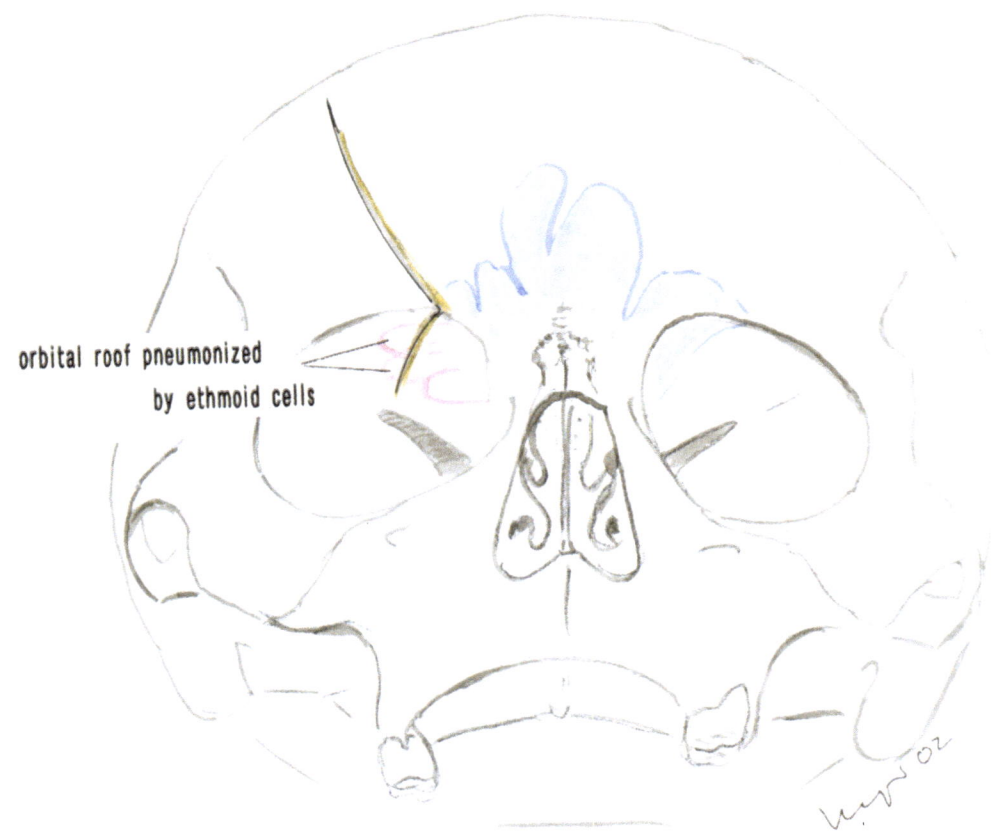

A
Orbita involved
by a skull fracture
orbital roof pneumonized
by ethmoid cells

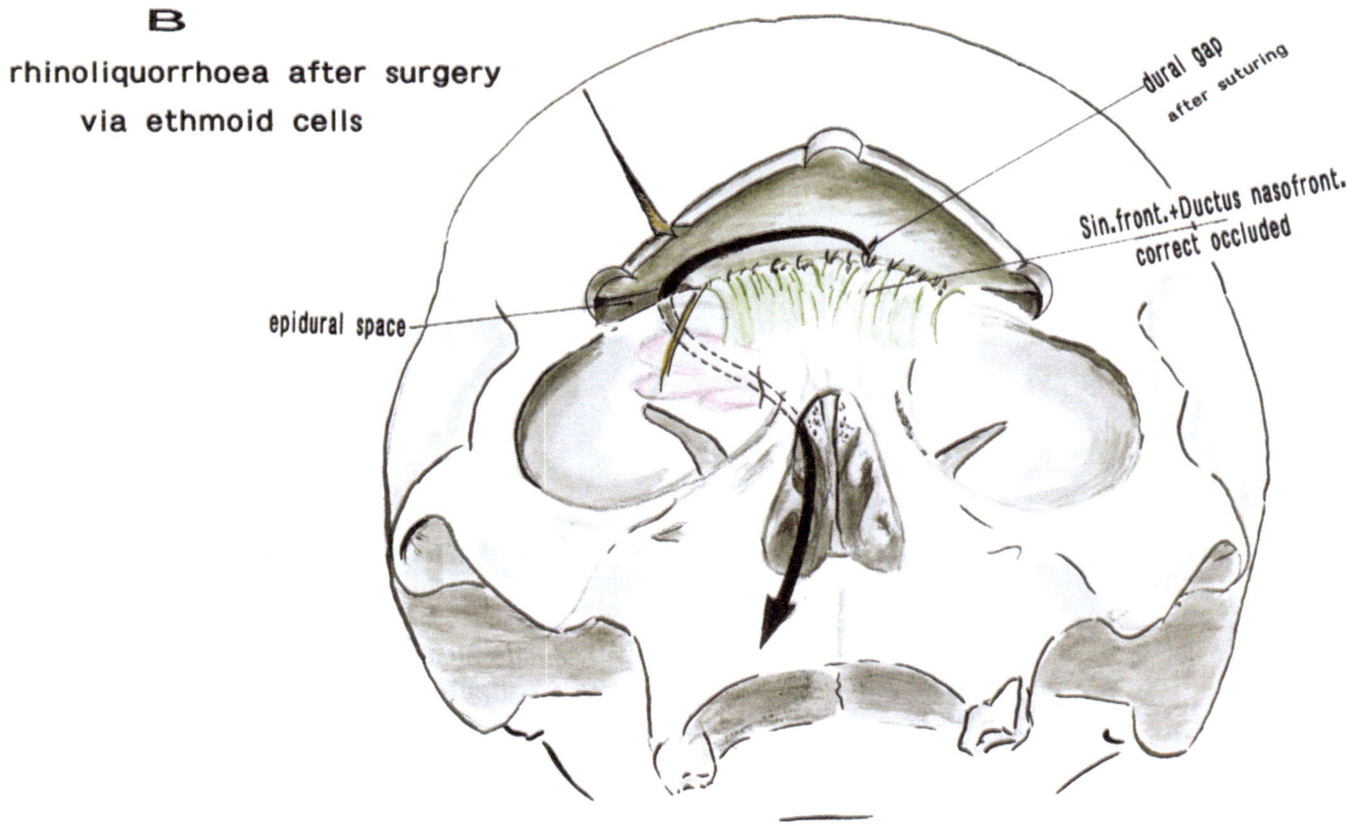

B
rhinoliquorrhoea after surgery
via ethmoid cells
dural gap
after suturing
Sin.front.+Ductus nasofront.
correct occluded
epidural space

Fig. 11. **Defect of the basal wall of Sinus frontalis (rare)**

A   Anatomical dissection
B   **Surgical aspect** (model)
      Loosening of Mucosa. **Close relationship with the contents of Orbita must be taken into consideration.**

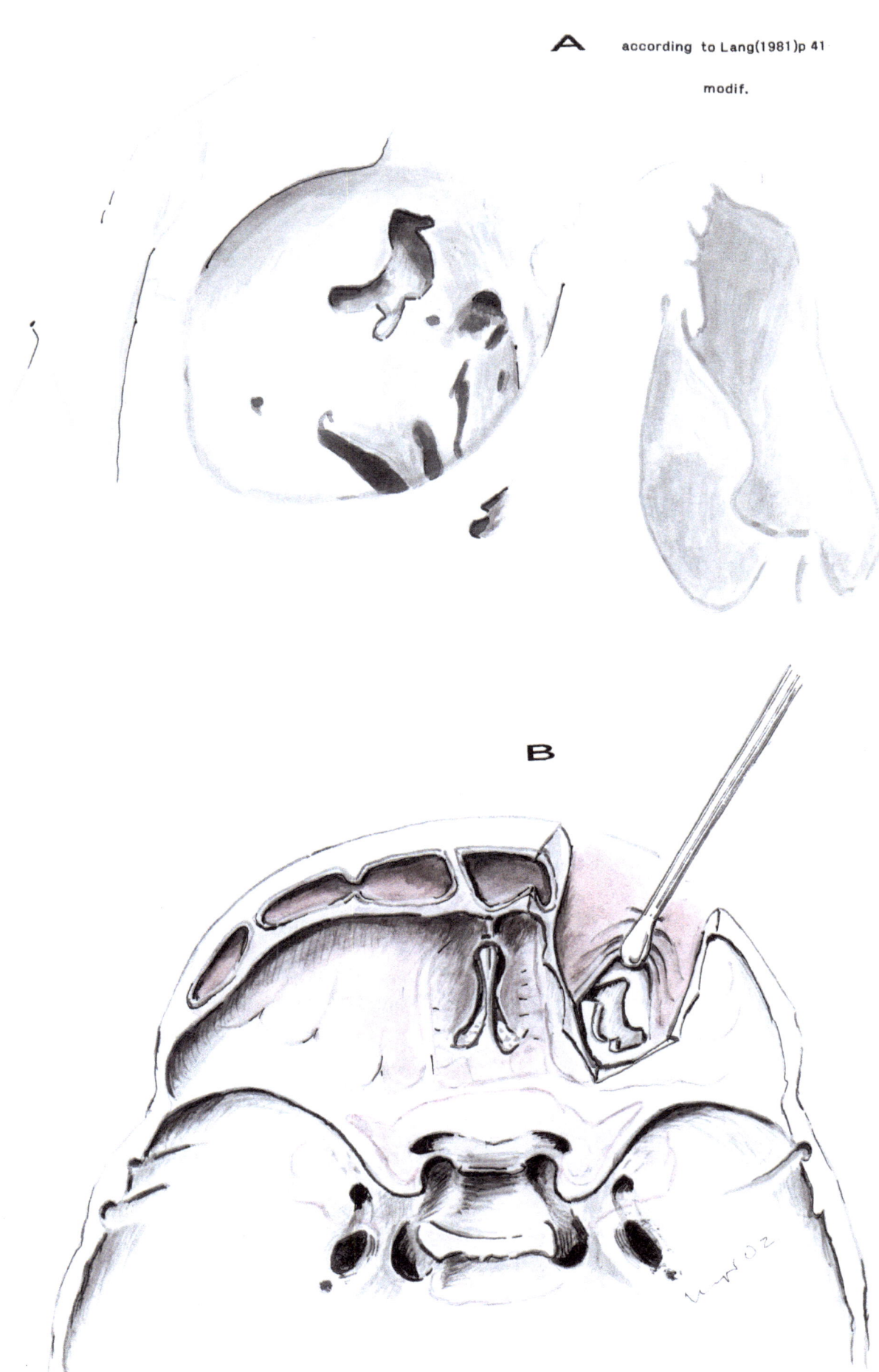

A   according to Lang(1981)p 41

modif.

B

Fig. 12. **Foramen caecum connected with Cavum nasi (rare variant)**

Usually, this foramen and Cavum nasi are separated by a thickwalled bony layer*.

A and **A'**	Drawing according to a photograph
B and **B'**	Intraoperative CSF leak following extensive electrocoagulation, hypothetical case, analogue
C	To the well known events after coagulation of Lamina cribrosa

Surgical aspect: Usually, this area is highly vascularized. Danger for postoperative rhinoliquorrhoea after a too extensive electrocoagulation in this variant.

---

* see C

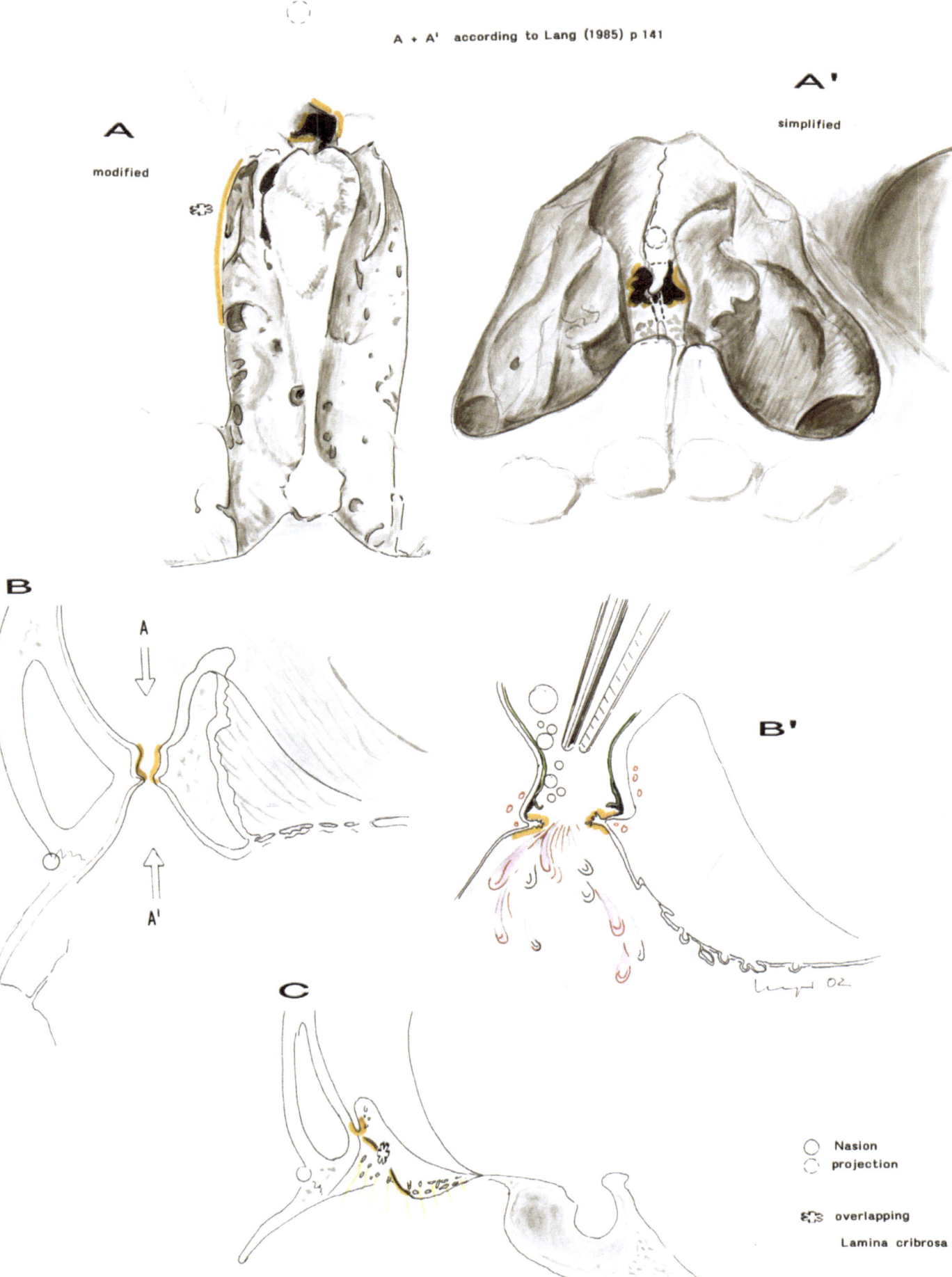

A + A' according to Lang (1985) p 141

A'
simplified

A
modified

B

A
A'

B'

C

◯ Nasion
◌ projection

✿ overlapping

Lamina cribrosa

Fig. 13. Further rare preconditions for traumatic CSF leak

A   **Pneumosinus of Crista galli**
B   **Circumscript aplasia of the dura close to Os ethmoidale** (Seeger)
C   **General widening of the pneumosinuses**

Abbreviations

a   Christa galli (variant)
b   Cellulae ethmoidales
c   Sin. frontalis
d   Sin. sphenoidalis
e   wall of the orbit
f   as a, normal finding
g   uplasia of the dura

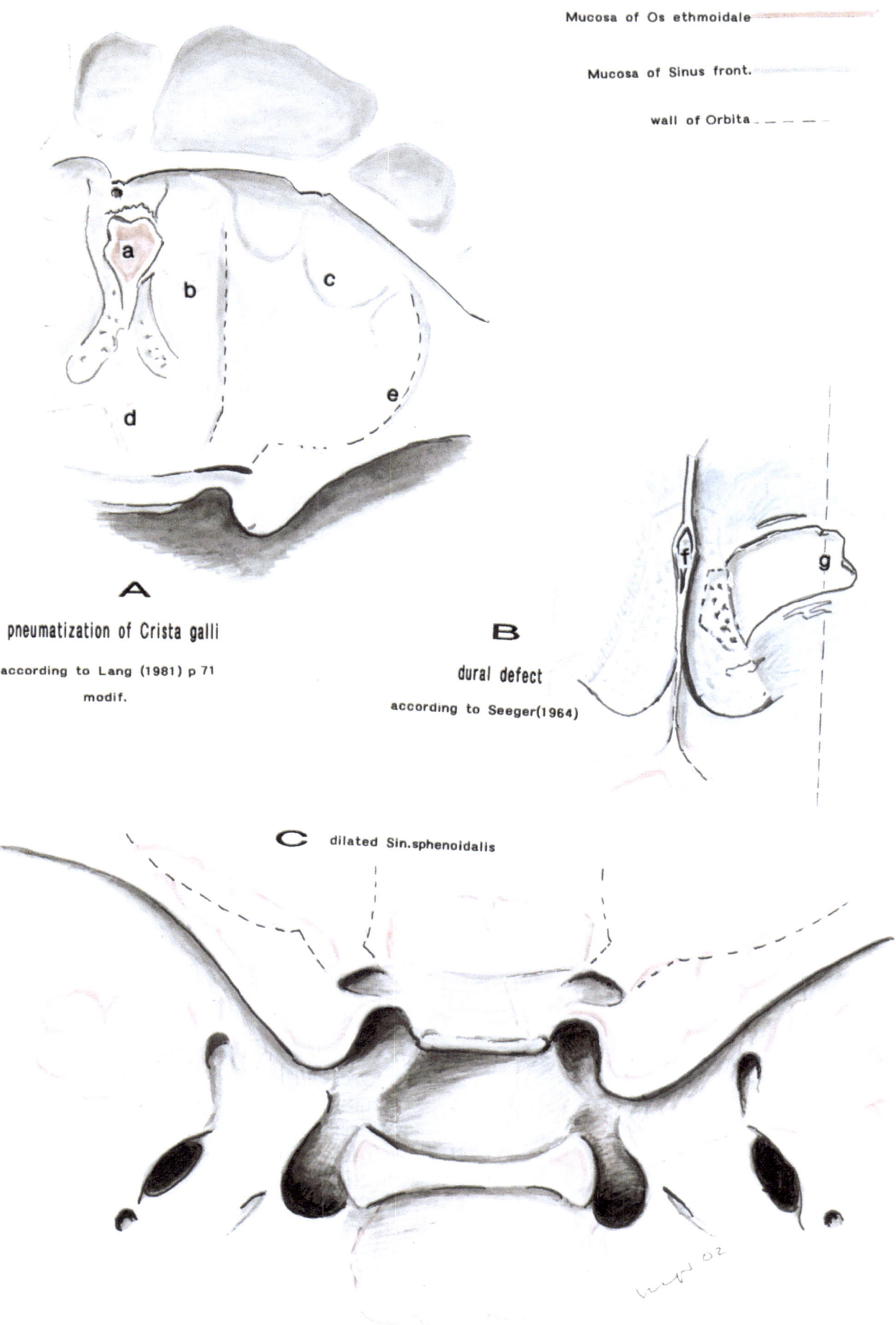

Mucosa of Os ethmoidale

Mucosa of Sinus front.

wall of Orbita

A

pneumatization of Crista galli

according to Lang (1981) p 71
modif.

B

dural defect

according to Seeger(1964)

C dilated Sin.sphenoidalis

Fig. 14. Addendum for Figs. 12 and 13

A Crista frontalis, bulging considerably into the intracranial space
B Well known microsurgical procedure, e.g. in Olfactory groove meningiomas
C See Fig. 13

Microneurosurgical aspects of basal approaches close to the midline

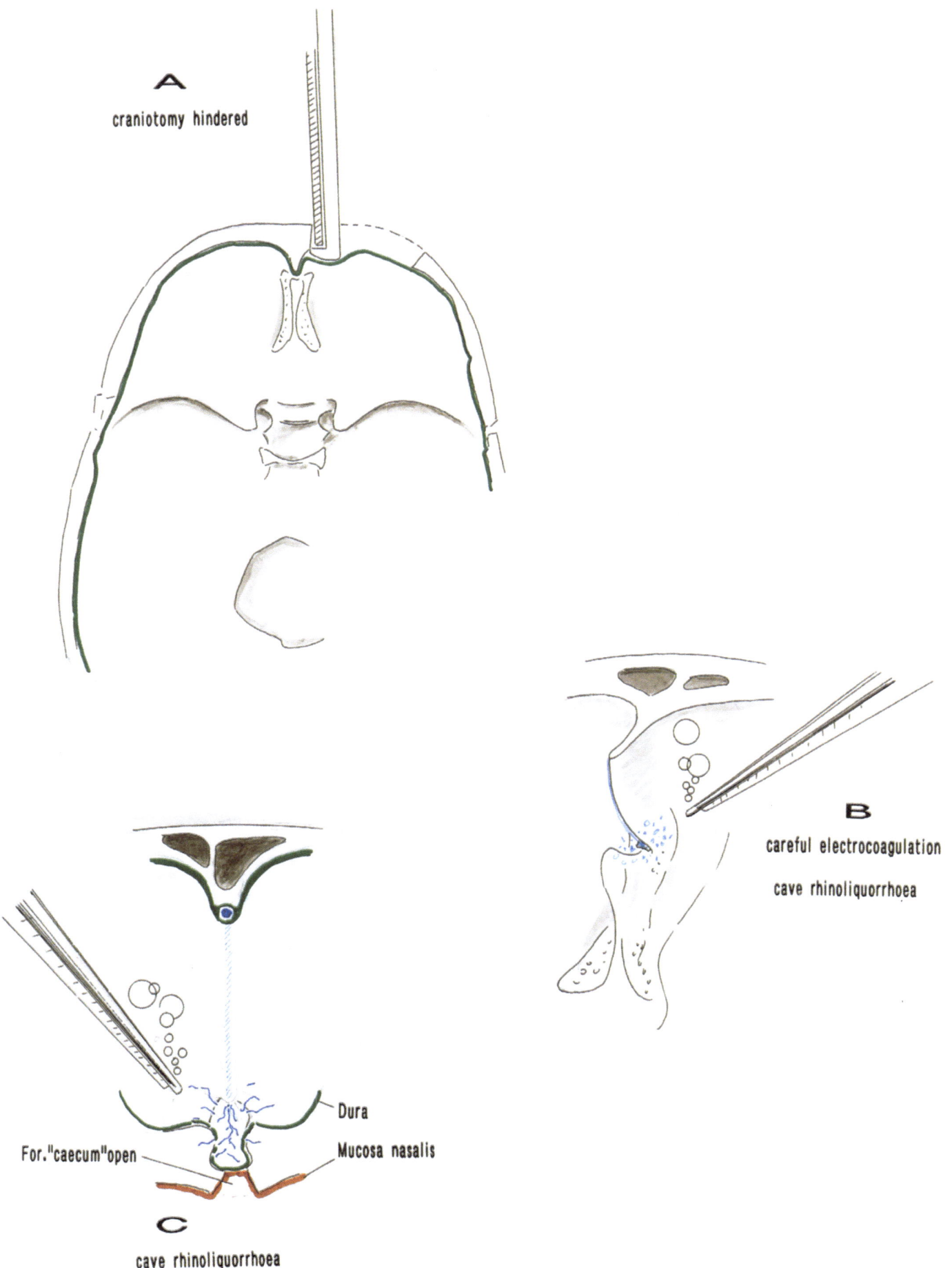

A craniotomy hindered

B careful electrocoagulation

cave rhinoliquorrhoea

For."caecum"open

Dura

Mucosa nasalis

C cave rhinoliquorrhoea

Fig. 15. **Os frontale and Os ethmoidale,** separated (arrows), survey

A	View from an anterior direction
B	From a lateral direction
	* upper surface of Os ethmoidale and lower surface of Os frontale from the superior row of Cellulae ethmoidales

These morphological aspects should be known for understanding the relationships of the ethmoid cells and Canales ethmoidales (ant. and post.) with the frontal and ethmoid bone

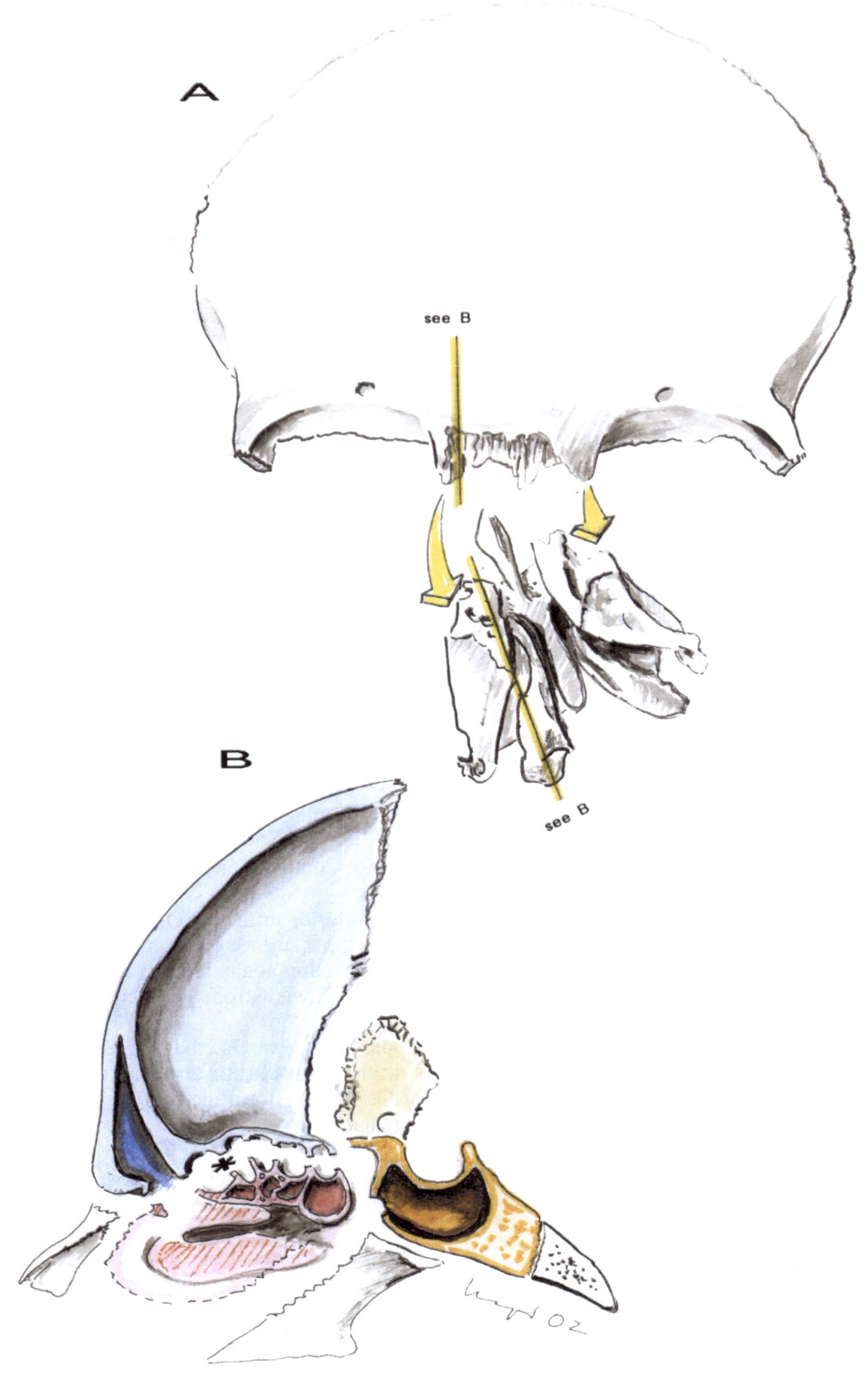

A

see B

see B

B

Fig. 16. Continuation of Fig. 15. Usual findings

A   Inferior surface of Os frontale
B   Upper surface of Os ethmoidale
C   Os ethmoidale, view from an anterior direction
D   As C, view from a lateral direction

Upper surface of Os ethmoidale form the bottom of **Canalis ethmoidalis ant.** and **post.** Inferior surface of Os frontale form the roof of Canalis ethmoidalis ant. and post.

These channels enclose Nn., Aa., and Vv. ethmoidales anteriores and posteriores.

Clinical aspects: The vessels of Canalis ethmoidalis ant. and post. are feeding some tumors (e.g. olfactory groove meningiomas) and some AVM's

FIG. 16

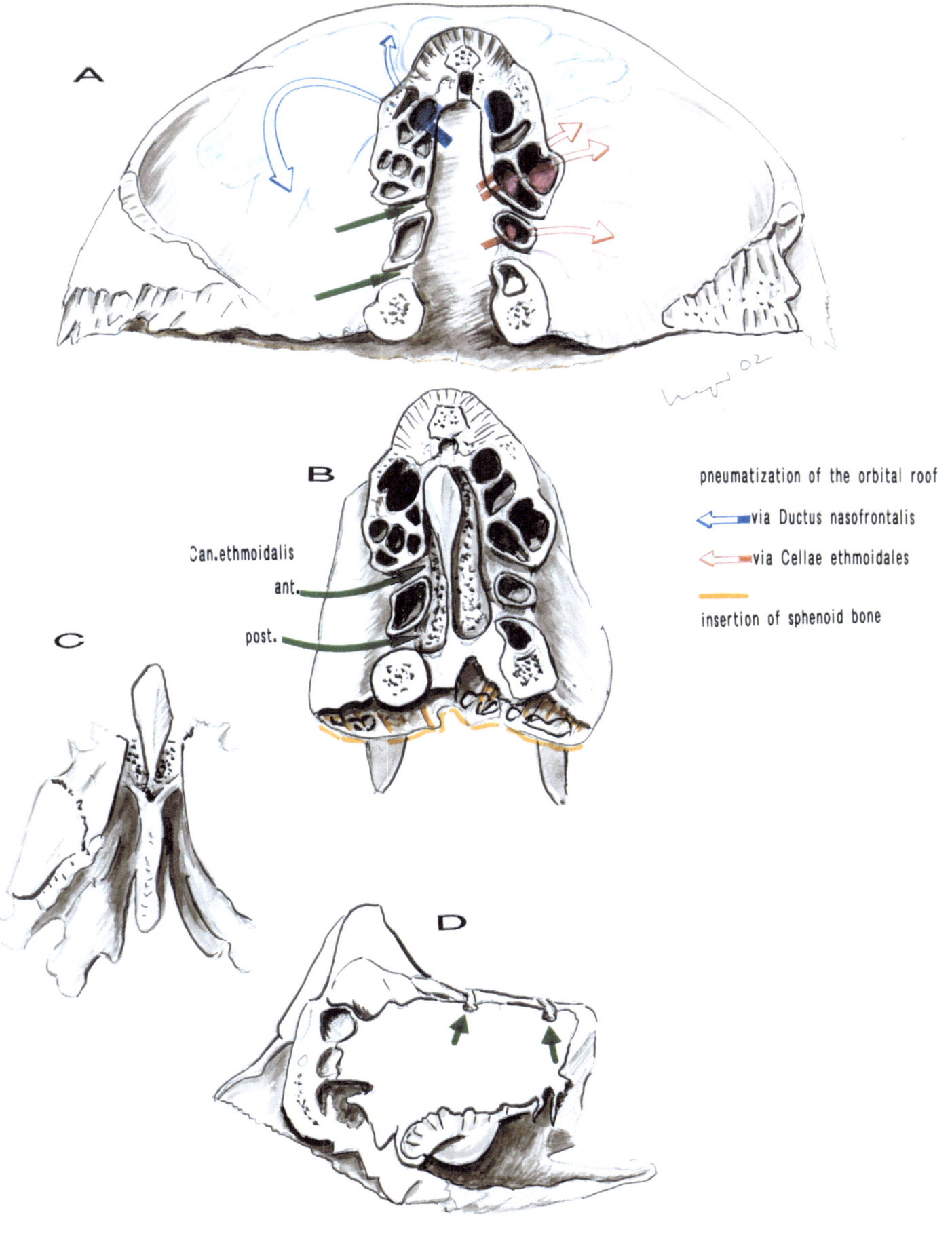

A

B

C

D

Can.ethmoidalis

ant.

post.

pneumatization of the orbital roof

via Ductus nasofrontalis

via Cellae ethmoidales

insertion of sphenoid bone

Fig. 17. Continuation of Fig. 16

A Survey
B Variable short distance of Canalis ethmoidalis post. and Canalis opticus (common finding)
C Horizontal transectional level close to Lamina cribrosa (see topogram)

Microsurgical aspects:
Danger for N. opticus, if normal or dilated (AVM) A. ethmoidalis post. must be electro-coagulated. Danger for CSF leak, if Canalis opticus and the sheets of N.II are opened

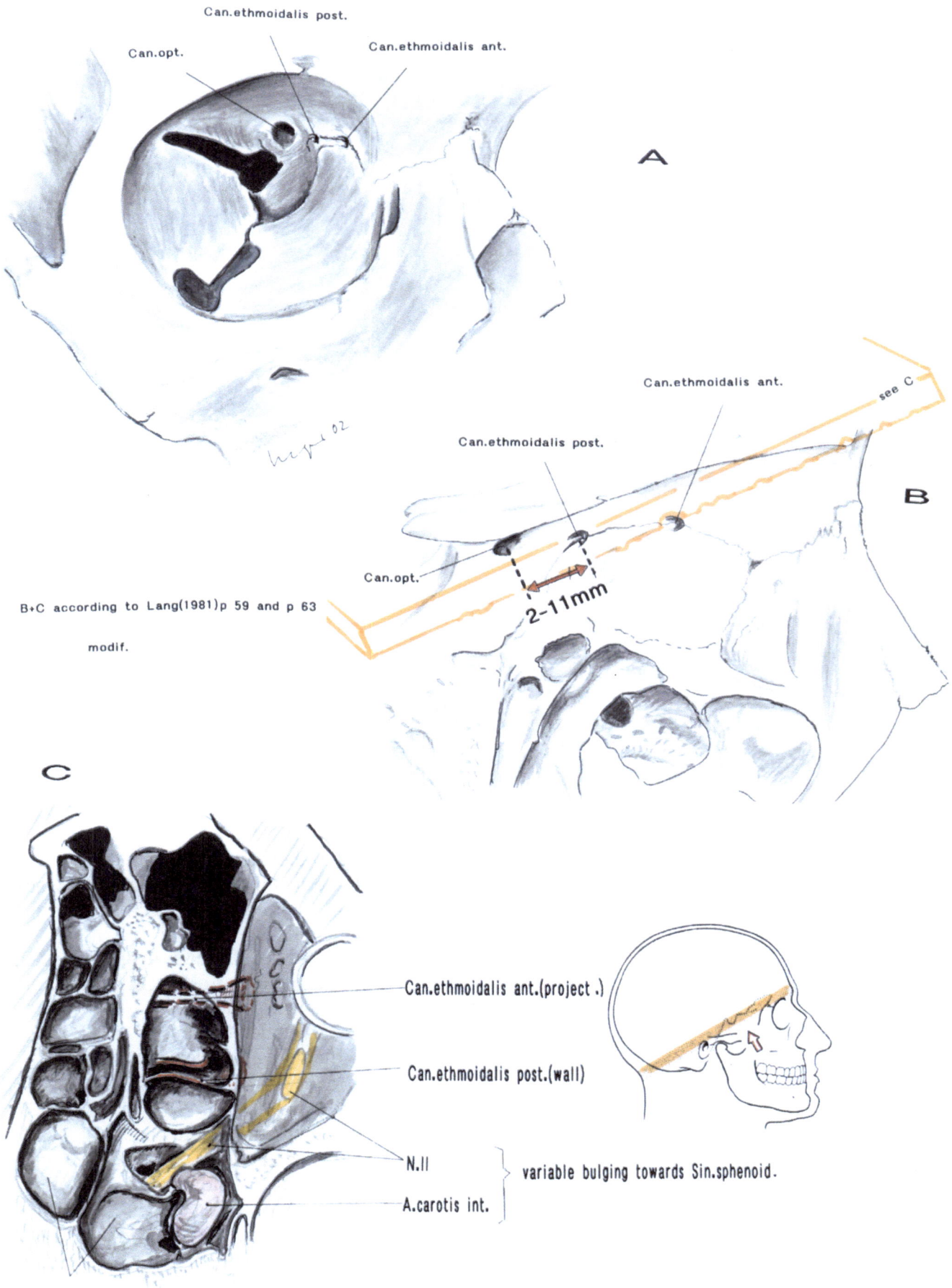

Can.ethmoidalis post.

Can.opt.

Can.ethmoidalis ant.

A

Can.ethmoidalis ant.

see C

Can.ethmoidalis post.

B

Can.opt.

2-11mm

B+C according to Lang(1981)p 59 and p 63

modif.

C

Can.ethmoidalis ant.(project.)

Can.ethmoidalis post.(wall)

N.II

A.carotis int.

variable bulging towards Sin.sphenoid.

Sin.sphenoidalis

Fig. 18. Addendum of Figs. 16 and 17

Canalis ethmoidalis ant. and post.
Its vessels are feeding the dural and bony structures
of the area of Lamina cribrosa and anterior Falx.

A	**Short distance between Canalis opticus and Canalis ethmoidalis post.**
B	Long distance
C and **D**	For understanding A and B. Probes are mapping Canales ethmoidales

Abbreviations

a	Foramen ethmoidale post.
b	intraorbital end of Canalis opt.
c	Foramen ethmoidale ant.
d	distance measurements of a and b
e	A. carotis int.
f	briging of A. carotis int. against Sin.sphen.
a' to *d'*	(projections) as a to d

FIG. 18

A

B

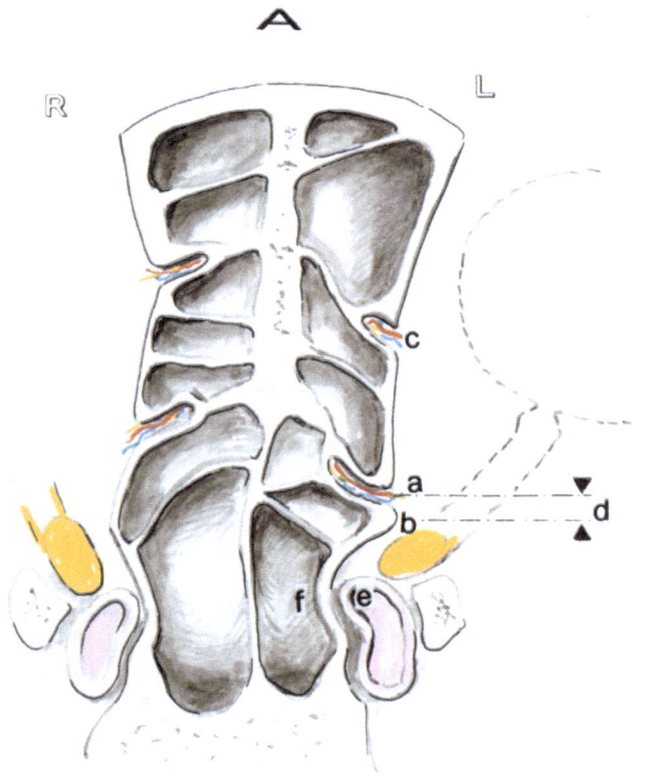

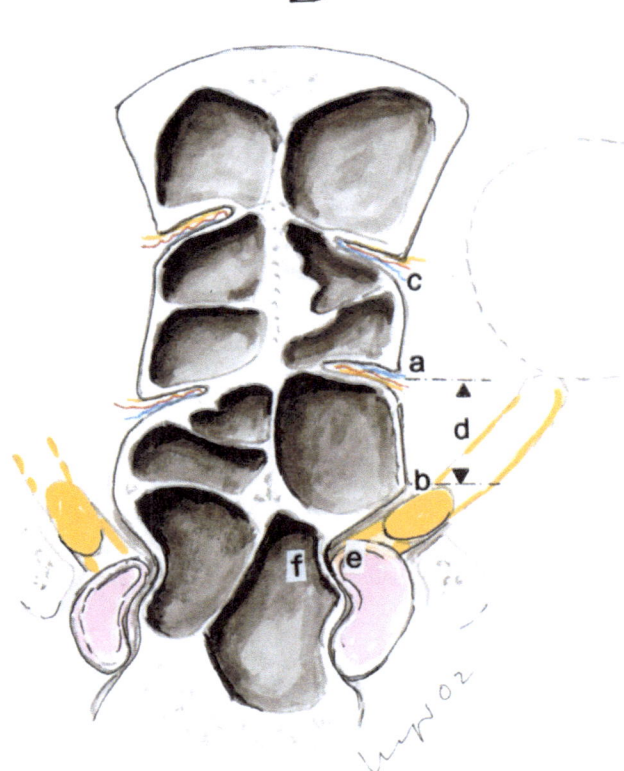

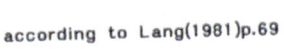
according to Lang(1981)p.69

modif.

C

D

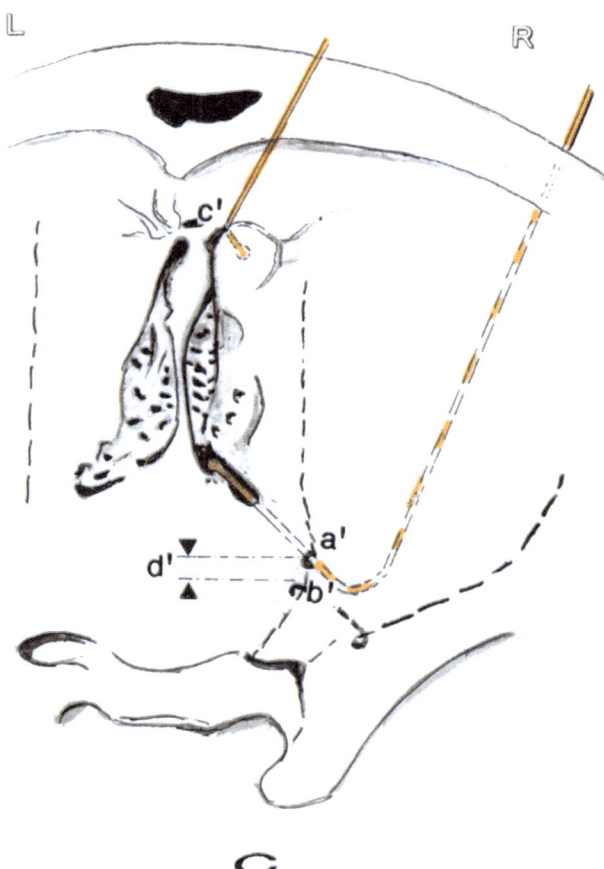

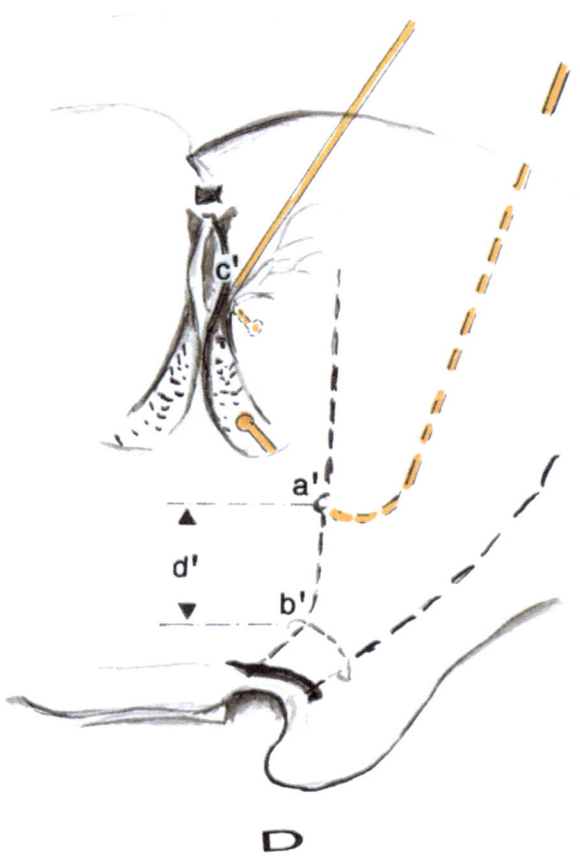

Fig. 19. **Asymmetric skull base and narrow Sinus frontalis**

Anatomical dissection (elderly female)

A	Basal view (photograph). Sinus frontalis not yet transsected
B	According to A, C, and C'
C and C'	Leftsided basal structures resected. Note asymmetrical insertion of Vomer at Alae vomeris

Abbreviations

a	Pterion
b	Sutura lambdoidea
c	**Crista frontalis,** common variant
d	Lamina mediana nasi, transsected
e	Lamina cribrosa
f	**Septum sphenoidale,** asymmetric, viewed through Apertura sphenoidalis dextra
g	Apertura sphenoidalis sinistra
h	**Vomer,** running from a right basal position to a dorso-medial position
i	Ala vomeris
j	Canalis opticus
k	Fissura orbitalis sup.
l	**Sinus maxillaris, narrow variant**
l'	**Thickened anterior wall of Sinus maxillaris**

Microneurosurgical aspects for pituitary surgery:
Asymmetric midline structures may produce problems for definition of the midline (danger for the carotid arteries, and for Nn. V/2). For hypoplasia of Sinus sphenoidalis and its unfavorable relationships with the lumen of Sella see Fig.20

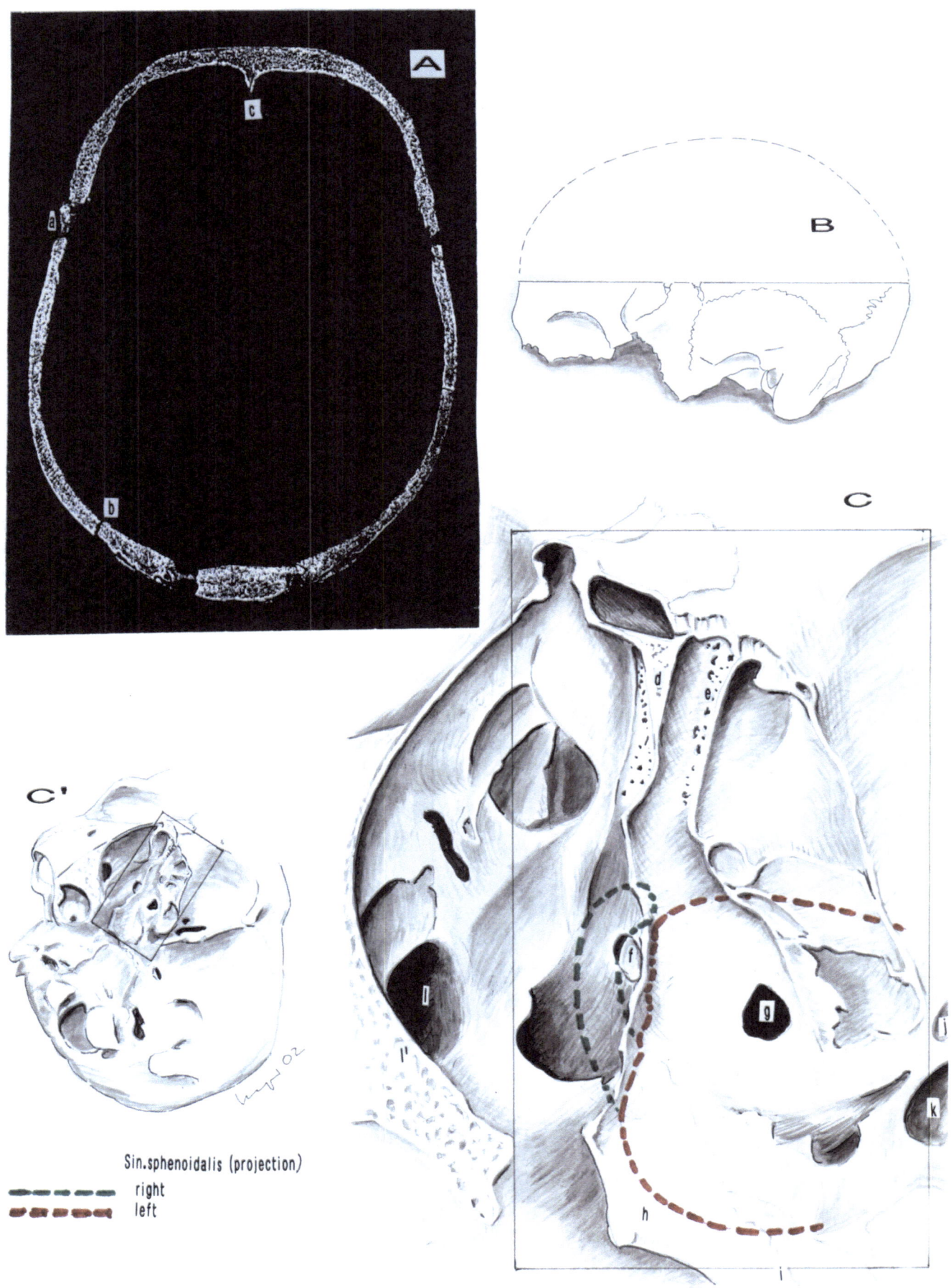

Sin.sphenoidalis (projection)
right
left

Fig. 20. **Combination of rare variants.** CT scan* of skull dissection Fig. 19

A — Left side close to the midline

A1 to **A3** — Sinus frontalis small, but it extends to the orbital roof –a-. This combination is rare

B to **C1** — Sinus sphenoidalis is located anterior to the floor of Sella turcica

Abbreviations

a — Sinus frontalis
b — left Sinus sphenoidalis (wider than right)
c — Corpus sphenoidale
d — Cristà frontalis, wide variant
e — Septum sphenoidale
f — Canalis opt.
g — Processus pterygoideus
g' — **Vomer, asymmetric variant**
h — Tuberculum sellae and posterior end of Sinus sphenoidalis
i — Lamina cribrosa
j — Planum sphenoidale
k — Sella
l — Dorsum sellae
m — Crista galli
n — Bulla ethmoidalis wide, Sinus sphenoidalis narrow
o — Dorsum sellae

Microneurosurgical aspects:
A transnasal-transsphenoidal approach to the sella would be problematic in a similar anatomical situation

---

* Section Neuroradiology (Prof.Schumacher and coworkers), University Freiburg/Br.

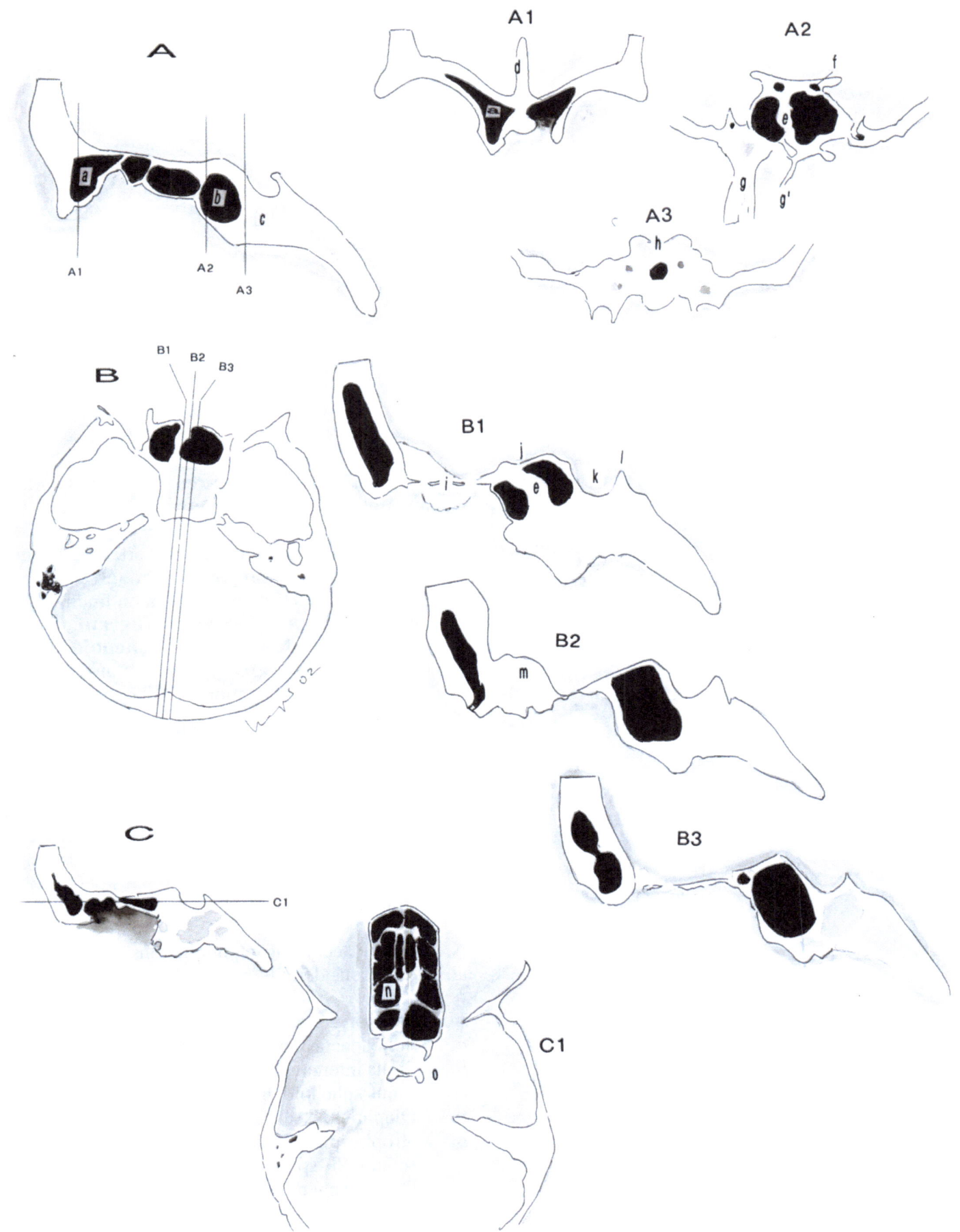

Fig. 21. **Variants of the area of Sella turcica, – microneurosurgical aspects –**

A **Dolichoectasy** of the carotid arteries.**Danger for transnasal surgical approach**

B For comparison with A: Common finding

C **Large Sinus intercavernosus ant. Small pneumatization of the sphenoid bone: Transnasal surgical approach obstructed**
Note the basal location of Diaphragma sellae (common variant)

D Senile atrophic pituitary, Diaphragma normal (according to an MRT of the Department of neurosurgery Freiburg)

E **No pneumatization of the sphenoid bone: Transnasal surgical approach obstructed Large Sinus intercavernosus ant. and post: Transnasal surgical approach obstructed**

F Sin. intercavernosi wide

A, B, C, E, F according to photographs of Bergland et al. (1968), considerably modified

Abbreviations – variants –
a dolichoectasy of carotid artery
b Sinus intercavernosus ant.
c Sinus sphenoidalis
d Diaphragma sellae
e atrophic pituitary and typical widening of intra-sellar CSF-space
f Corpus sphenoidale (no pneumatization)
g Sinus intercavernosus ant. and post.

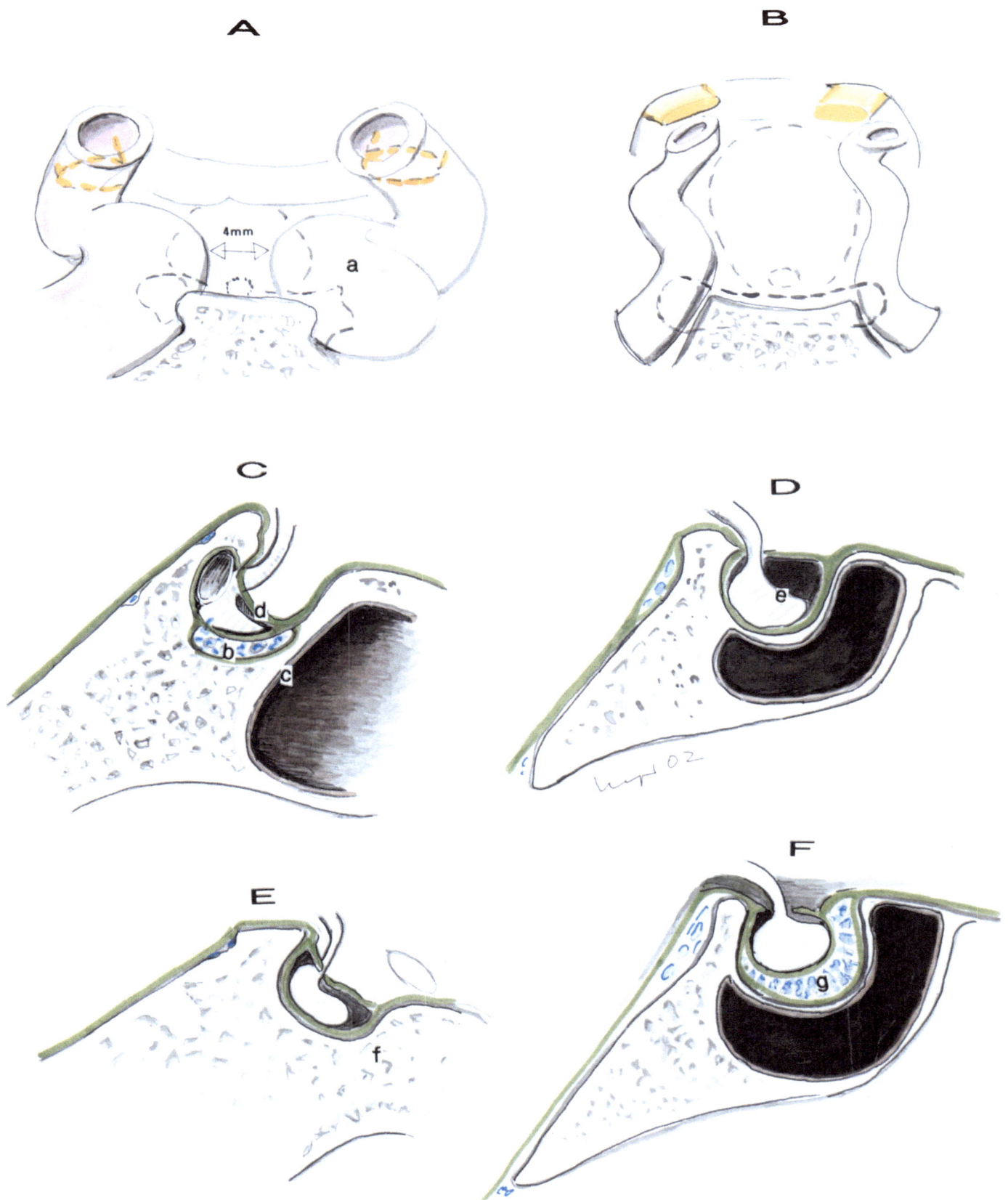

Fig. 22. **Further variants of Diaphragma sellae – microsurgical aspects –**

A to **C**   Narrow opening for the pituitary stalk (normal findings)

D   Pituitary tumor enclosed by a diaphragm with narrow opening: **Favorable for microneurosurgery**

E and **F**   Wide opening for the pituitary stalk (common variant)

G   Pituitary tumor not enclosed by protecting diaphragm: **Less favorable for microneurosurgery**

Green colored: Diaphragma sellae and other dural structures

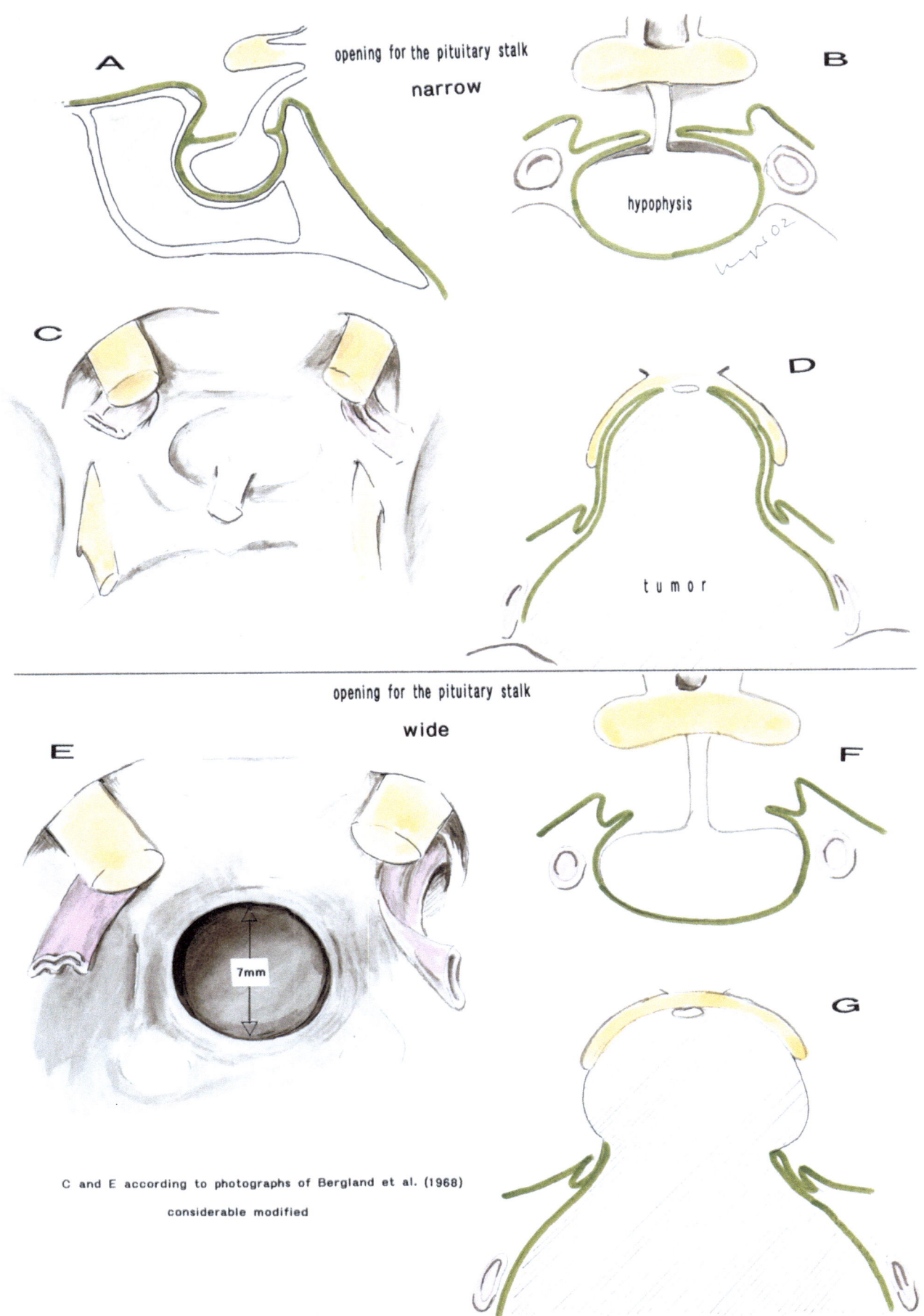

opening for the pituitary stalk
narrow

hypophysis

tumor

opening for the pituitary stalk
wide

7mm

C and E according to photographs of Bergland et al. (1968)

considerable modified

Figs. 23 to 28.

Variants of A. ophthalmica and its connections

For further variants see Chapter 2 (Figs. 51 to 63)

Fig. 23. **Primitive arteries**
Which may persist in children and adults

Neuroradiological aspects: Problems of interventional embolization

FIG. 23

A.trigemina persistens

A.otica(acustica)persistens

A.hypoglossica

A.intersegmentalis (proatlantica),
    originating from A. carotis int.or ext.

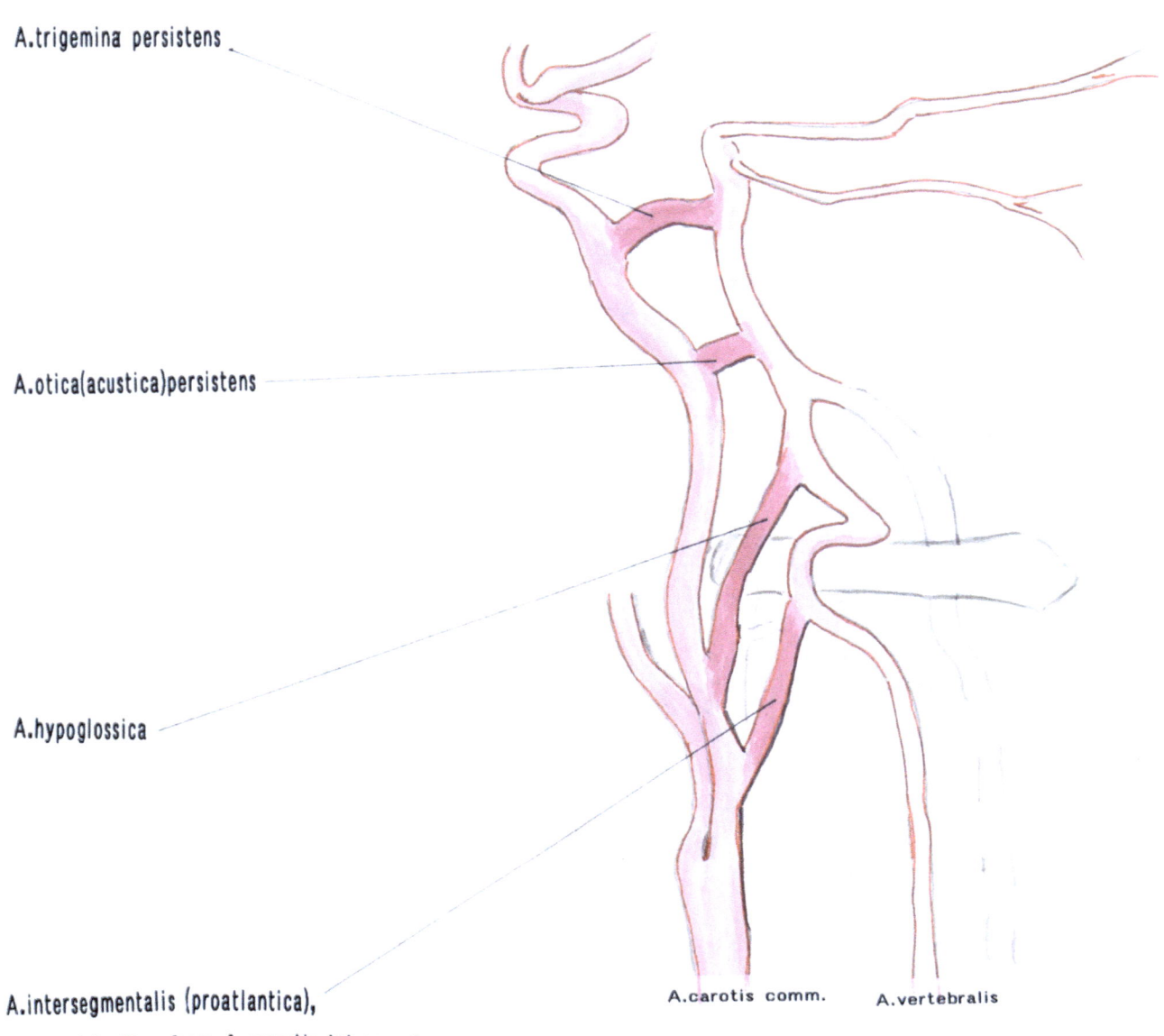

A.carotis comm.   A.vertebralis

according to Krayenbühl and Yasargil (1957)

Huber (1979)

Fig. 24. Continuation of Fig. 23

Fetal development of primitive anastomoses according to Padget (1948), modified

A   Early stage: A. trigemina primitiva –b- located close to A.ophthalmica primitiva –c-

B   Later on: Definitive presentation of A. communicans post. –f-.
No connection of A. ophthalmica primitiva as well as A. hypoglossica and the primitive basilar artery. Stage: 5mm

C   Stage: 7–12 mm

Neuroradiological aspects:
The knowledge of the development of arteries is necessary for understanding atypical connections or aplasia of arteries, to avoid fatal complications by endovascular treatment

Abbreviations

a   A.carotis int.
b   A.trigemina primitiva
c   A. ophthalmica primitiva
d   A. hypoglossica primitiva
e   A. intersegmentalis
f   A. communicans post.
g   A. basilaris
h   A. vertebralis
i   A. subclavia

FIG. 24

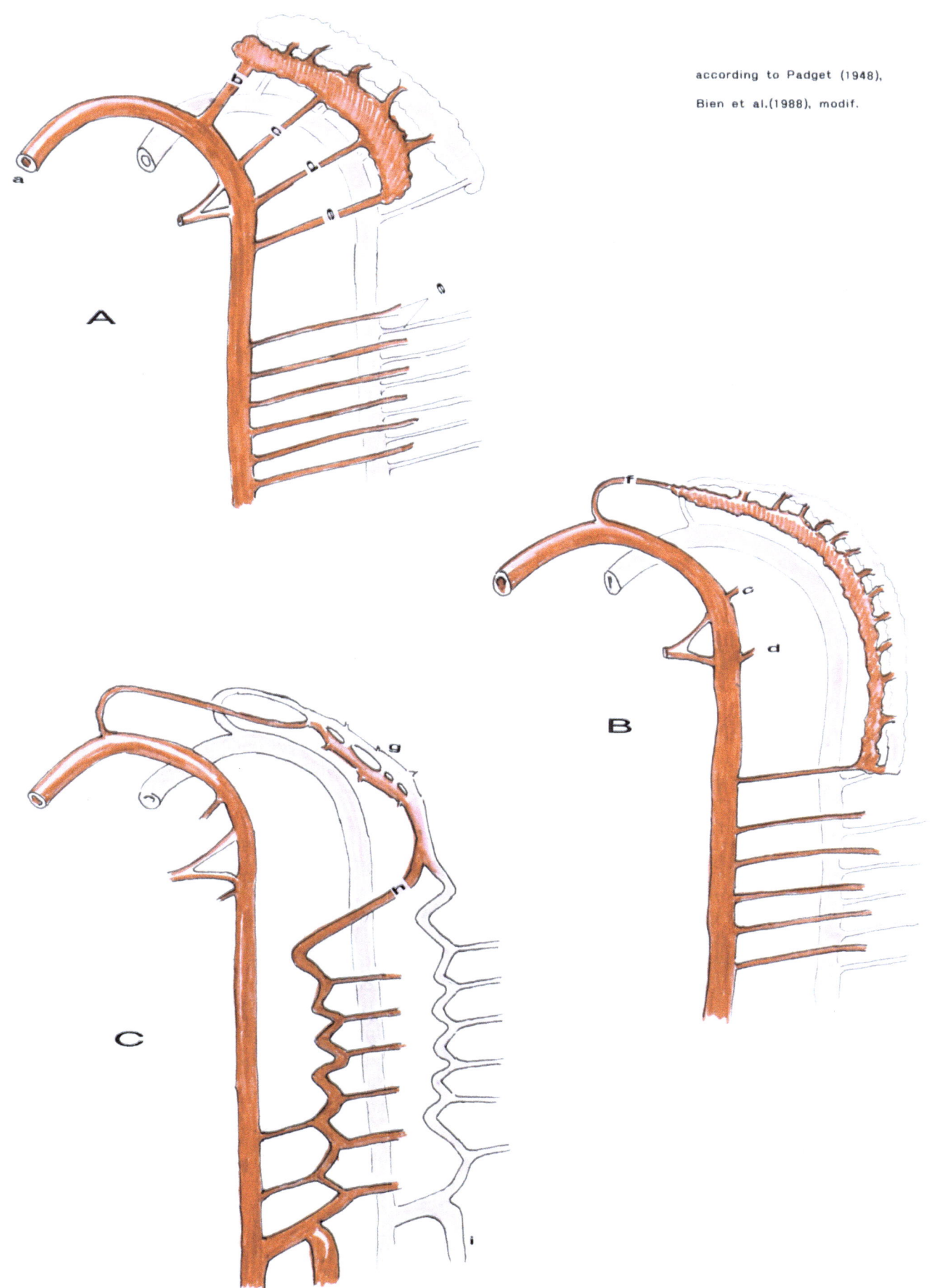

according to Padget (1948),

Bien et al.(1988), modif.

A

B

C

Fig. 25. Continuation of Fig. Fr 24

A Stage of development when the trigeminal and stapedial arteries still exist (modified by Schumacher and Wakloo, 1994, according to Padget)

B Trigeminal type in an adult. According to an angiogram*

Abbreviations

a A. carotis int.
b A. communicans post.
c A. cerebelli sup.
d A. basilaris
e primitive dorsal ophthalmic artery
f primitive ventral ophthalmic artery
g retinal vesicle
h stapedial artery
i pouch of Rathke
j Pharynx
k A. cerebri post.

---

* Section Neuroradiology (Prof.Schumacher and coworkers) University Freiburg/Br.

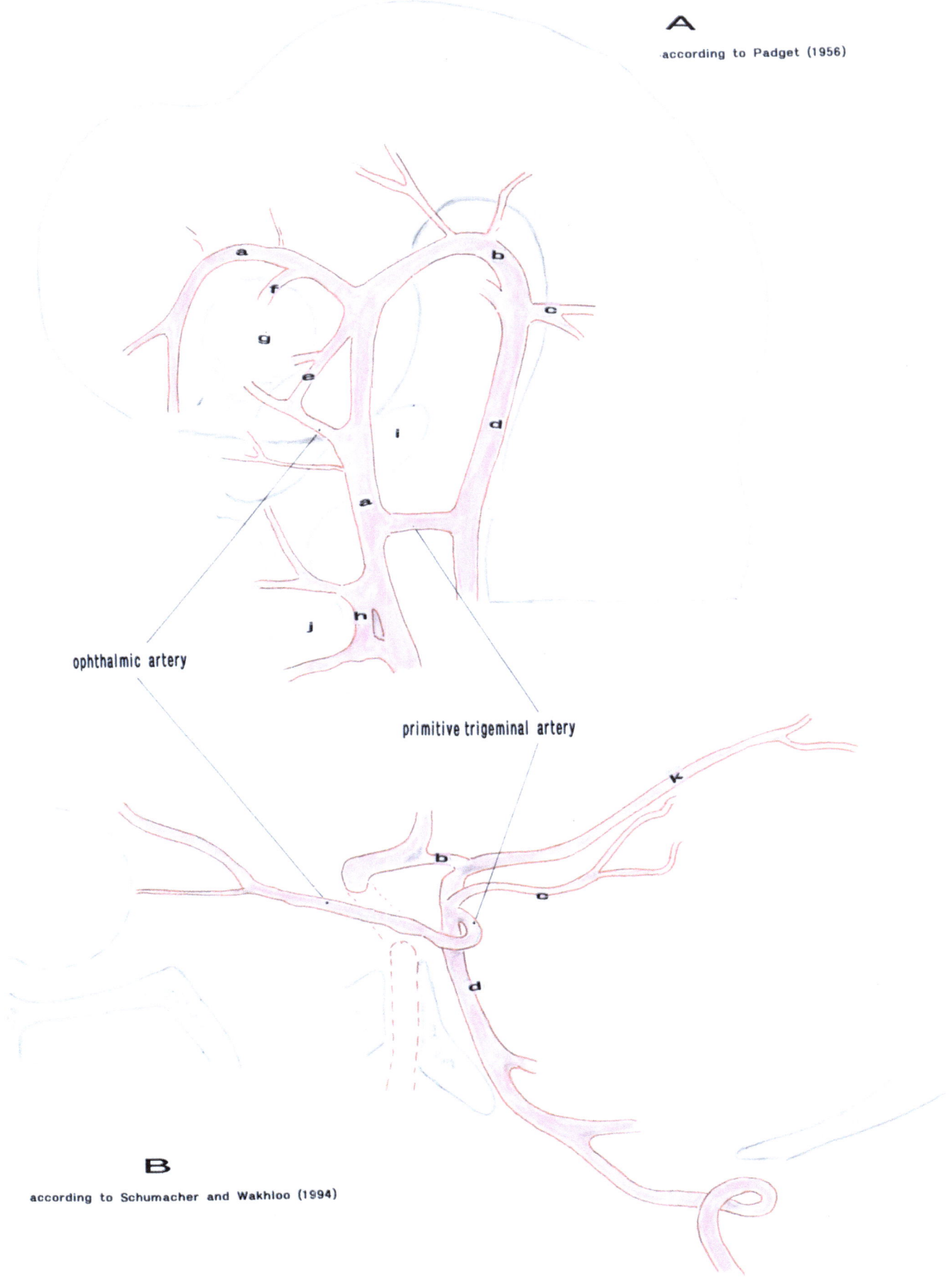

A

according to Padget (1956)

ophthalmic artery

primitive trigeminal artery

B

according to Schumacher and Wakhloo (1994)

Fig. 26.

A   **A. ophthalmica** and its branches, normal findings, with adjacent structures
B   As A, bony structures omitted

Abbreviations
a   main trunk
b   A. centralis retinae
c   A. lacrimalis
d   anastomosis with A. meningea media, anterior branch
e   A. ethmoidalis post.
f   A. ethmoidalis ant.
g   R. (Rr.) front., connected with A. angularis from A. facialis
h   R. (Rr.) supraorbit.

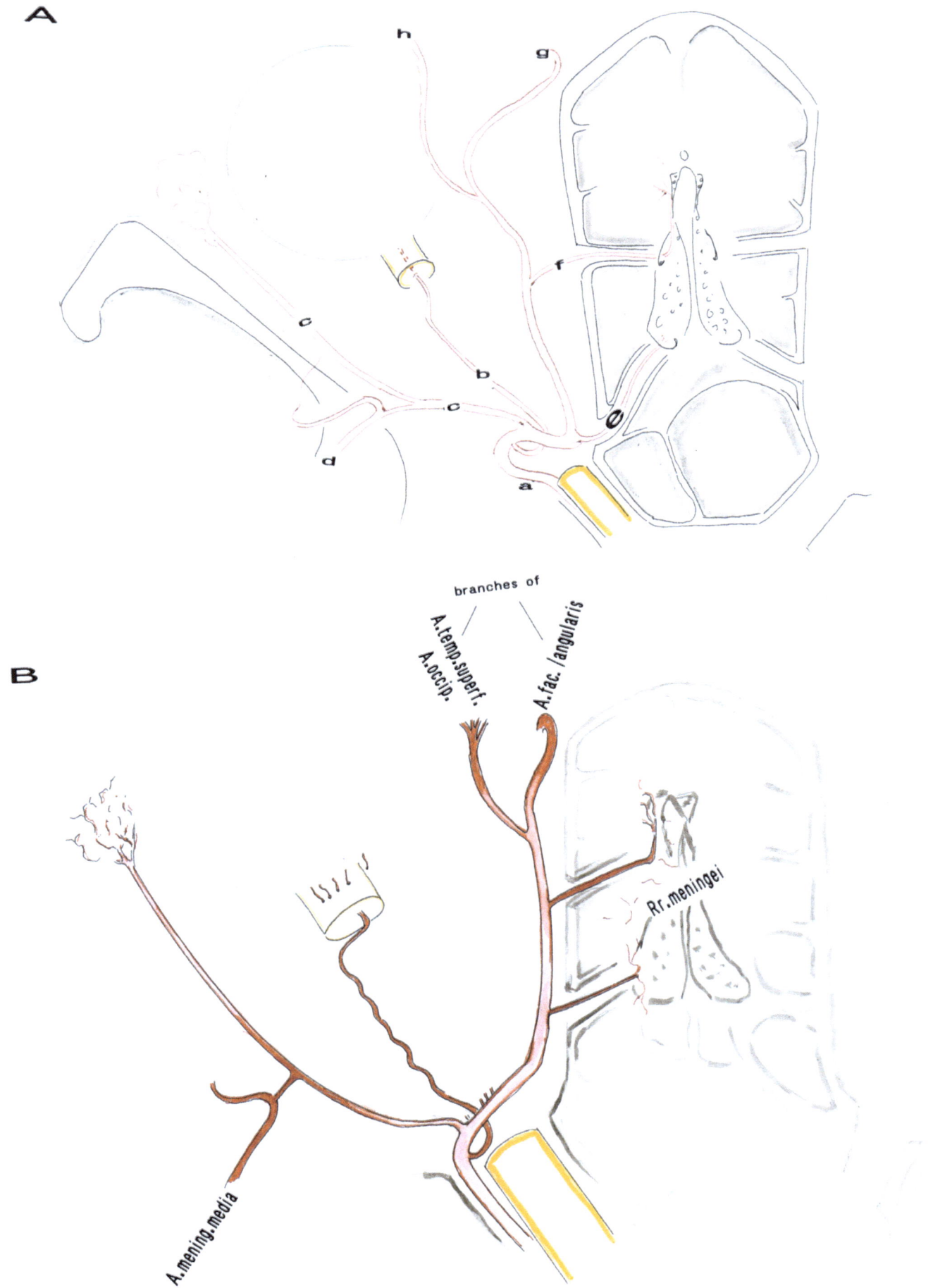

Fig. 27.

A   **A. ophthalmica** and its branches, normal finding, for comparison with B and C

B   Main trunk **undercrossing N. II, common variant**
Aplasia of the normal anastomosis between A. ophthalmica and A. ethmoidalis, see C

C   **Aplasia of R. anastomoticus with A. lacrimalis,** rare variant (according to Jazuta, 1905, cit. Lang, 1979, p. 531)

Microneurosurgical aspects of orientation see B and C

Neuroradiological aspects for interventional embolization of AVM's and richly vascularized tumors see C

Abbreviations for Figs. 26 and 27

a   origin of A. ophthalmica from A. carotis int.
a'   aplasia of a (see A Fig. 27)
b   Vagina ext. of N. II
c   N II
d   A. lacrimalis
e   anastomosis between d and f
f   anterior branch of A. meningea media
g   A. lacrimalis

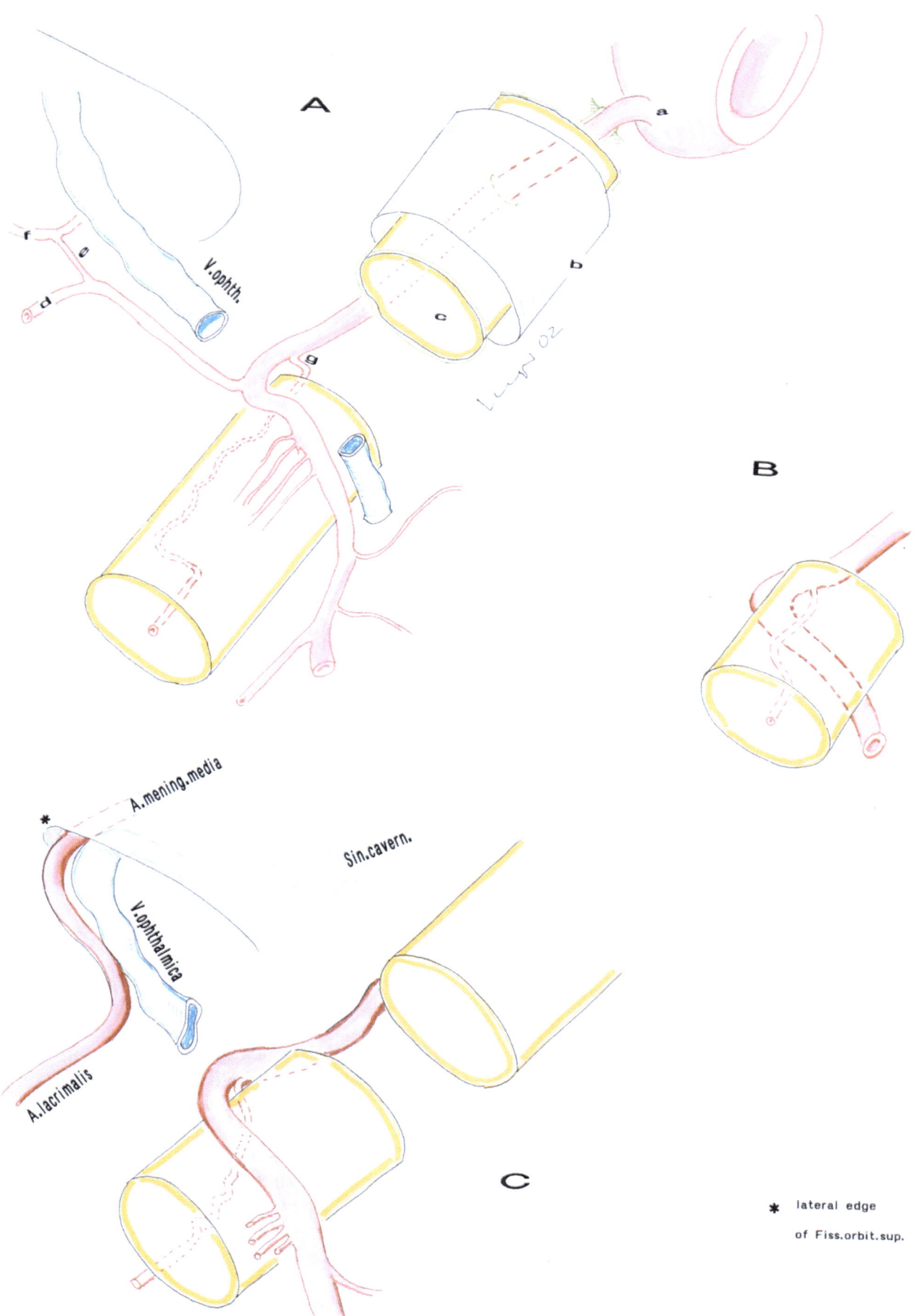

A

V.ophth.

B

A.mening.media

Sin.cavern.

V.ophthalmica

A.lacrimalis

C

* lateral edge
of Fiss.orbit.sup.

Fig. 28. Continuation of Fig. 27

Further variants of A. ophthalmica and its branches

D A. ophthalmica **originating from A. meningea media** (Singh and Dass, 1960, cit. Lang, 1979, p. 529) rare variant
E Wide connection of A. ophthalmica with A. meningea media
F **A. centralis retinae originating distal** from this connection

For clinical aspects see Figs. 24 and 27

For abbreviations see Fig. 27

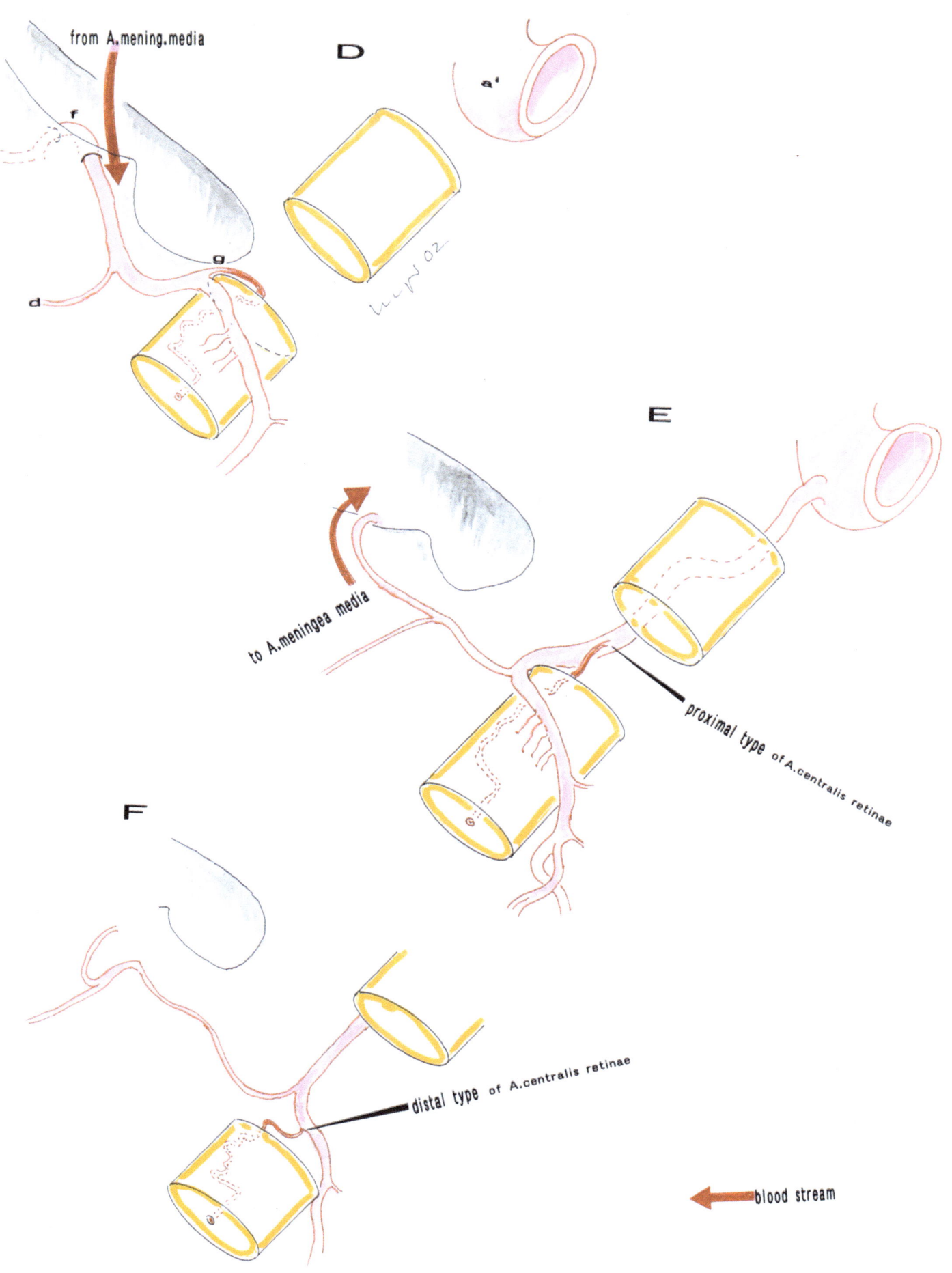

from A.mening.media

D

to A.meningea media

E

proximal type of A.centralis retinae

F

distal type of A.centralis retinae

blood stream

Figs. 29 to 47. **Frontodorsal-lateral and medial variants**

Fig. 29

A   **Sutura coronalis and sagittalis,** normal
B   **Persistent Sutura frontalis** (normally: synostotic after the early second year of life) **– common variant –**
      Synostosis of Sutura frontalis (normal finding in elderly individuals)

Clinical aspects of B:
Definition of Bregma is more difficult

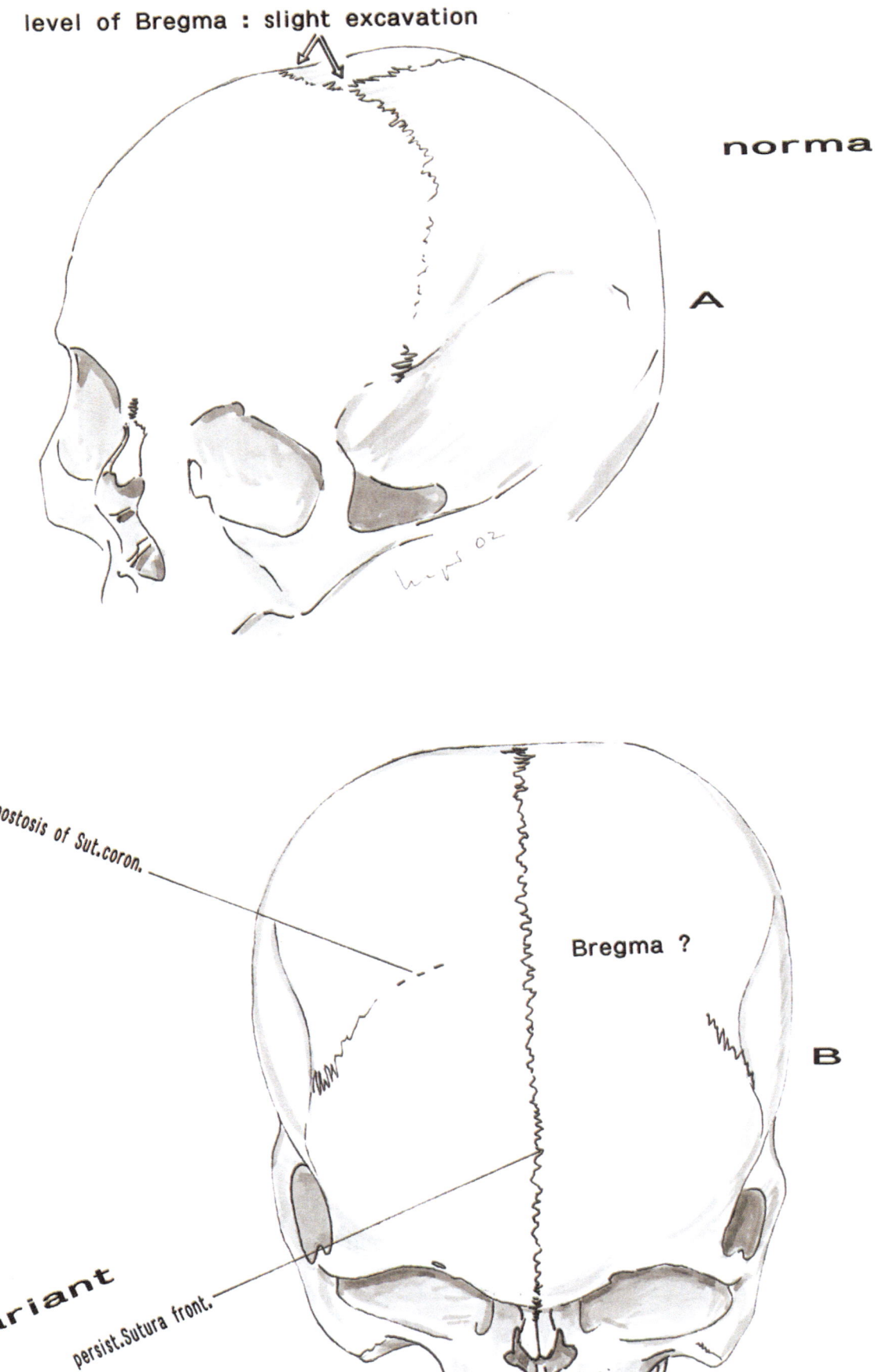

level of Bregma : slight excavation

normal

A

synostosis of Sut.coron.

Bregma ?

B

variant

persist.Sutura front.

Fig. 30. **Enostosis frontalis interna (Morgagni)**

A	**Normal finding**
B and **C**	Hyperplasia and deformation of Tabula interna, combined with thin-walled and defect dural and arachnoidal structures
C'	**Danger for cortical damage by surgical procedures**

Microsurgical aspects: see C'

FIG. 30

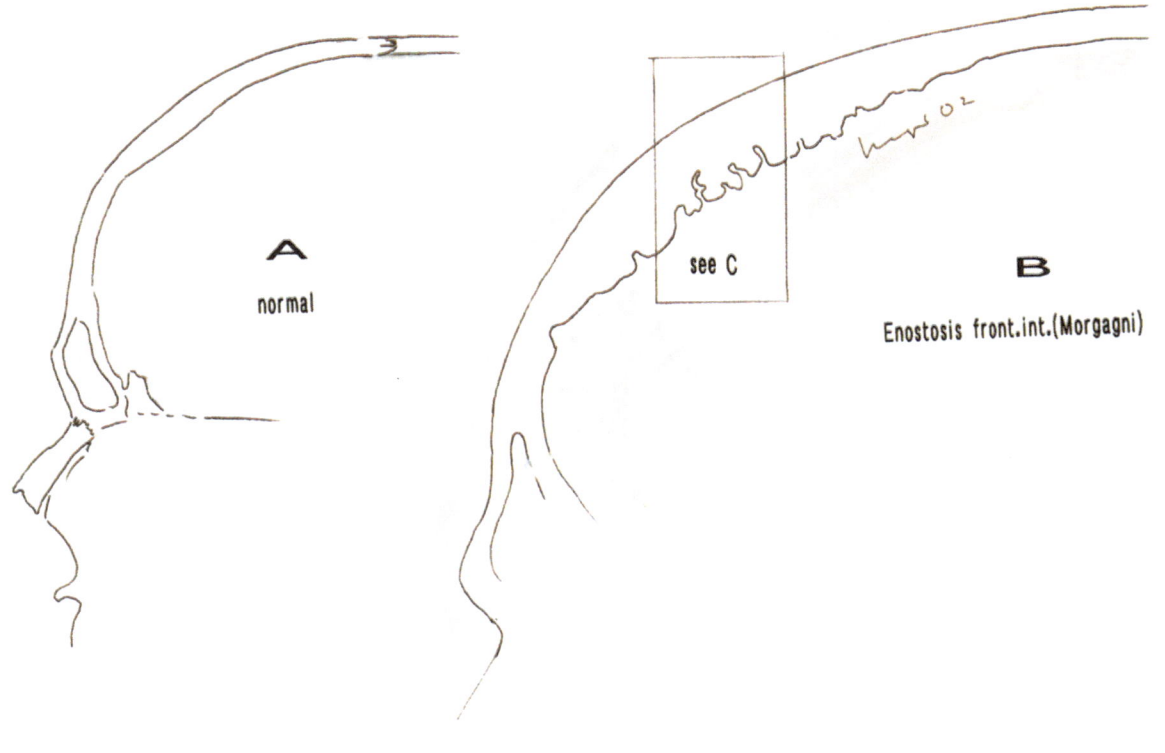

A

normal

see C

B

Enostosis front.int.(Morgagni)

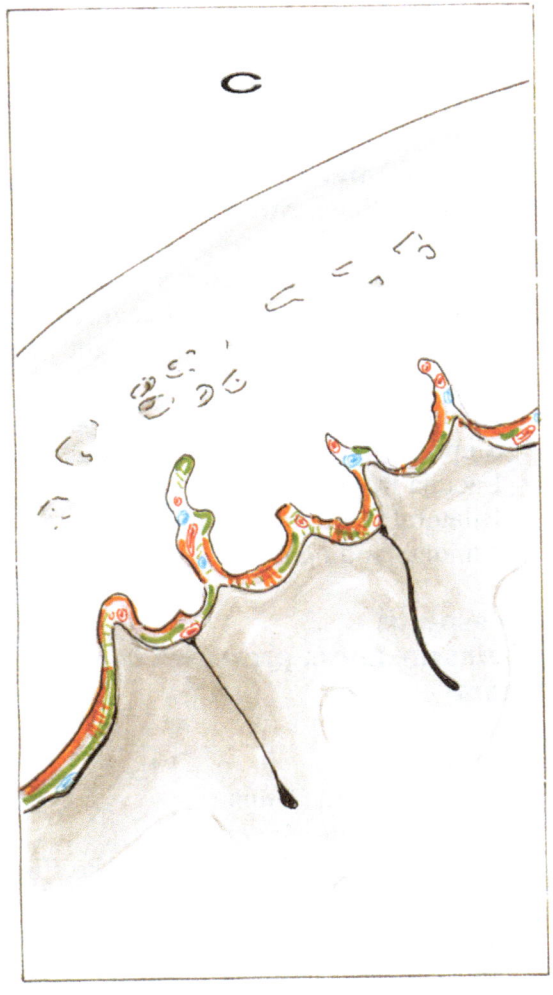

C

C' postop.

Cortex damaged

Fig. 31. **Frontopolar superficial cerebral veins, variants (simplified). Survey**

A   Veins and adjacent structures
  • Obliteration of the rostral segment of Sinus sagittalis sup., common variant (Browder and Kaplan, 1976, p 49)
  • Frontopolar crossing veins (Vesalius, 1542), Stephens and Stilwell, 1969, p 135)
  • Lacuna frontalis (common variant)
  • Bilateral sagittal bridging vein (rare variant)
B   For understanding A

Clinical aspects:
These veins are better presented in MRI than in angiogram

Abbreviations
a   obliteration of the sinus
b   frontopolar bridging veins
c   bilateral sagittal vein, parallel to the sinus
d   Lacuna frontalis (normal: located parietal)
e   Granulationes arachnoidales (Pacchioni)

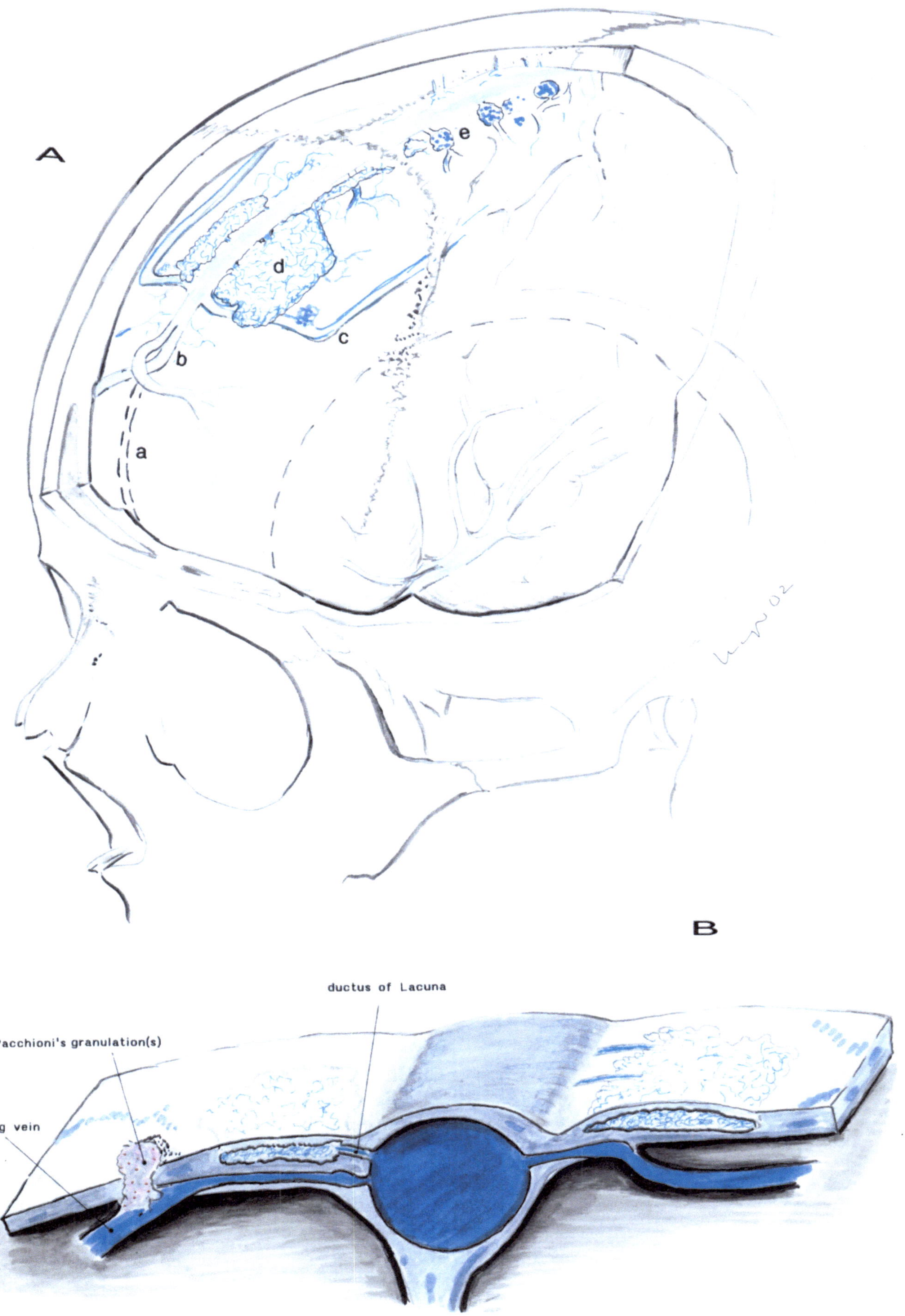

FIG. 31

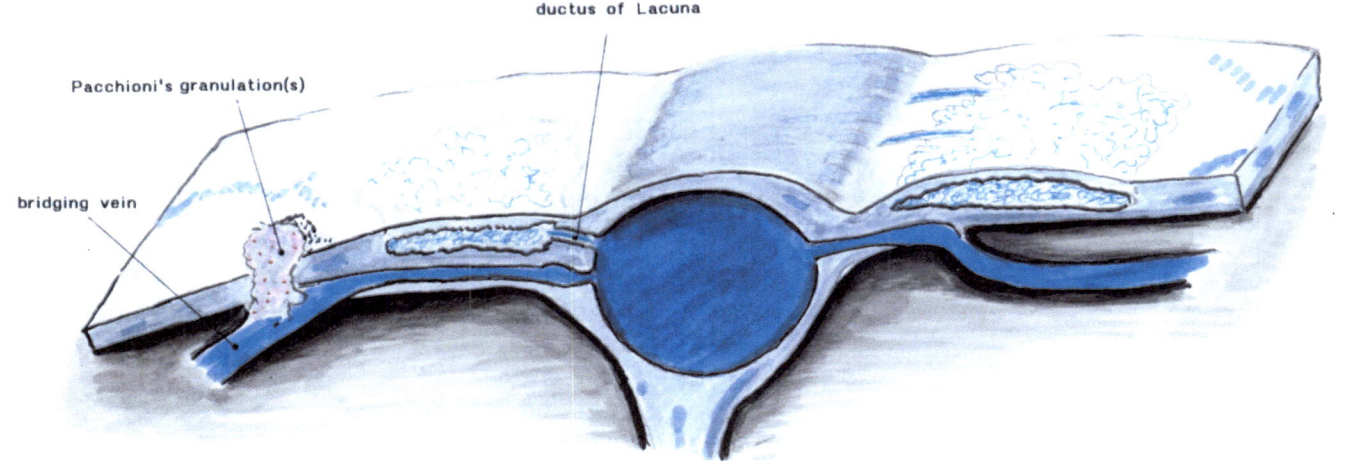

A

e

d

c

b

a

B

ductus of Lacuna

Pacchioni's granulation(s)

bridging vein

Fig. 32. **Frontal and parietal bridging veins, common variants**

Vinylite casts, colored drawing according to photographs of Browder and Kaplan (1976), pp 47 to 52. Skull drawn in*

A Irregular course of the frontal segment of Sinus sagittalis sup.
Normal parietal Lacunae
B Hypoplasia of the frontal segment of the sinus
C Aplasia of the frontal segment of the sinus. Note fork-like configuration and widening of frontopolar bridging veins. The course of the sinus is not located exact in the midline.

Clinical aspects:
Midline structures, e.g. Sutura sagittalis, are not congruent with the sinus. This is important for microneurosurgical midline approaches (problems for definition of the lateral margin of the sinus and mixing up with Lacuna are possible)

---

* Corresponding points of drawings and simplified topographs

FIG. 32

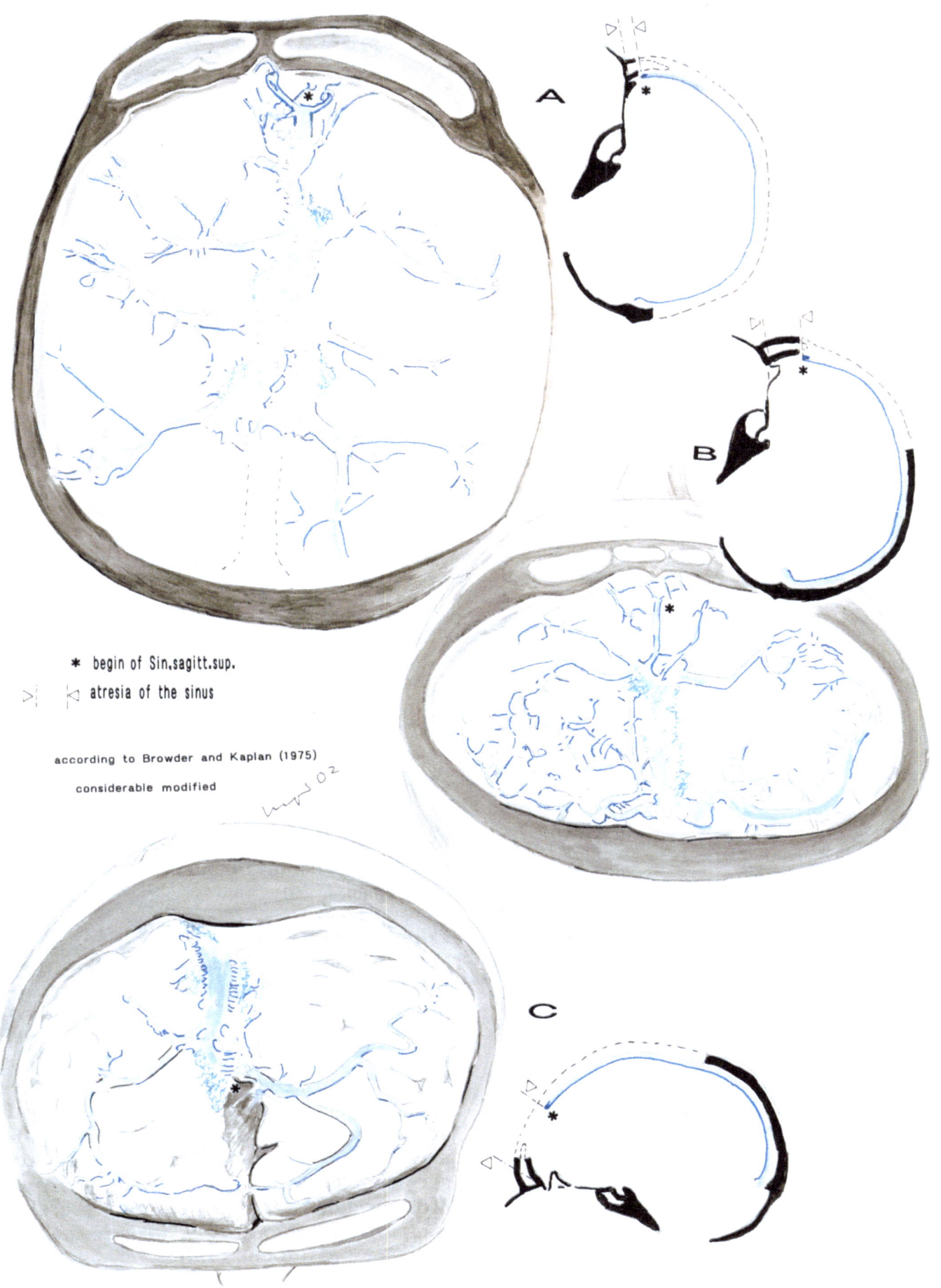

* begin of Sin.sagitt.sup.

⊳⃒ ⊲ atresia of the sinus

according to Browder and Kaplan (1975)

considerable modified

Fig. 33. **Sinus duplicated**

A By a sagittal septum (upper drawing)
B By horizontal septum (inferior drawing)

Most bridging veins are connected with the i n f e r i o r compartment of the sinus (Browder and Kaplan, 1976)

Clinical aspects:
Septa are often favorable for Microsurgery of meningiomas, which are penetrating the superior compartment of a sinus with a horizontal septum. Bridging veins of the inferior compartment are preserved. The inferior compartment of the sinus is mostly smaller than the superior compartment. If the sinus is divided by a sagittal septum, then tumors of the wider compartment are often more favorable for microsurgery than tumors of the smaller compartments: Most bridging veins are entering the smaller compartment (Lang, 1979, in agreement with Browder and Kaplan, 1976)

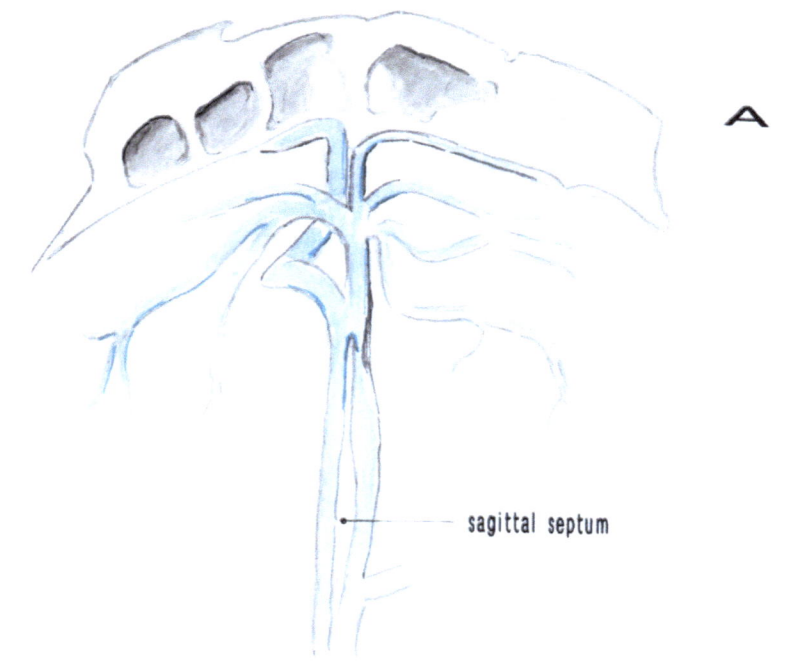

A

sagittal septum

B

horizontal septum

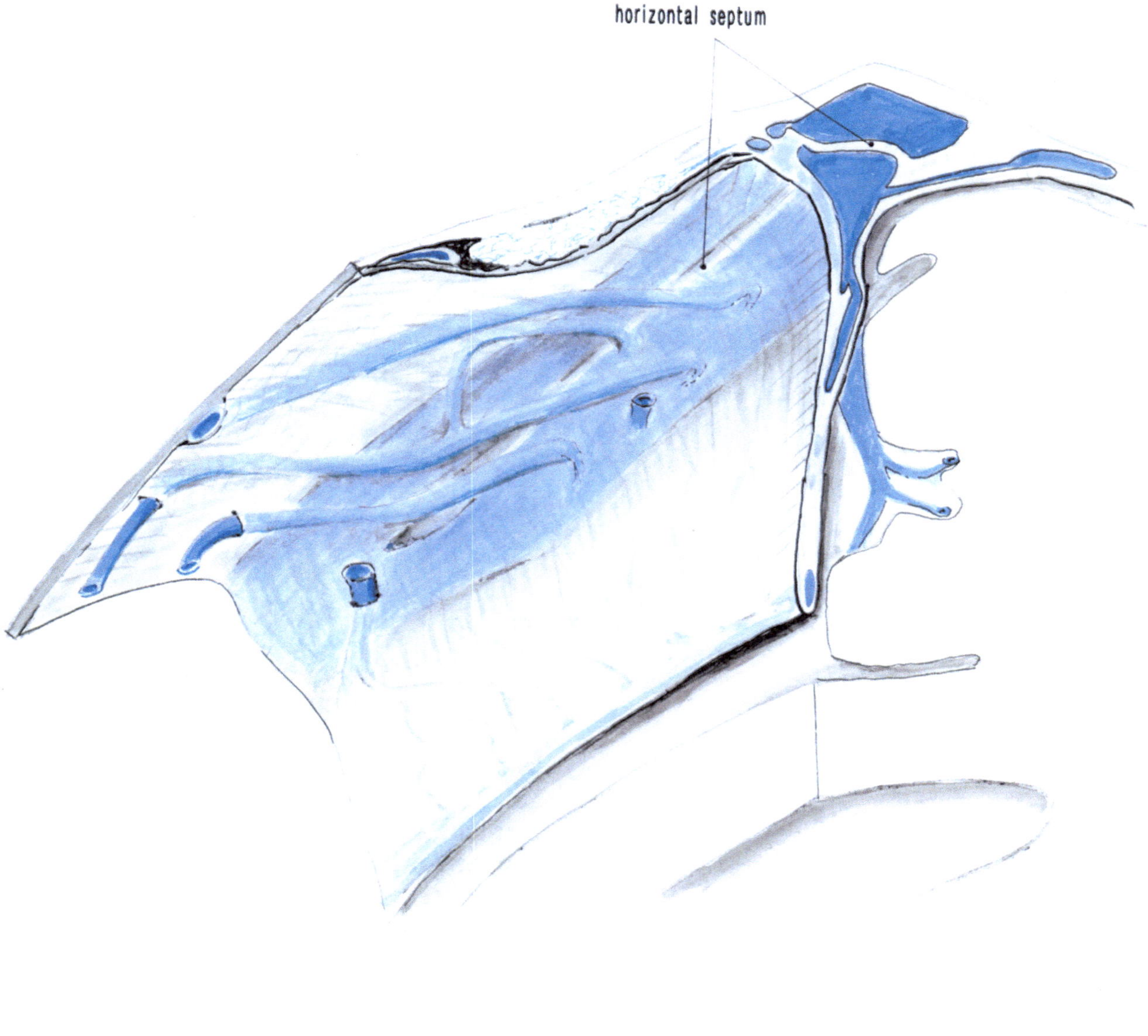

Fig. 34. **Atresia of the rostral segment of Sinus sagittalis sup., and widening of the fork-like configurated bridging veins** (according to Browder and Kaplan, 1976, considerably modified)

Surgical aspects:
Danger for brain damage after ligation of bridging veins (yellow colored)

FIG. 34

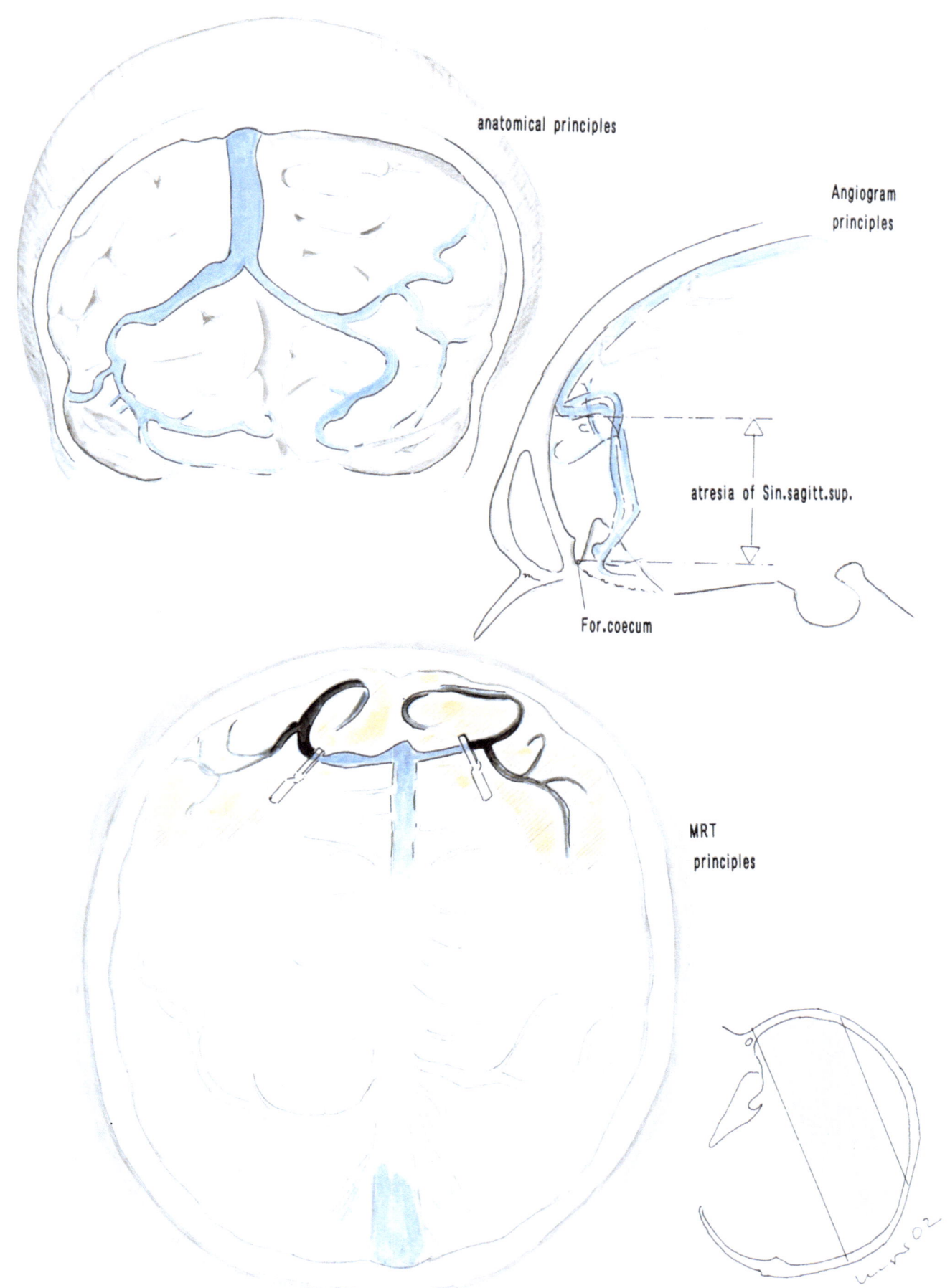

anatomical principles

Angiogram
principles

atresia of Sin.sagitt.sup.

For.coecum

MRT
principles

Fig. 35. **Duplication of the sinus by a sagittal septum. Bridging veins crossing the midline**

Microneurosurgical aspects

Clipping of small compartments dangerous, if bloodstream is accelerated

Abbreviations

a	wide compartment of the sinus
b	small compartment
c	wide briding veins entering b
d to *g*	bridging veins yellow colored: areas endangered by interruption of veins or sinus

FIG. 35

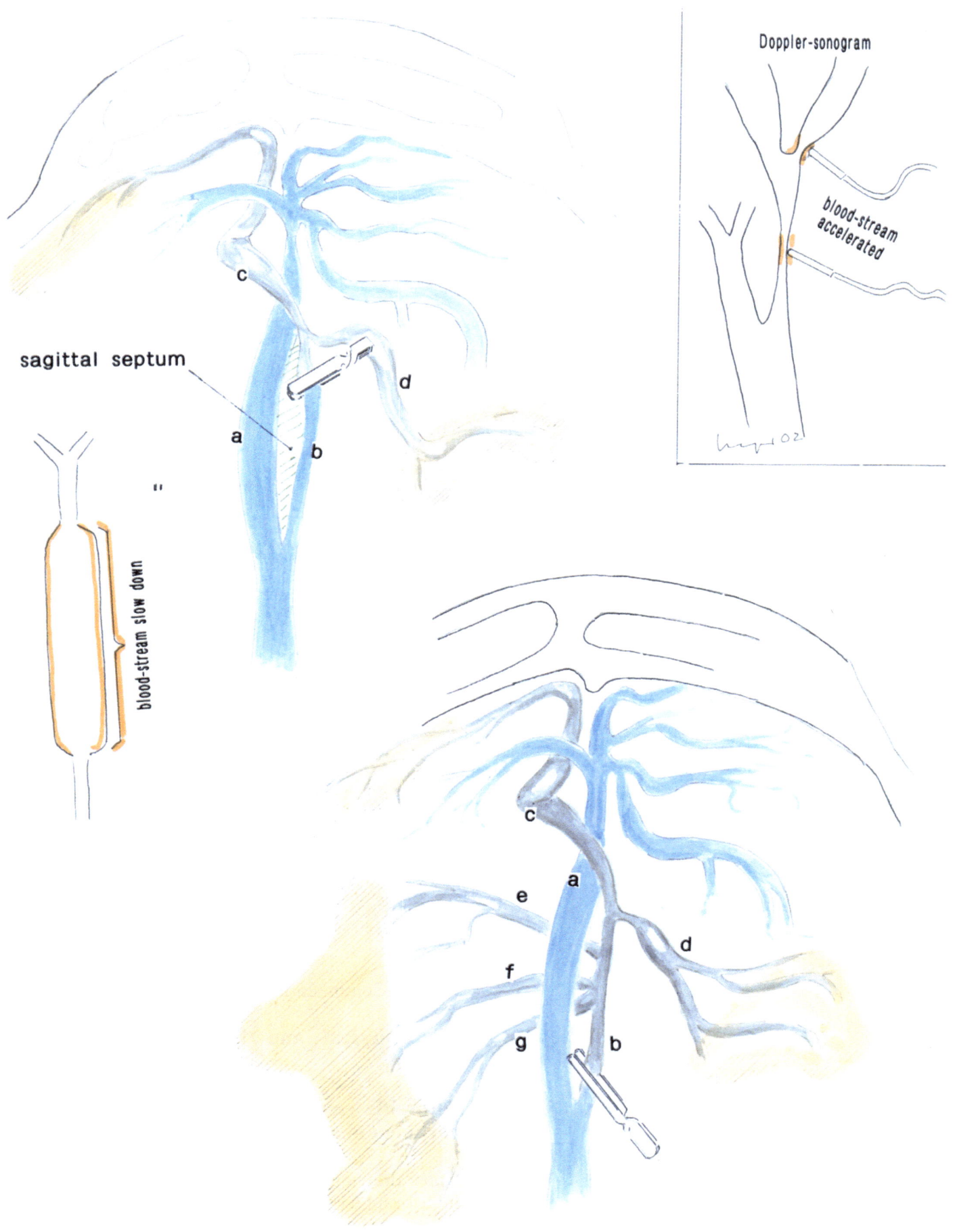

Doppler-sonogram

blood-stream accelerated

sagittal septum

blood-stream slow down

a

b

c

d

a

b

c

d

e

f

g

Fig. 36. Continuation of Fig. Fr 35

Microsurgical and Dopplersonographic aspects:

Favorable conditions for ligation of a wide compartment of a duplicated sinus

Upper drawing: Sagittal septum
Inferior drawing: Horizontal septum

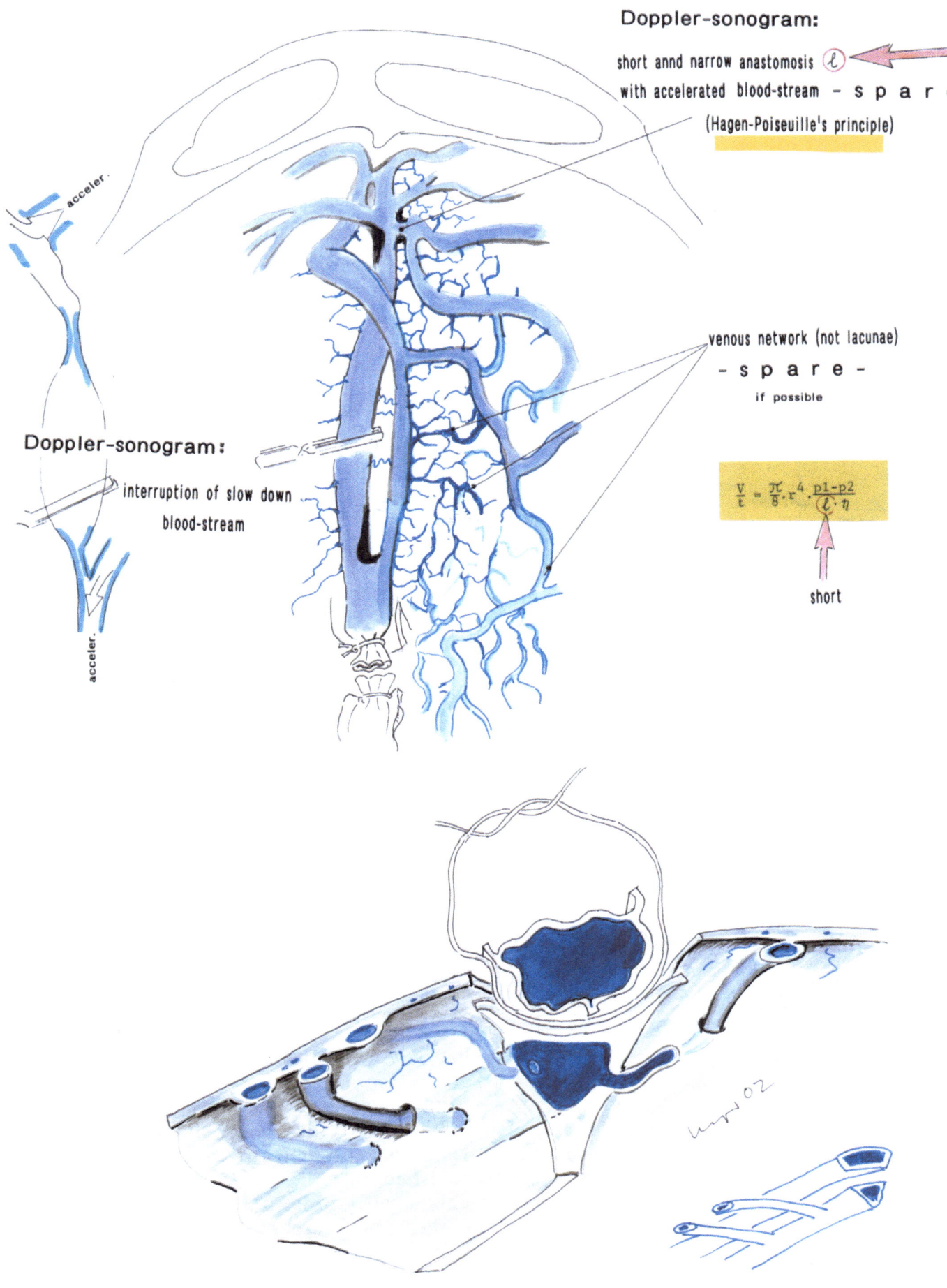

Doppler-sonogram:

short annd narrow anastomosis ⓛ ⟵

with accelerated blood-stream – s p a r e –

(Hagen-Poiseuille's principle)

acceler.

Doppler-sonogram:

interruption of slow down
blood-stream

acceler.

venous network (not lacunae)
– s p a r e –
if possible

$$\frac{V}{t} = \frac{\pi}{8} \cdot r^4 \cdot \frac{p1-p2}{\ell \cdot \eta}$$

short

Fig. 37. Anterior diameter of Falx, variants

Microneurosurgical aspects:
Falx wide – favorable for subdural route along
Falx, e.g. approaches to Area subcallosa and 3rd
ventricle
Falx small – unfavorable, because the route is
longer and more problematic when crossing
CSF-space

FIG. 37

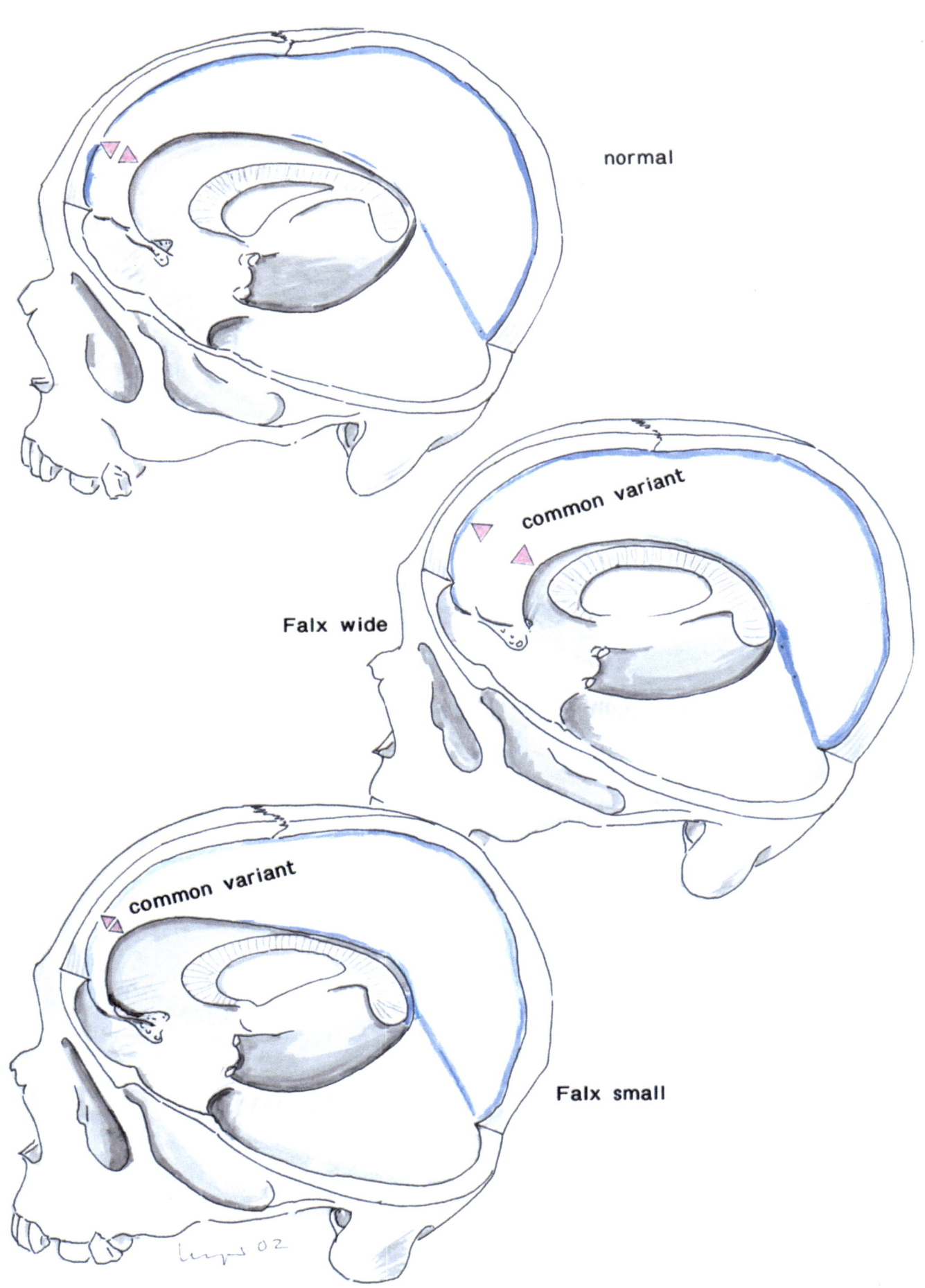

normal

Falx wide

common variant

common variant

Falx small

Fig. 38.

Microsurgical aspects:
Subdural midline approach is interrupted by
adhesions of both hemispheres in the areas of
the gaps

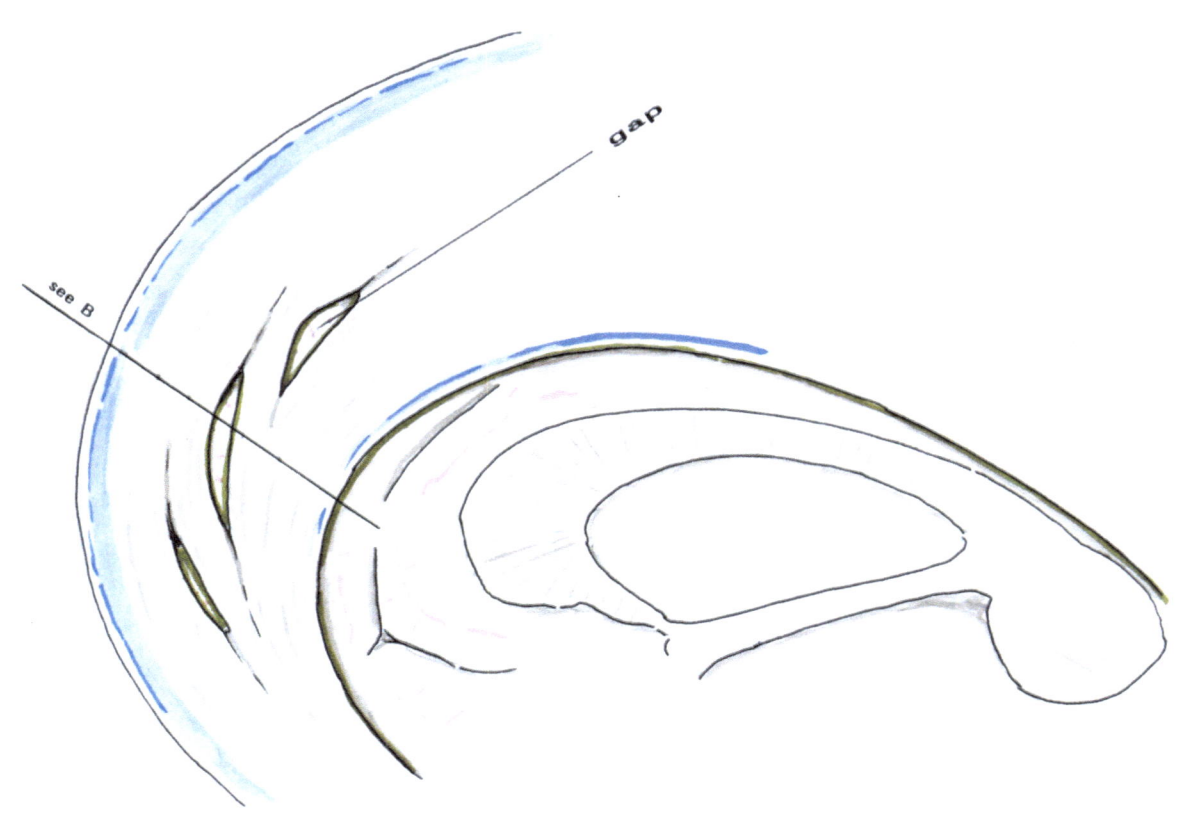

gap

see B

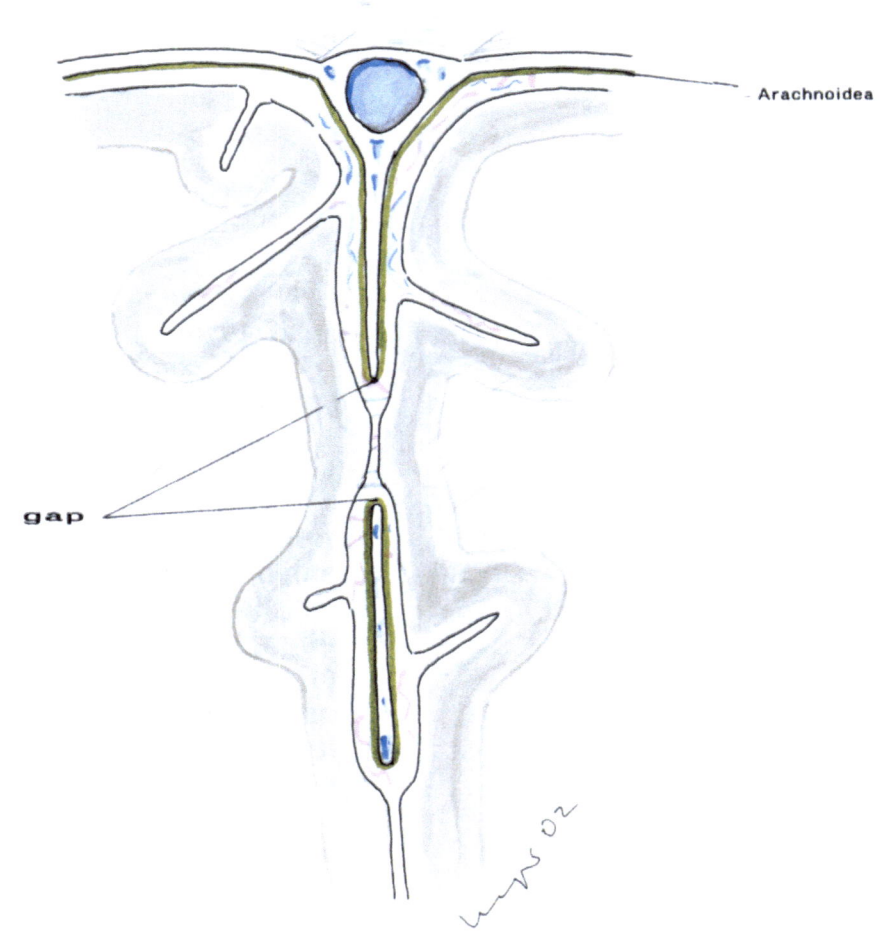

Arachnoidea

gap

Fig. 39. **A2 and distal branches of it.** According to Marino (1976)

Normal findings

Bilateral ramifications of both A2

Clinical aspects:
- **angiogram: Contralateral A1 and A.communicans ant. may be not filled with contrast medium, but distal contralateral branches close to the dorsal surface of the hemisphere are presented. This is a typical angiographic finding**
- **microneurosurgery: Interruption of the A2 on one side, e.g. during hemispherectomy or tumor surgery, may be followed by a fatal contralateral brain damage**

FIG. 39

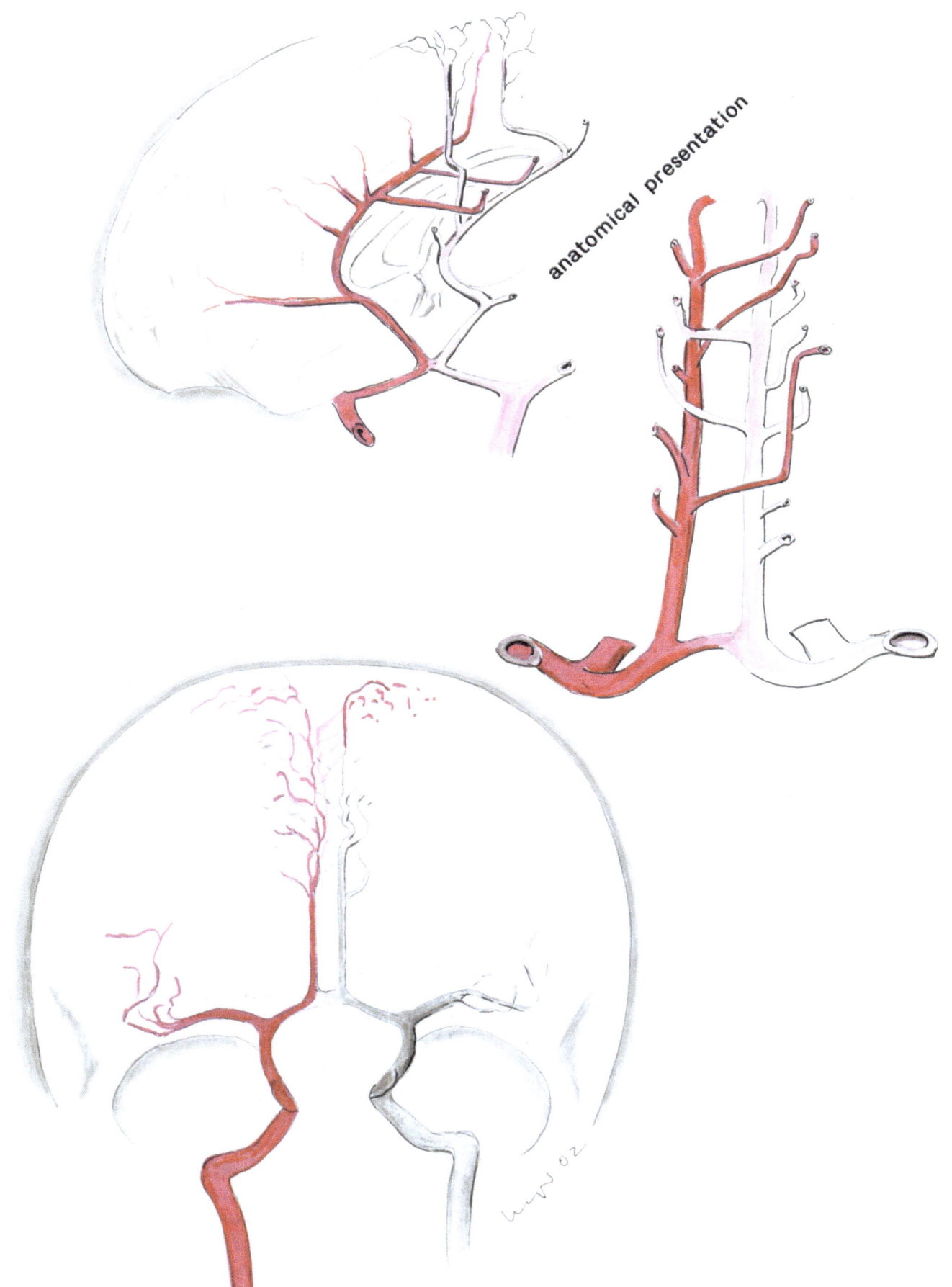

anatomical presentation

angiographic presentation

Fig. 40. **Bihemispheric branches of both Aa.cerebri ant. Different types, according to Marino** (1976) Marino's drawings modified. The „azygos type" called midline artery is identic with A. corporis callosi mediana, which was described by Stephens and Stilwell (1969), pp 62, 63, 68, and 69

Clinical aspects:
Preoperative diagnostic procedures (angiogram, MRT-angiogram), interventional embolizations and microsurgical procedures for aneurysms and angiomas of A. communicans ant., and A2, and its branches, and danger for complications

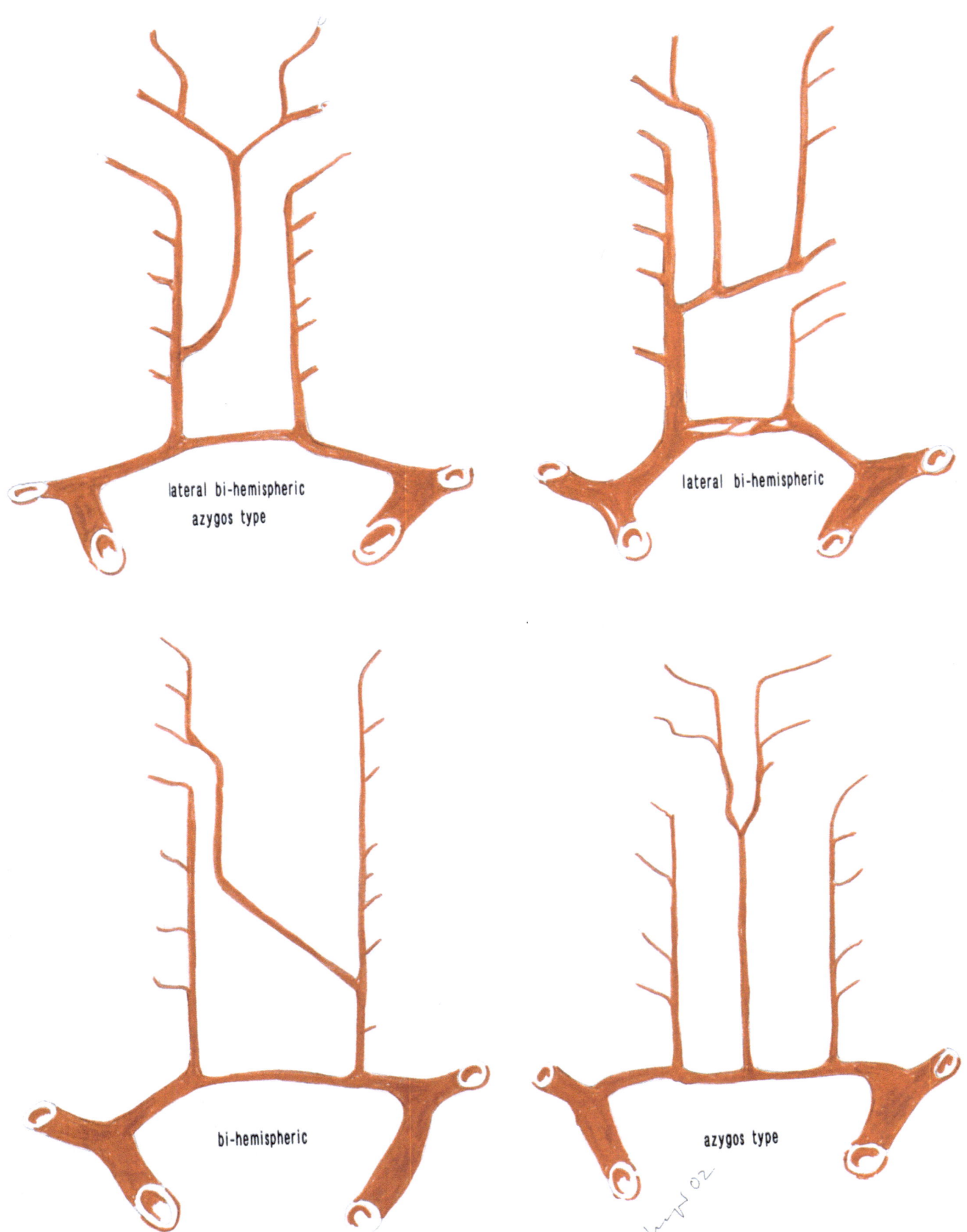

lateral bi-hemispheric
azygos type

lateral bi-hemispheric

bi-hemispheric

azygos type

Figs. 41 to 47. **Variants of the frontal lobes,** some clinical aspects

Fig. 41. **Rostrum corporis callosi thickened, common variant**

Note the close relationship of Rostrum with Commissura ant. and Fornices of normal finding and of the variant.

Aspects for epilepsy surgery:
Anterior commissurotomy must be stopped rostrally from Columnae fornicis and Commissura ant. Damage of these structures can be avoided by using neuronavigatory landmarks (Seeger and Zentner, 2002)

FIG. 41

normal finding

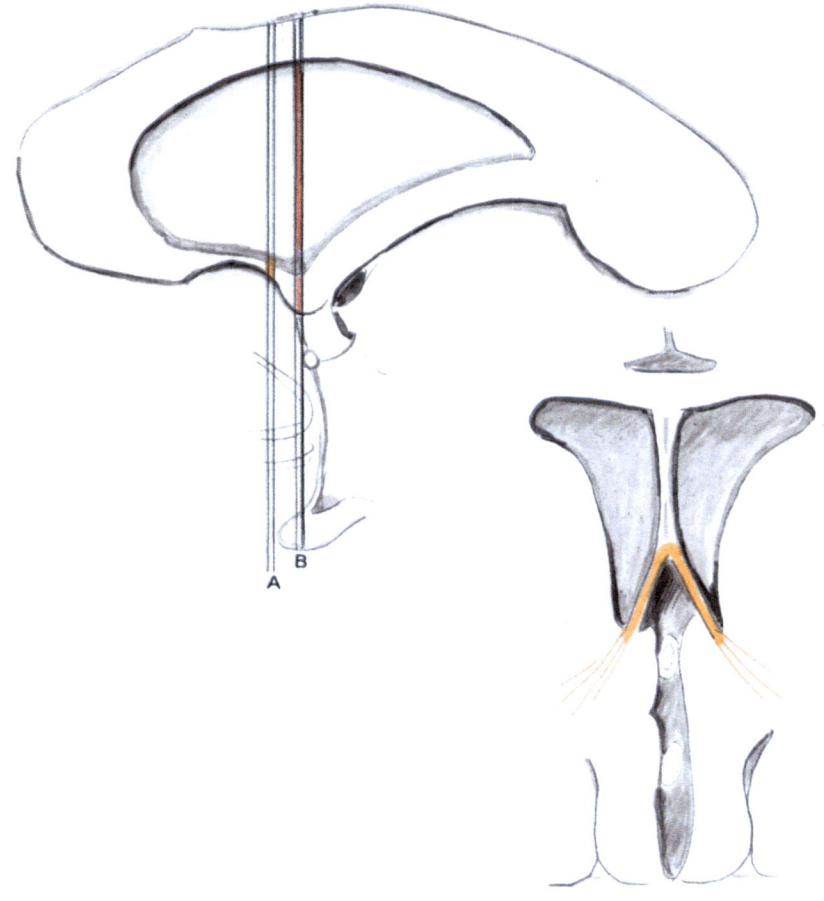

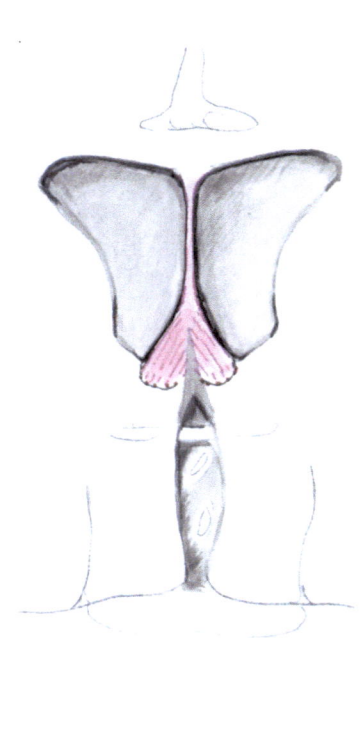

variant of Rostrum

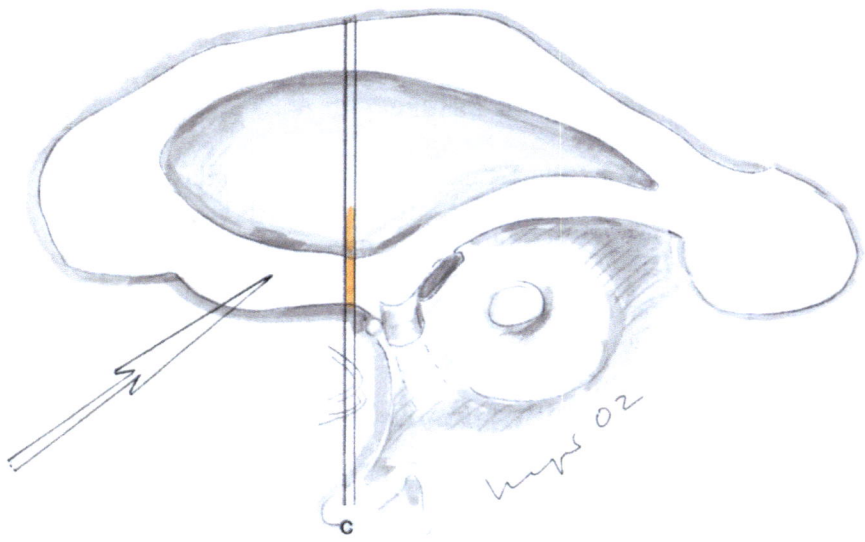

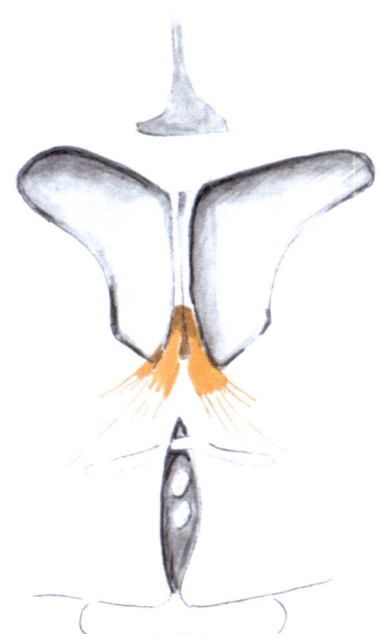

Fornix/Sept.pelluc.

Rostrum corporis callosi

Fig. 42. **Anterior insular area = anterior tip of the sylvian fissure**

A   Plastic casb (elastic) of a cadaver brain
B   Sketch according to a MRT
C   Sketch for comparison with A

Short distance measurements of the anterior tip of the sylvian fissure and its vessels and the anterior area of Cornu anterius
Variability: Approximately 5 to 10 mm, especially in elderly people

Clinical aspects:
After extirpation of the thin-layered brain tissue between Insula and Cornu ant. (by tumor resection, e.g.), the sylvian cistern and the ventricular system built one common CSF-space. Hemorrhagic sugillation of the numerous fine vessels of the cistern with following intraventricular bleeding is possible

FIG. 42

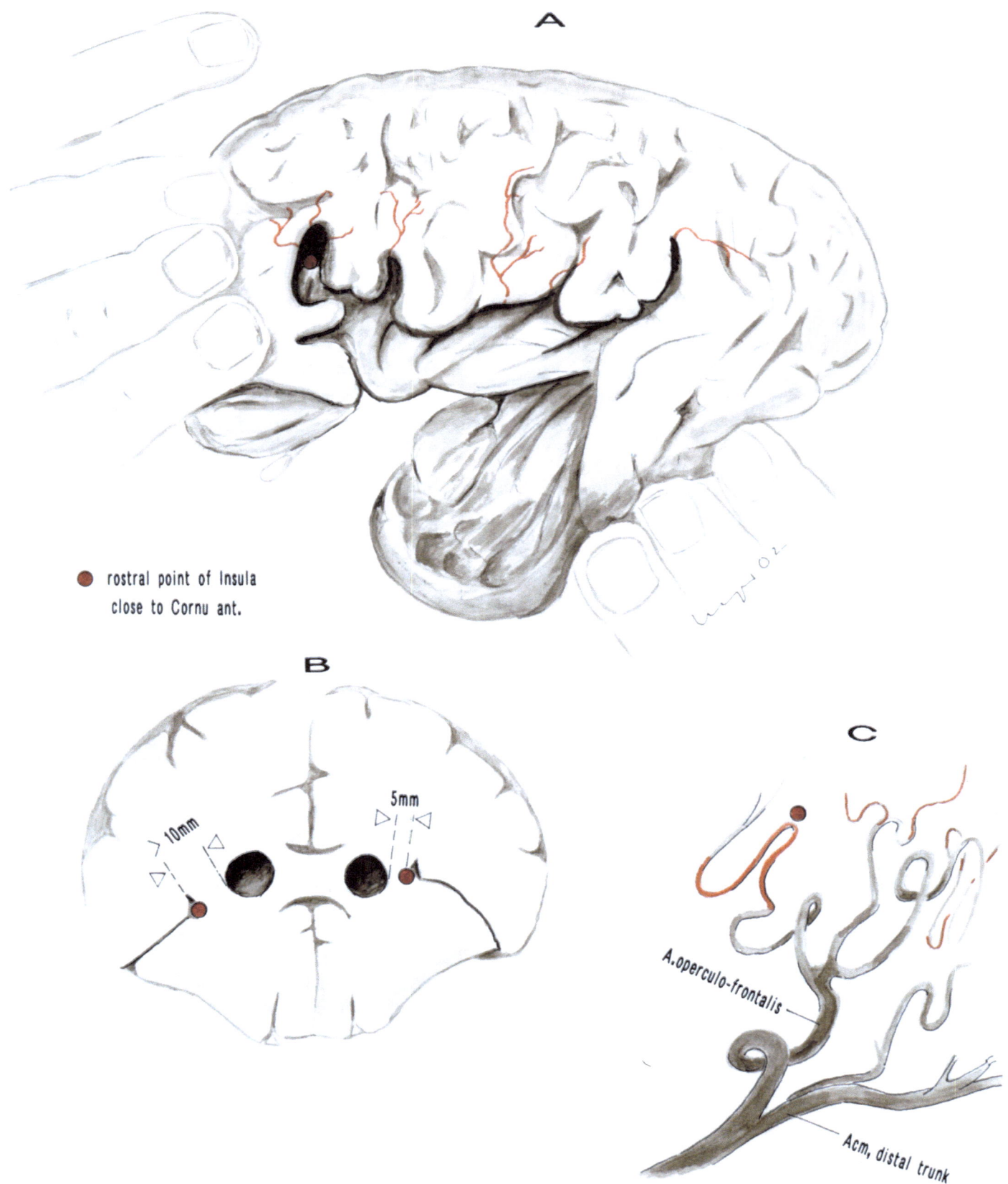

A

● rostral point of Insula
close to Cornu ant.

B

>10mm

5mm

C

A.operculo-frontalis

Acm, distal trunk

Fig. 43. **Motor cortex, common variant**

Bridging gyri interrupt the courses of Sulcus centralis and/or Sulcus praecentralis

These variants may mask the configuration of Gyrus praecentralis

Clinical aspects:
Errors in defining the motor cortex are to be excluded by fMRT, MEP, and evoked potentials

Abbreviations

a   Sulcus centralis
a'   interruption of a
b   Sulcus praecentralis
b'   interruption of b
c   Pars opercularis gyri praecentralis
       (base of motor gyrus). Note relationship with
       temporal pole
d   upper segment of Gyrus praecentralis
e   medial segment of Gyrus praecentralis

FIG. 43

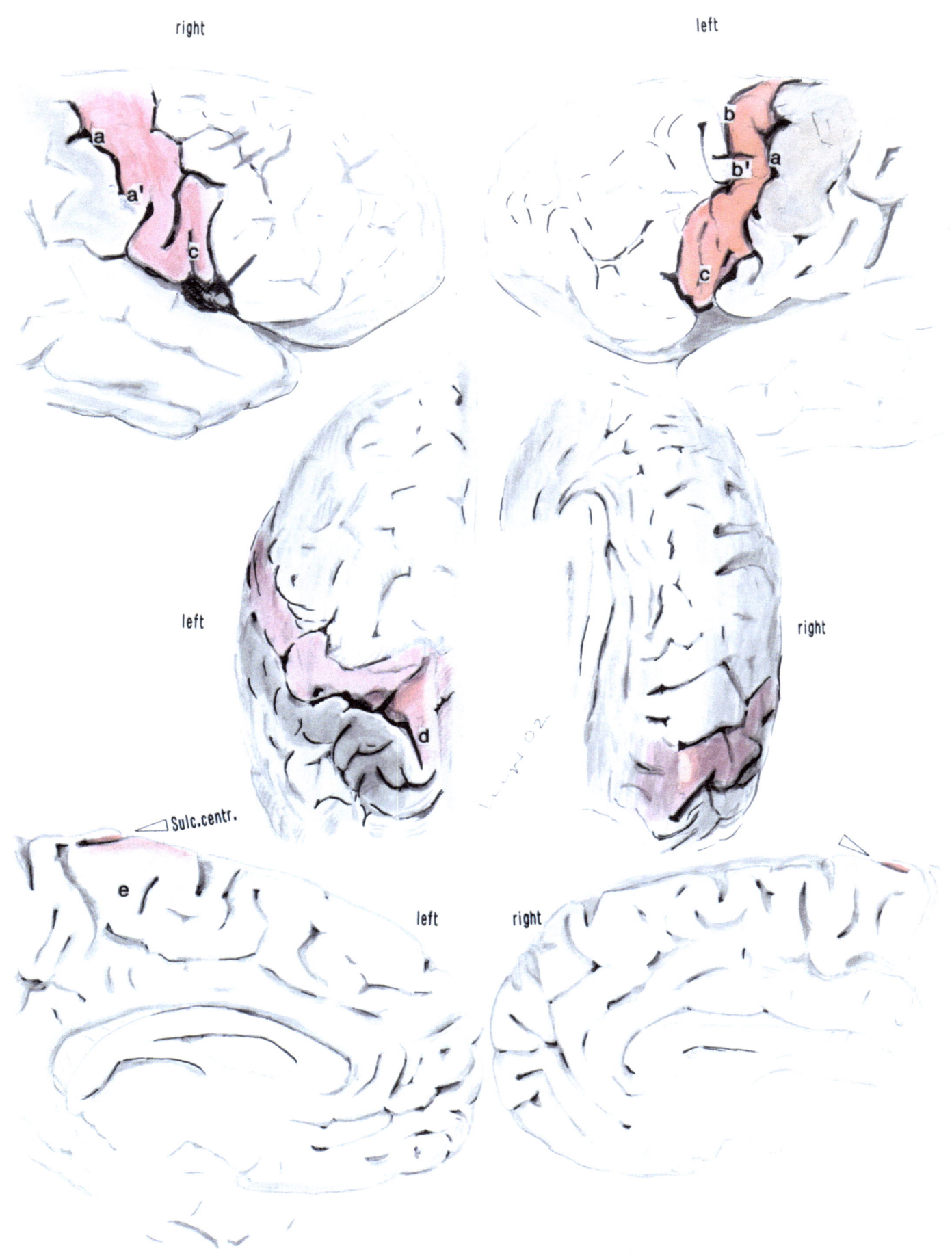

right

left

left

right

left

right

Sulc.centr.

Fig. 44. **Bihemispheric asymmetry of the motor cortex is a normal finding, no variant**

Corresponding points of the left hemisphere are located more in an occipital direction than the corresponding points of the right hemisphere (red colored arrows). Cause is unknown (motor speech area –ms- ?)

Clinical aspects:
See Fig. 43

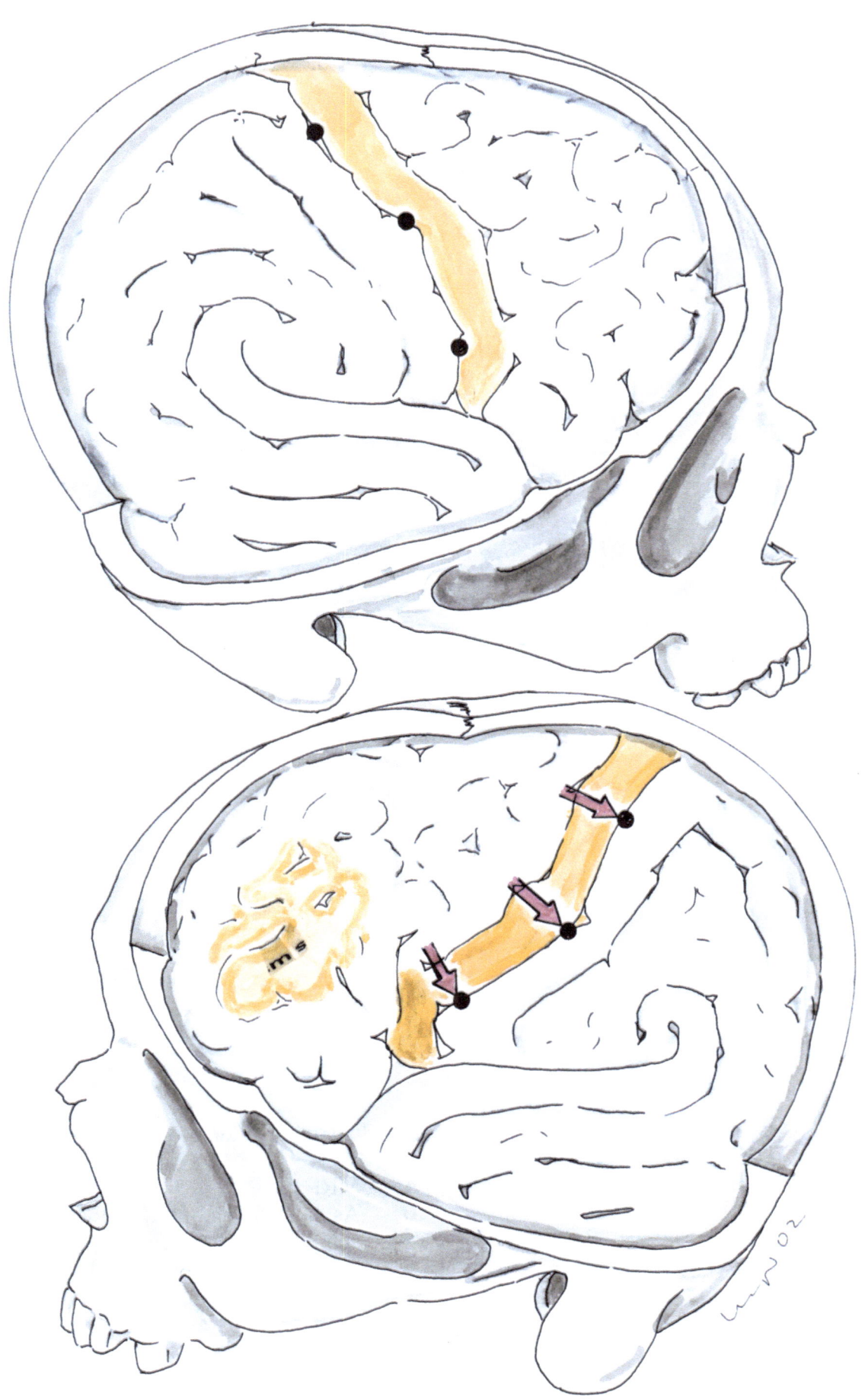

Fig. 45. Continuation of Fig. 44

A MRI
B As A, mapping of the motor area for compari-
       son with the right side. These findings are
       nearly congruent with the high resolution ma-
       gnetencephalogram (MEG)

Clinical aspects:
See Fig. 44

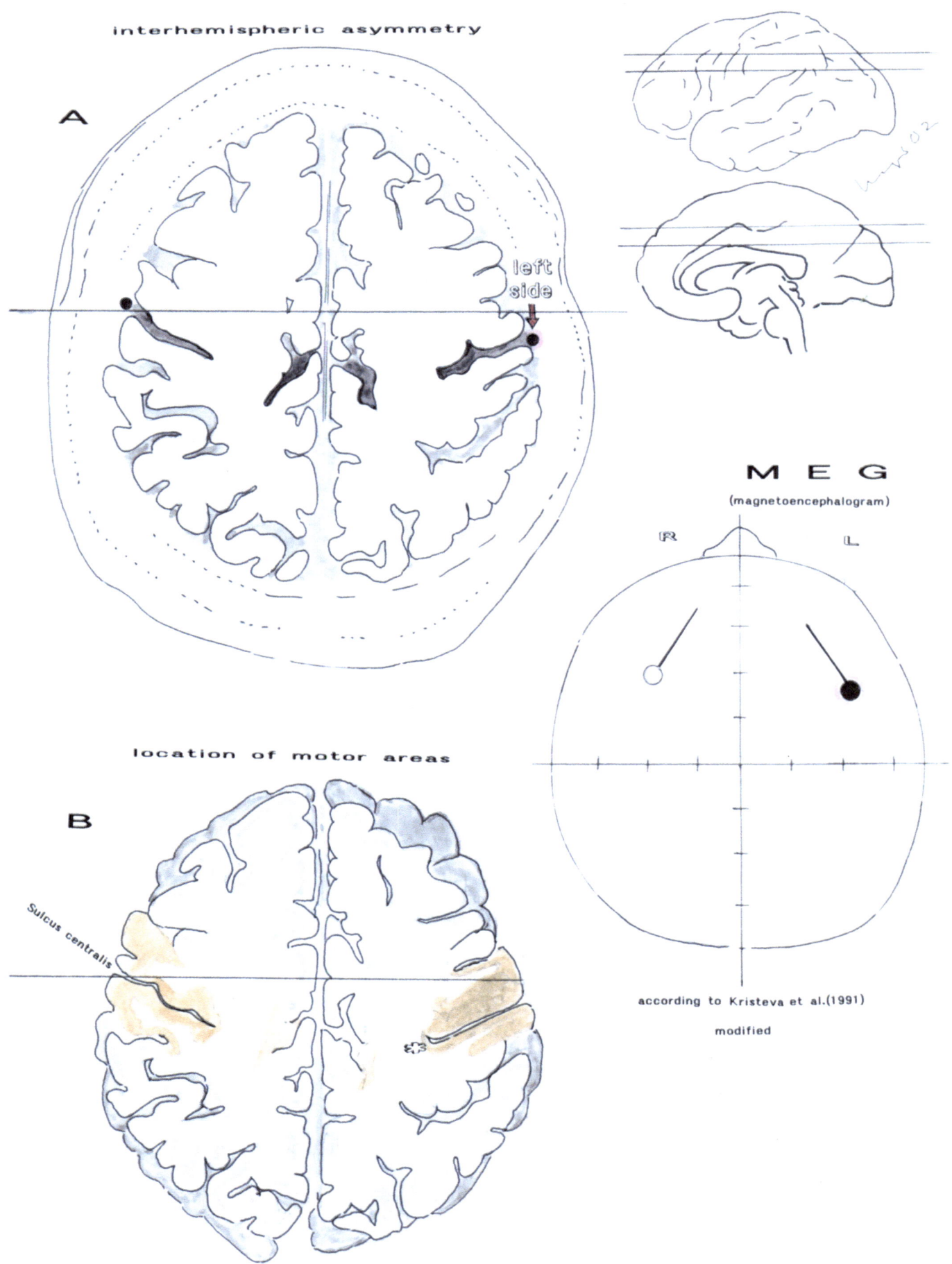

interhemispheric asymmetry

A

left side

location of motor areas

B

Sulcus centralis

M E G

(magnetoencephalogram)

R   L

according to Kristeva et al.(1991)

modified

Fig. 46. Continuation of Fig. 45

The modern findings of asymmetry of the motor area (area of Gyrus praecentralis) confirms with historical presentations

FIG. 46

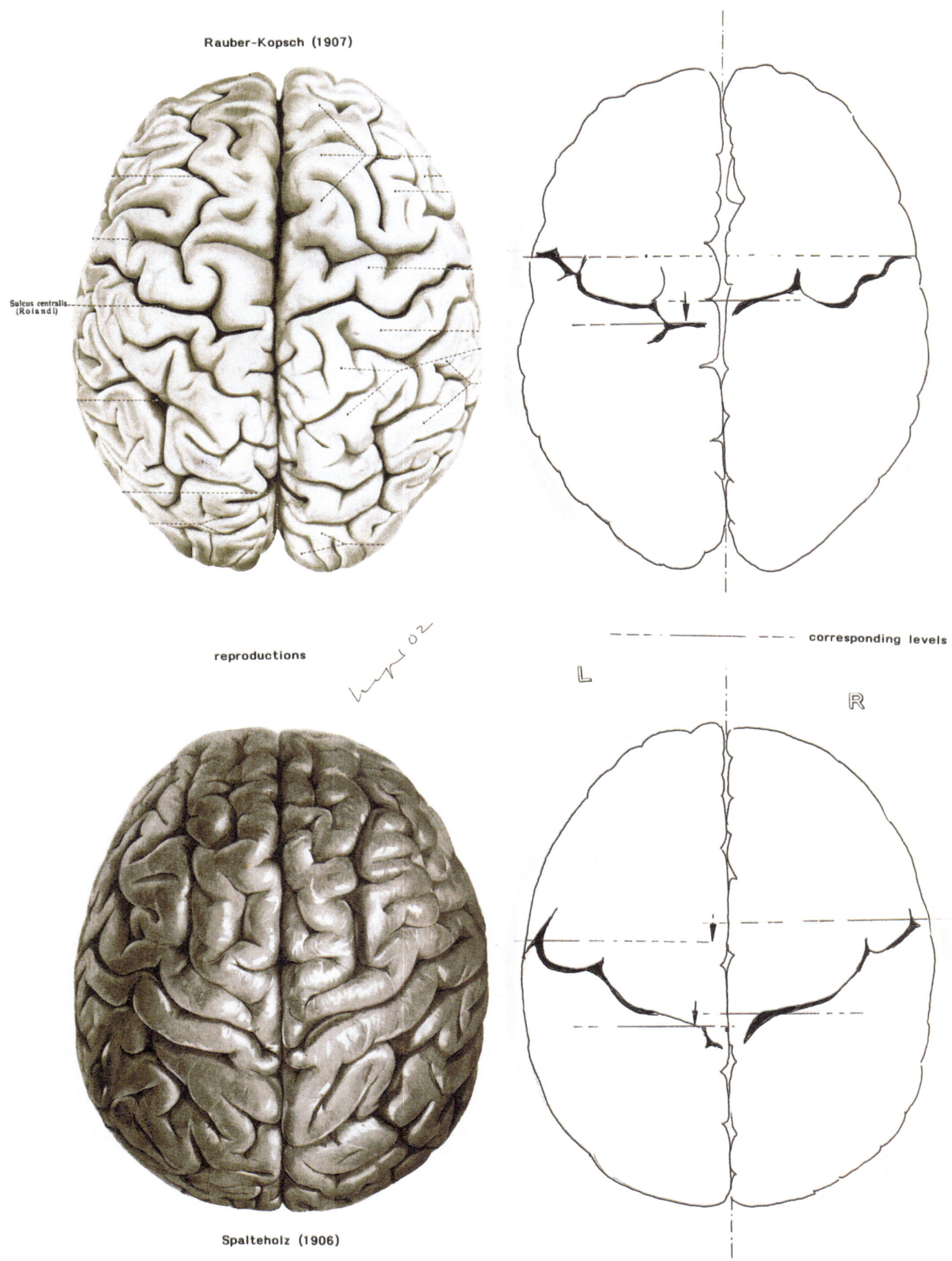

Rauber-Kopsch (1907)

Sulcus centralis
(Rolandi)

reproductions

corresponding levels

L

R

Spalteholz (1906)

Fig. 47. No significant difference between right and left hemisphere in apes and monkeys

Copy of Bischoff's presentation of a human brain from 1868 added

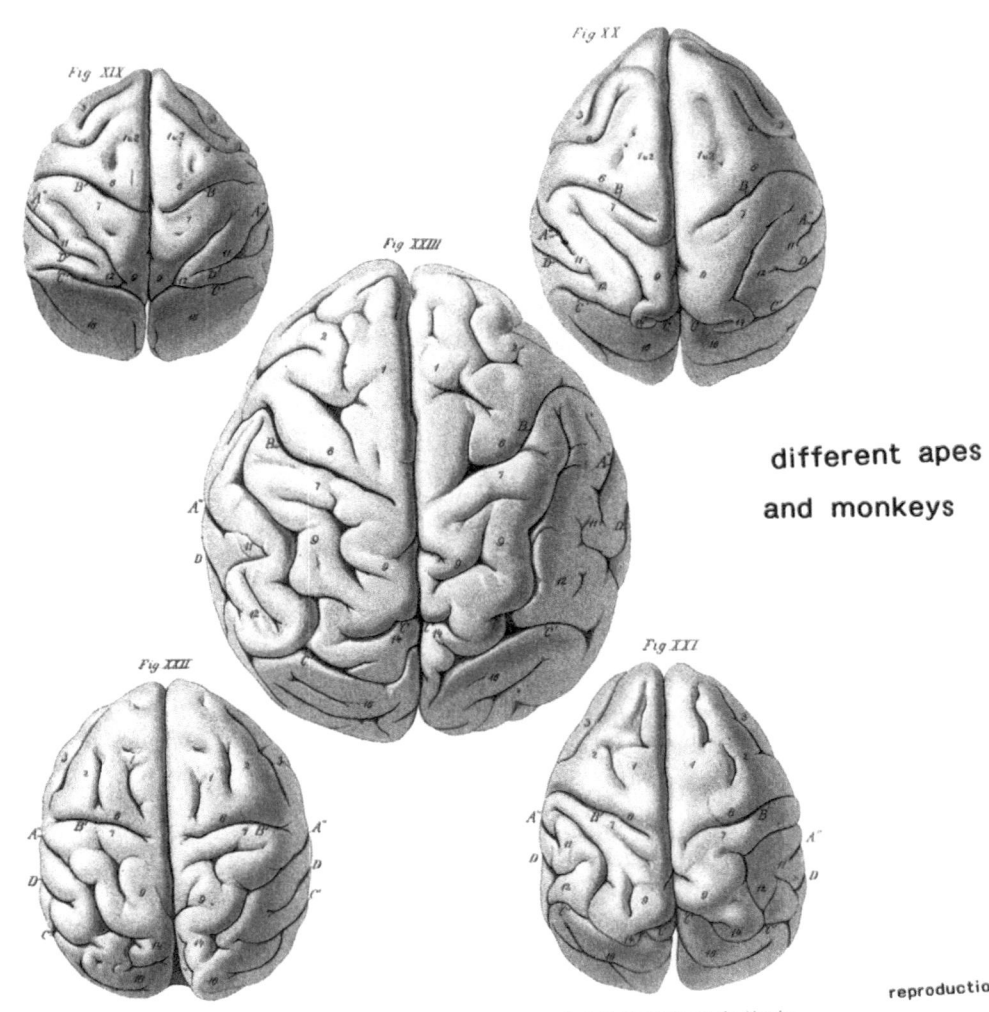

copy

2001

human

Die

Grosshirnwindungen des Menschen

mit Berücksichtigung ihrer Entwicklung bei dem Fötus
und ihrer Anordnung bei den Affen.

Neu untersucht und beschrieben

von

Dr. Th. L. W. Bischoff,

Professor der Anatomie und Physiologie, ordentlichem und correspondirendem Mitgliede der
Akademien der Wissenschaften zu München, Berlin, Wien, St. Petersburg und der Royal Society
zu London etc. etc.

Mit sieben Tafeln.

Aus den Abhandlungen der k. bayer. Akademie der W. II. Cl. X. Bd. II. Abth.

München 1868.

Verlag der k. Akademie,
in Commission bei G. Franz.
Akademische Buchdruckerei von F. Straub.

Taf VI

Fig XX

Fig XIX

Fig XXIII

different apes

and monkeys

Fig XXII

Fig XXI

reproduction

Math. phys. Cl. X. II.

A. Mermann zu München

Bischoff Hirnwindungen des Menschen

Chapter 2
Variants of the fronto – temporo – basal (pterional) areas
(Figs. 48 to 85)

Fig. 48. Pterional microsurgical approaches

Schematic presentation

* Bony landmarks

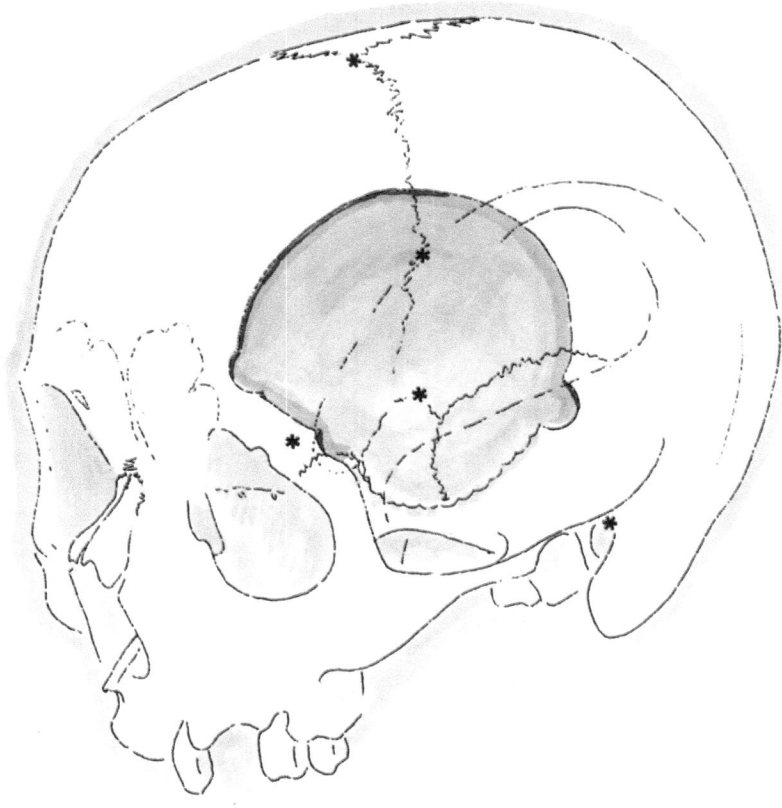

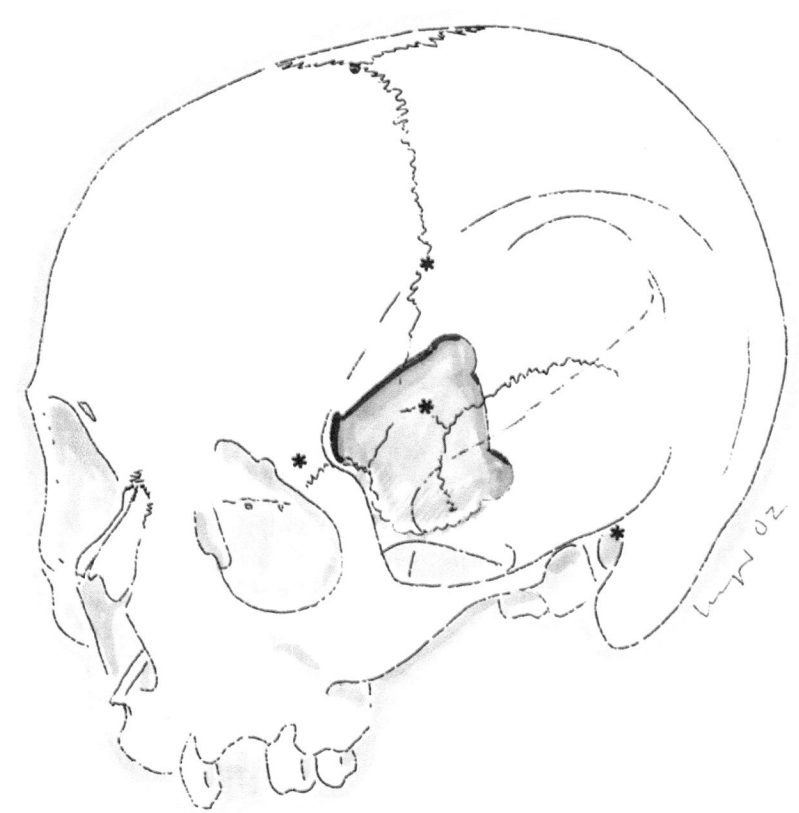

Fig. 49. **Gap of Ala major = fenestration of Facies temp. and Facies orbitalis of Ala major, rare variant**

A   Gap (arrow)
     * further variant: Exostosis of Processus frontalis of Os zygomaticum (rare)

B   **Surgical (microsurgical) aspects:**
     Danger for lesioning of intraorbital structures by drill, rongeur or craniotome, because Periosteum and Dura may be adherent

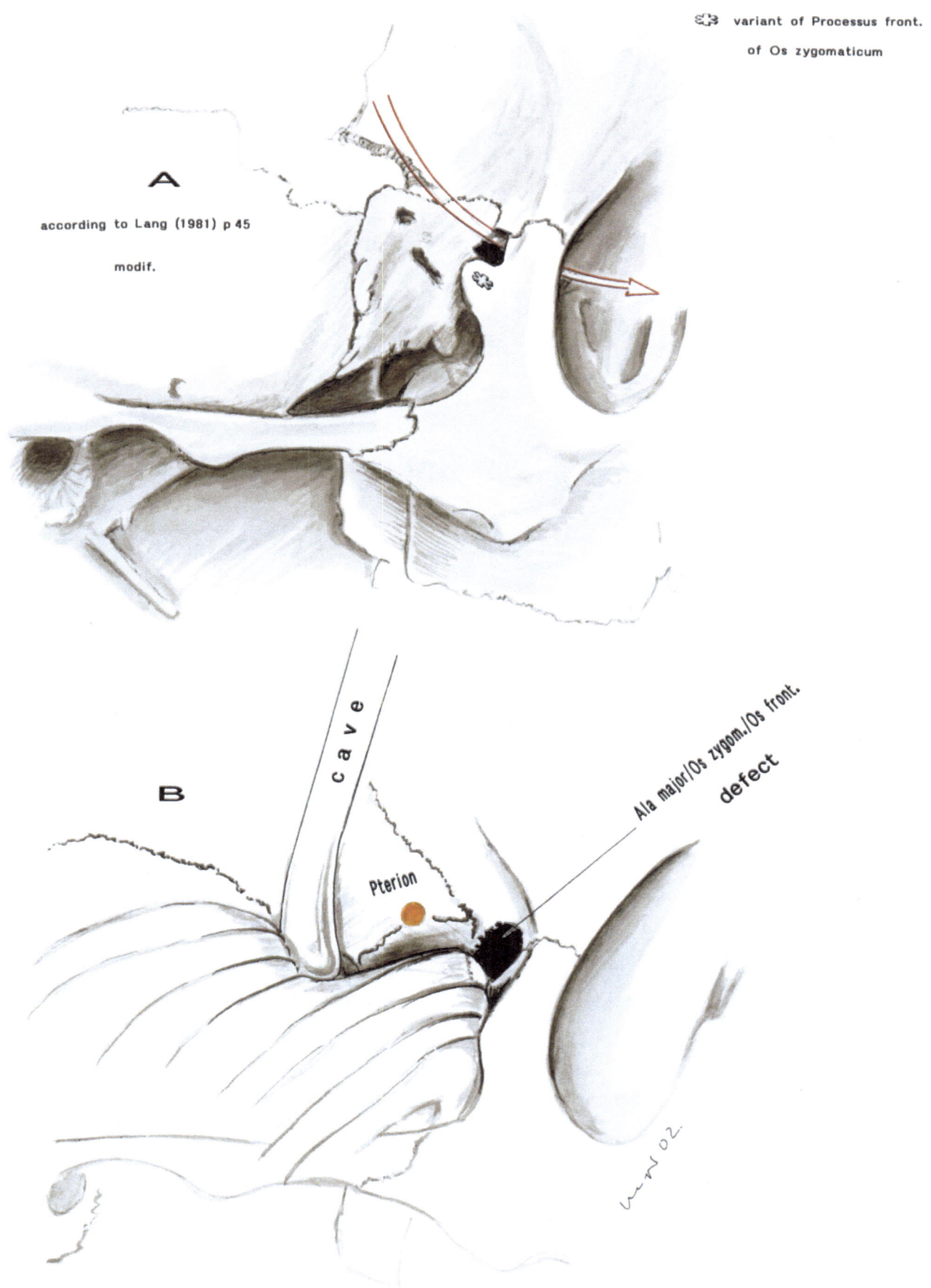

variant of Processus front. of Os zygomaticum

A

according to Lang (1981) p 45

modif.

B

cave

Ala major/Os zygom./Os front. defect

Pterion

Fig. 50. **Edge of Ala minor everted in a dorsal direction, rare variant**

A   Survey
B   Normal finding (left)
C   Variant, bed of Gyrus front. III

Clinical aspects:
This variant should be considered, if the fronto-temporal dura must be dissected from the bone, as it is necessary for Yasargil's pterional trepanation.

Note: A. cerebri media and its branches, as well as its aneurysms may be located close (approximately 5 – 10 mm) to the edge of Ala minor

FIG. 50

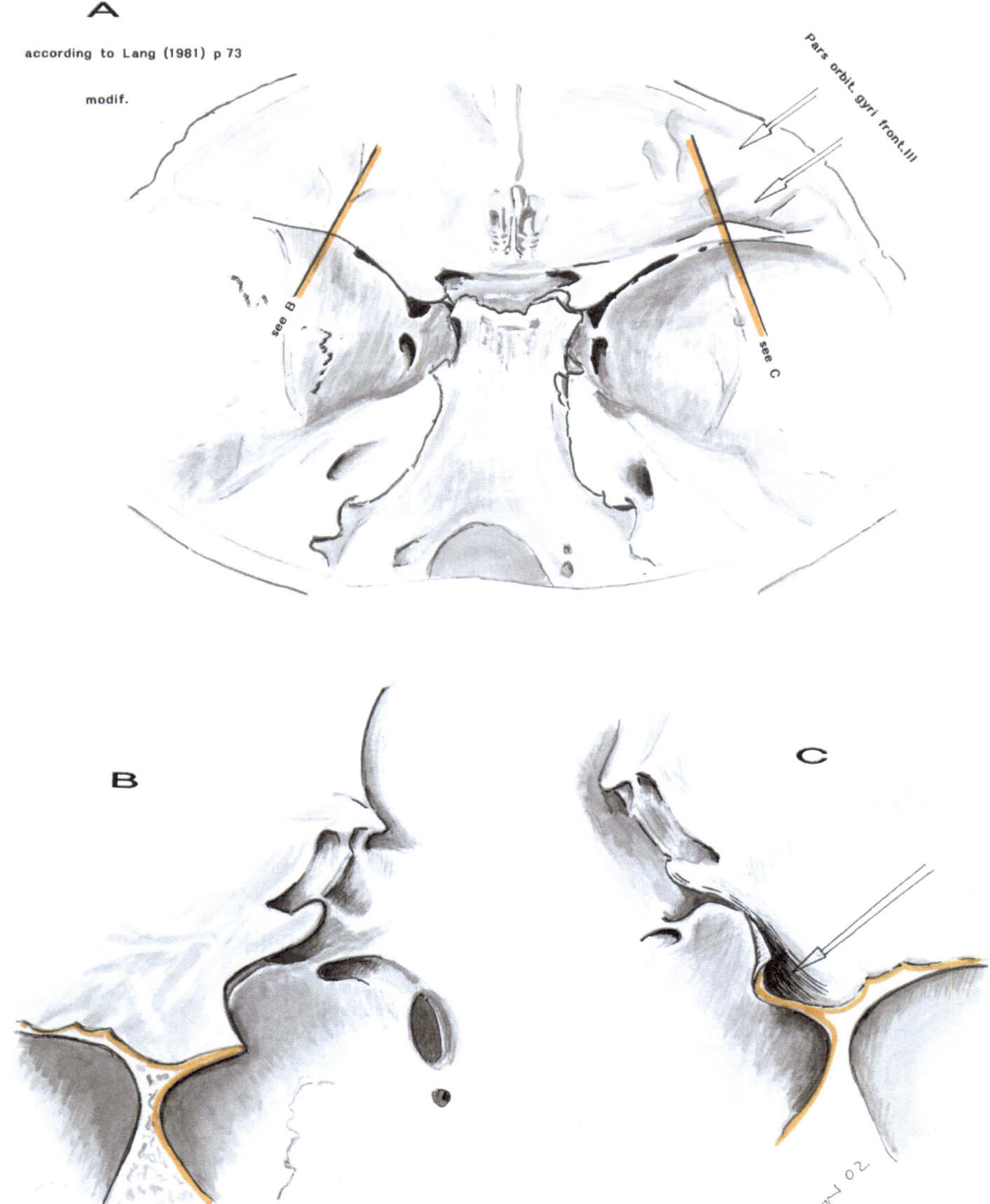

A

according to Lang (1981) p 73

modif.

Pars orbit. gyri front. III

see B

see C

B

C

Figs. 51 to 63

Variants of A. ophthalmica and its branches, and surrounding structures

For further variants see Chapter 1 (Figs. 23 to 28)

Fig. 51. Distal origin of A. centralis retinae from A. ophthalmica, common variant, combined with origin of A. ophthalmica from A. meningea media, rare variant (Schumacher et al, 1995)

A Normal finding, for comparison with B
B A. ophthalmica with A. centralis retinae are originating from A. meningea media. No or small connection with siphon of A. carotis int. (for literature see Lang, 1979 and 1981)

Clinical aspects:
• **Neuroradiological: Danger of blindness after interventional embolization (according to Goetz and Schumacher, 1995)**
• **Microneurosurgical: Blindness after ligation of the anterior branch of A. meningea media**

Abbreviations
a Fissura orbitalis sup.
b Anulus tendineus (Zinni)
c Can. opt.
d N. opt.
e A. ophthalmica, basal segment
f as e, dorsal segment
g A. nasoethmoidalis
h A. ethmoidalis post.
i A. ethmoidalis ant.
j as g
k A. supraorbitalis

FIG. 51

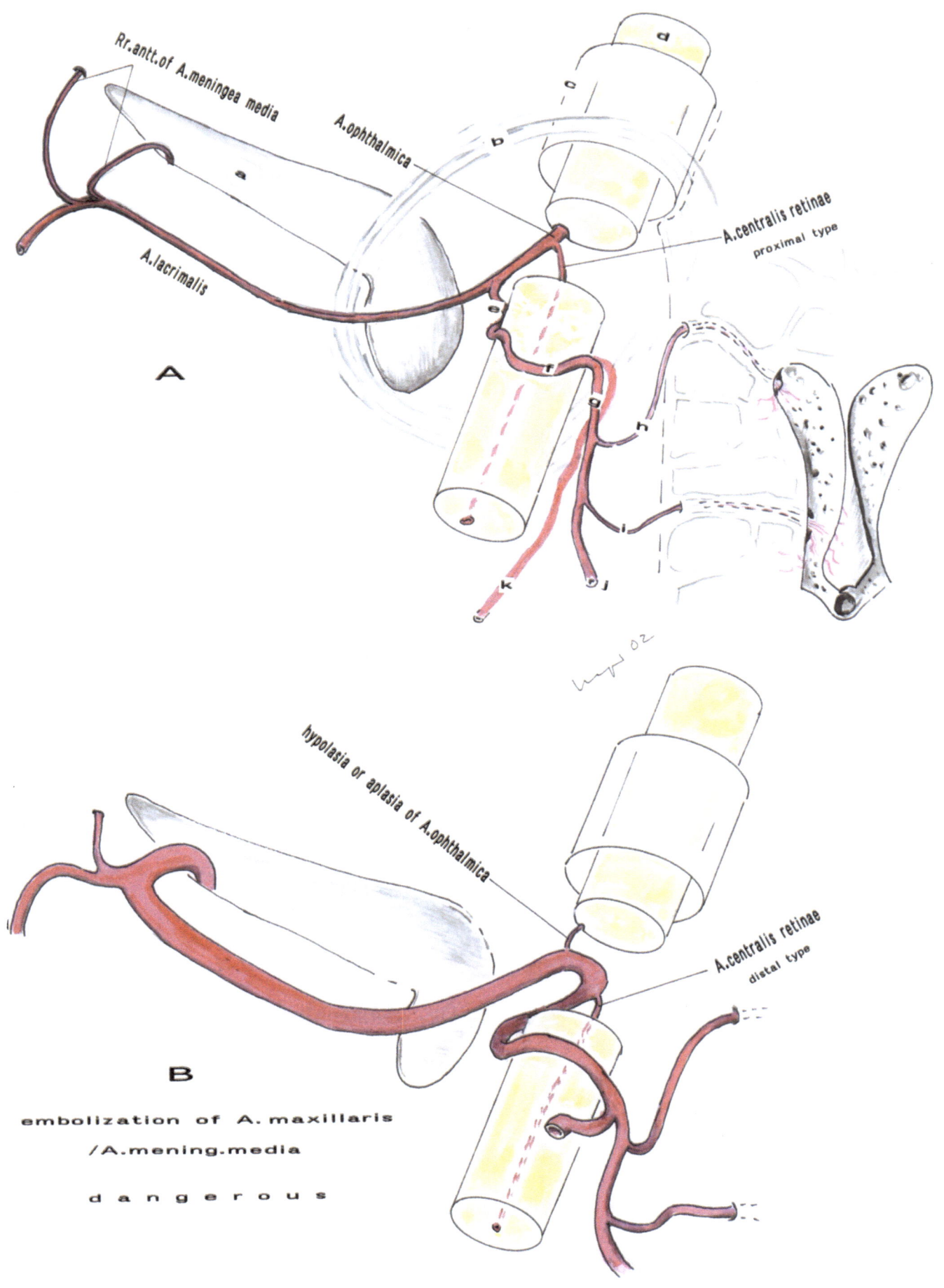

Rr.antt.of A.meningea media

A.ophthalmica

A.lacrimalis

A.centralis retinae
proximal type

a

b

c

d

e

f

g

h

i

j

k

A

hypolasia or aplasia of A.ophthalmica

A.centralis retinae
distal type

B

embolization of A.maxillaris
/A.mening.media

d a n g e r o u s

Fig. 52. **A. meningea media, normal finding**

Skull dissection and sectional enlargement from it (from author's collection)
Rami meningei (≙ sulci meningei and bony penetration points) added. Pterional trepanation and small sections of the dura added. Veins added.

A constant branch connects A. meningea media with A. lacrimalis. Other branches return to the Dura -*- (Lasjaunias, Berenstein, 1987)

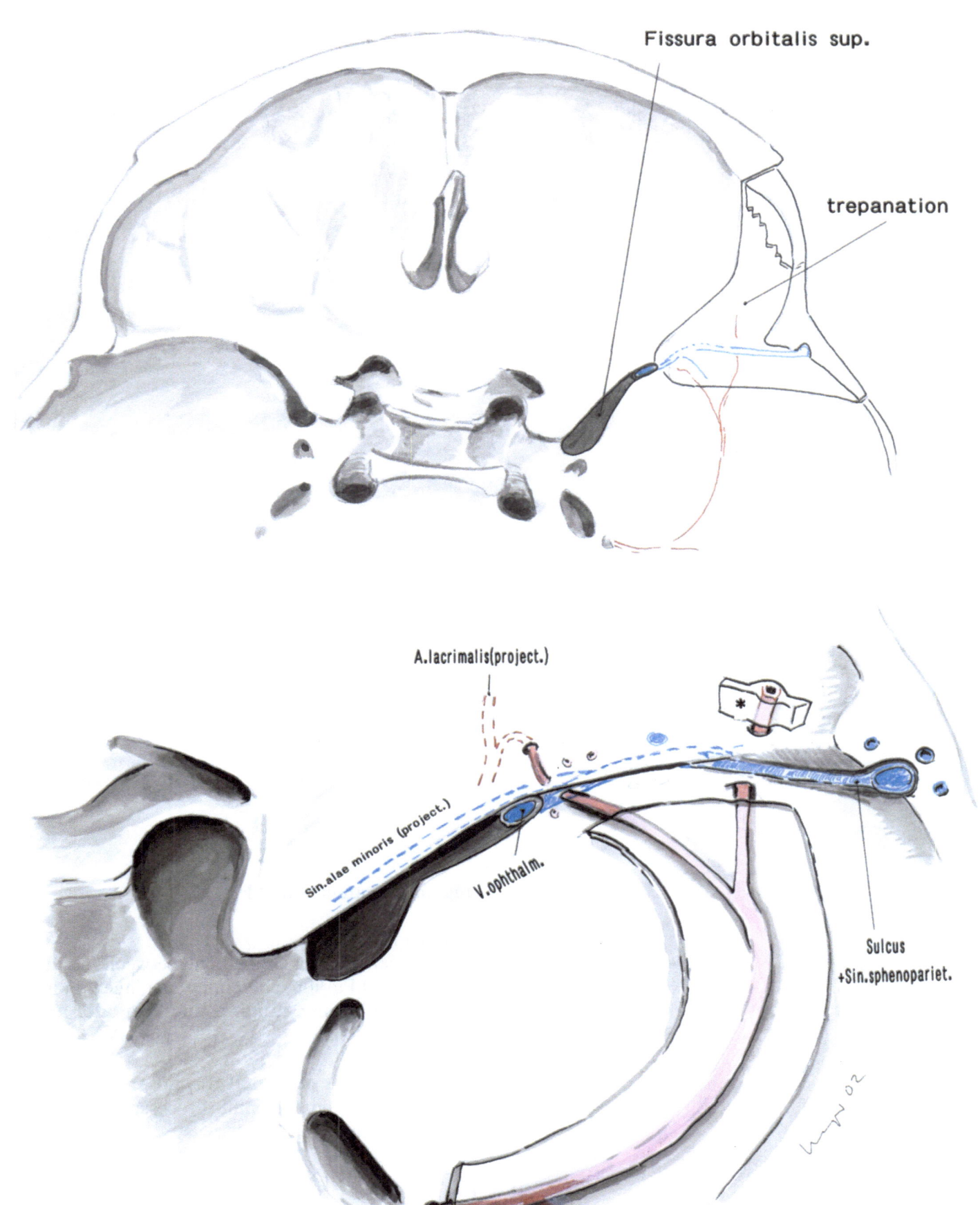

Fissura orbitalis sup.

trepanation

A.lacrimalis(project.)

Sin.alae minoris (project.)

V.ophthalm.

Sulcus
+Sin.sphenopariet.

For.spinosum

Fig. 53. Continuation of Fig. 52

Normal findings:
Typical penetrations of the bone between Fissura orbit. sup. and Pterion, containing meningeal branches (arteries and veins), **number and diameter are variable***

Landmarks for pterional trepanation
1 frontozygomatic point (Processus zygomaticus of Squama frontalis)
2 Pterion
3 Lateral edge of Fissura orbitalis sup.

---

* from the author's skull collection

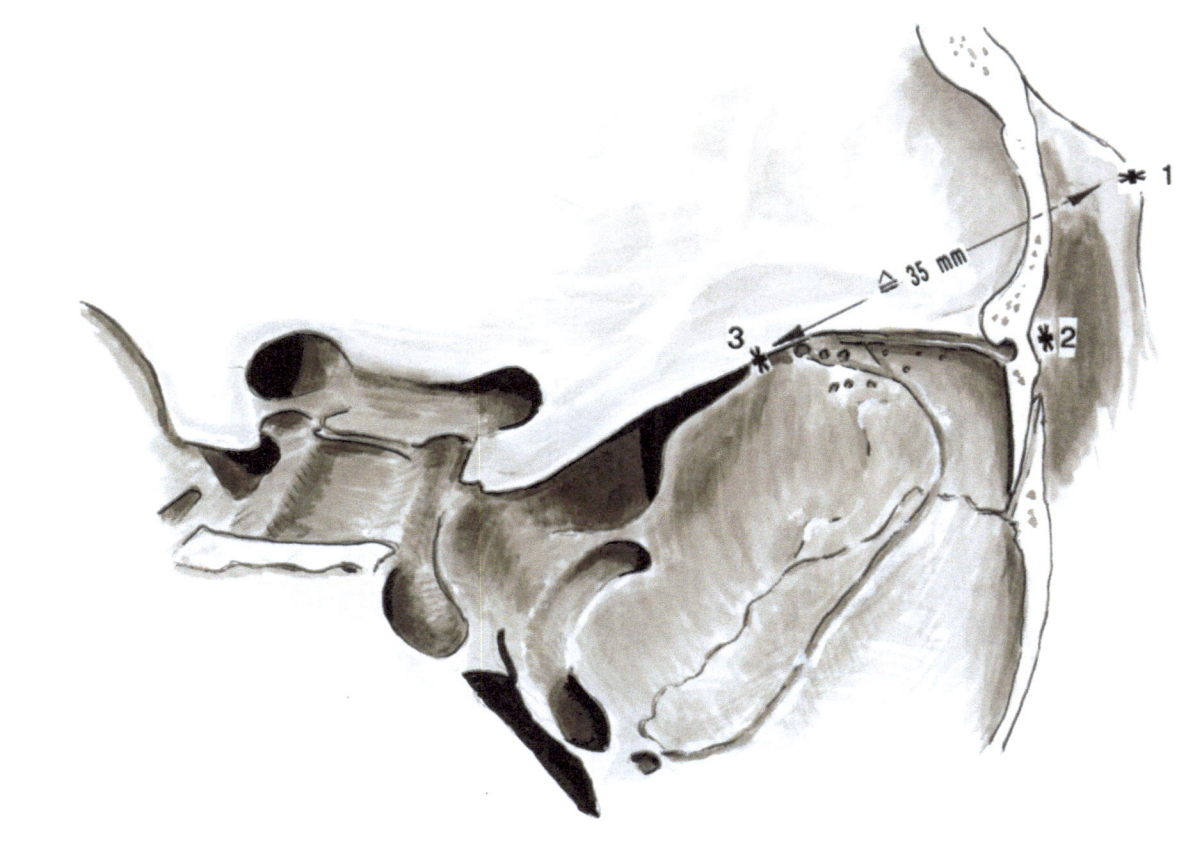

Fig. 54. Continuation of Fig. 53

Anterior branches of A. meningea media, normal findings*

---

* from the author's skull collection. Vessels and nerves added.

FIG. 54

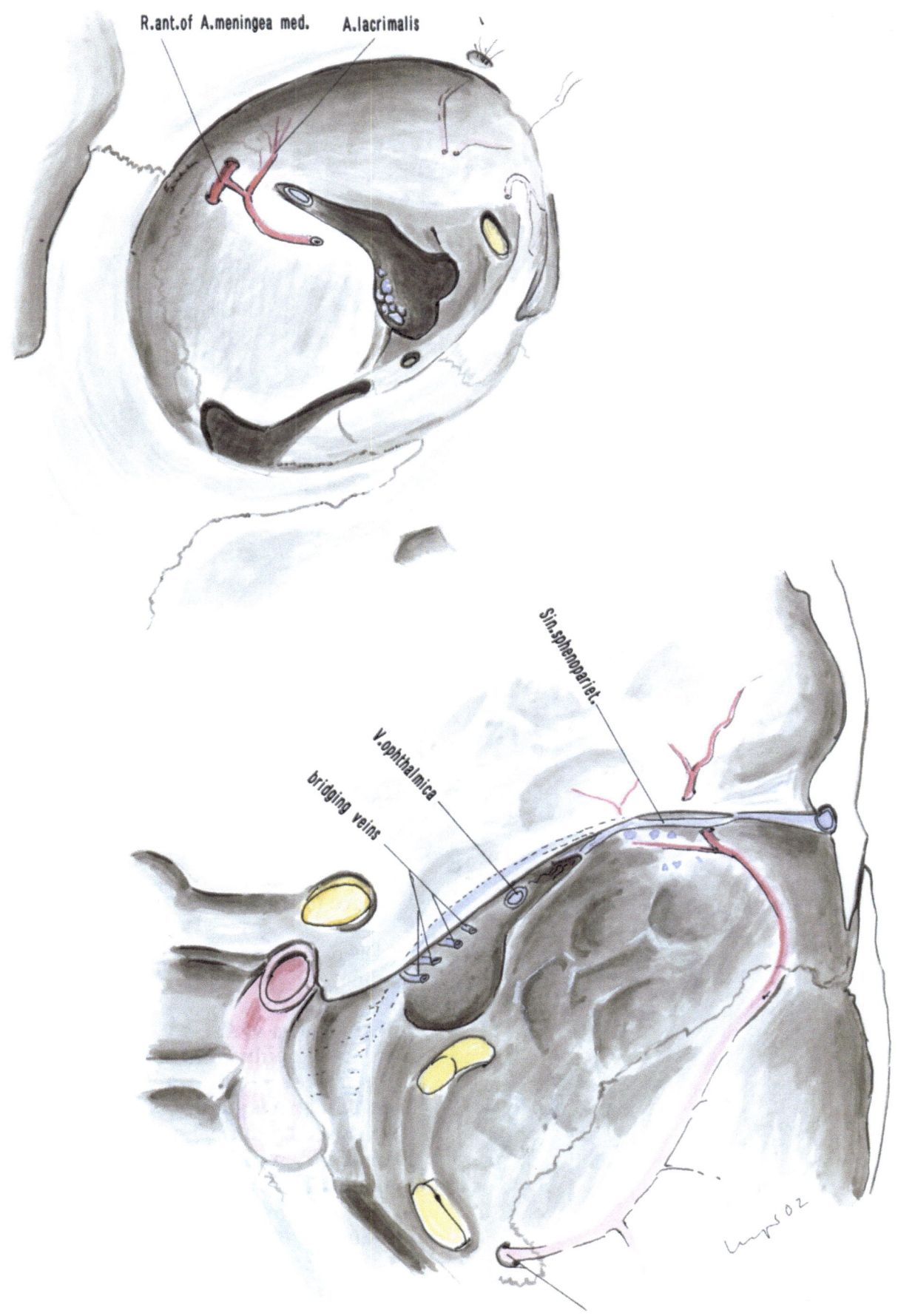

R.ant.of A.meningea med.   A.lacrimalis

Sin.sphenoopariet.

V.ophthalmica

bridging veins

For.spinosum/A.meningea media

Fig. 55.

A **Foramen spinosum missing. Atypical horizontal course of Sulcus meningeus. Rare (?) variant***

B Deep Sulci meningei in a child*, normal finding

Clinical aspects of A:
- **Surgery of epidural hematomas after trauma: A. meningea media would be missed in the area of Foramen spinosum. Foramen spinosum would be missed, too.**
- **Embolization of AVM's, e.g.: Embolization of A. maxillaris may include A. ophthalmica and A. centralis retinae – danger for unilateral blindness**

---

* from the author's skull collection

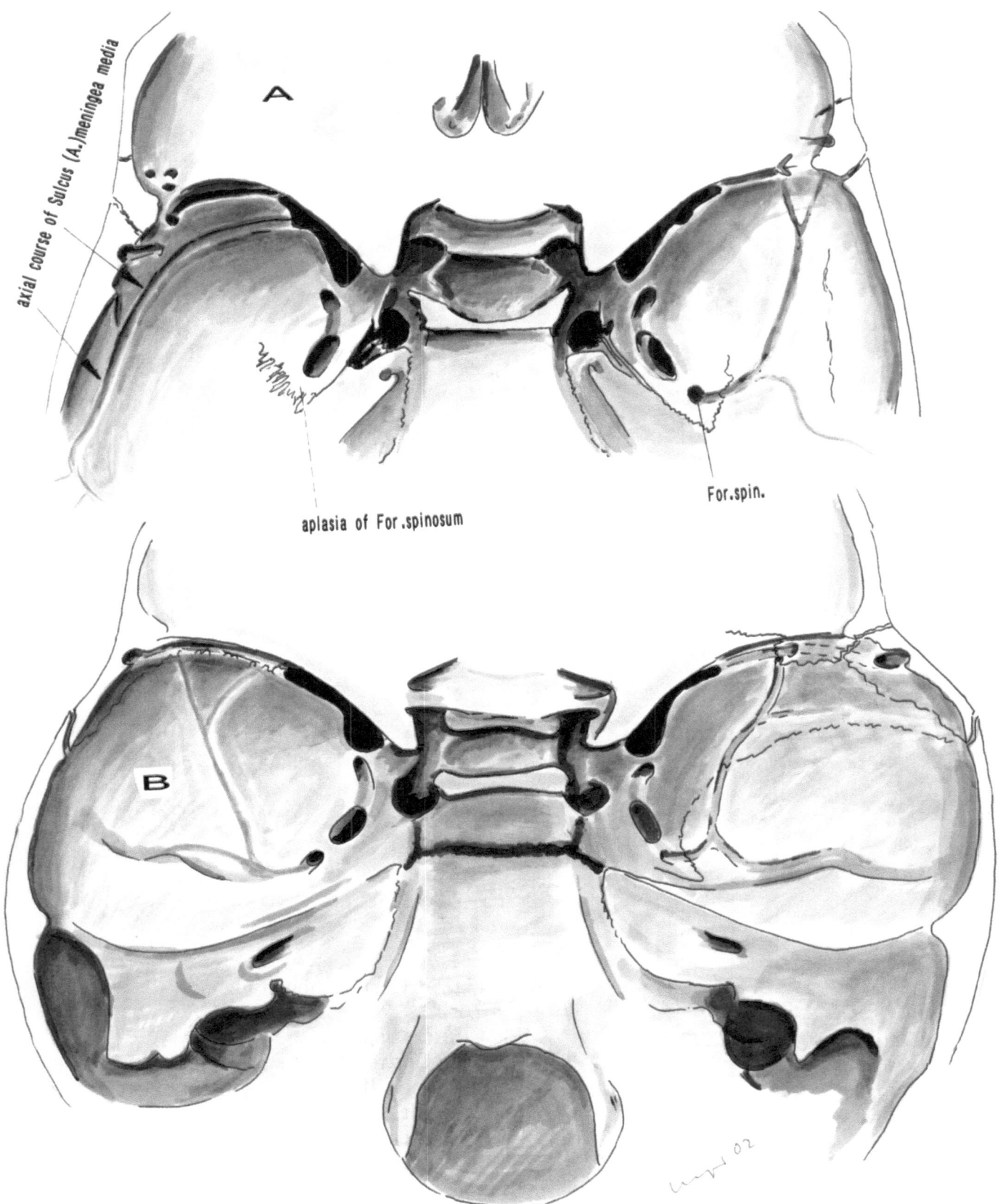

axial course of Sulcus (A.)meningea media

A

aplasia of For.spinosum

For.spin.

B

Fig. 56. Continuation of Fig. 55

Clinical aspects:

A   Conventional surgery of epidural hematomas, if Foramen spinosum is missing. Atypical course of A. meningea media. No surgical problems

B   A meningea media as in Fig. A.
Atypical problematic localization when using minimal invasive surgery

---

* usual anatomical landmarks

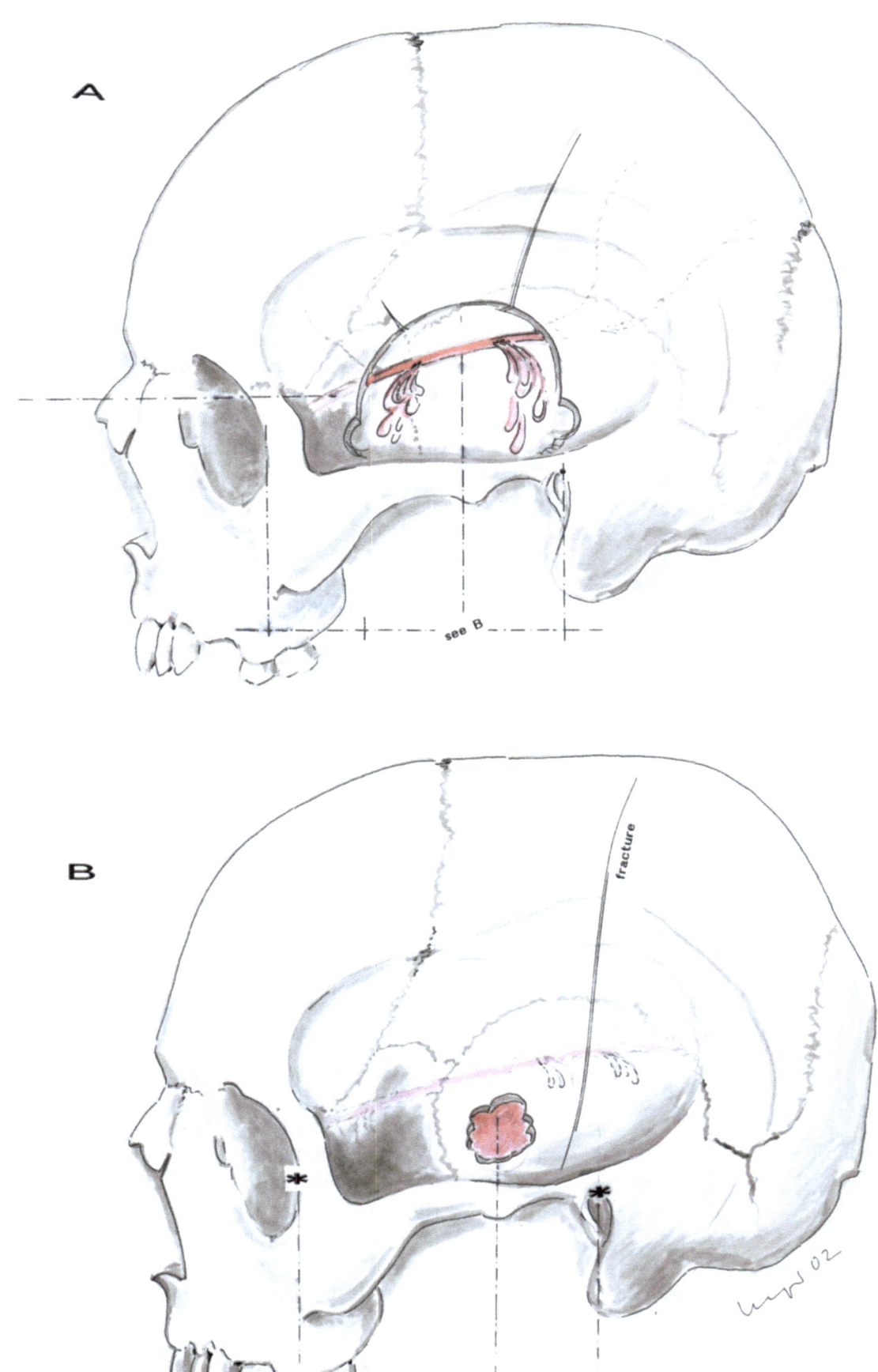

A

B

see B

fracture

Fig. 57. **Origin of A. meningea media from A. ophthalmica,** rare (?) variant
A.centralis retinae normal (originating at the proximal trunk of A.ophthalmica)
Additional drawing (inferior presentation) shows adjacent structures, according to Seeger (1978), modified

Clinical aspects:
• see Figs. 55 and 56

Abbreviations

a	For. opticum
b	A. ophthalmica, main trunc
c	A. centralis retinae (according to Lang, 1979, 1981)
d	Foramen rotundum and N. maxillaris (and vessels)
e	frontozygomatic point
f	A. lacrimalis
g	A. meningea media (anterior main branch)
h	N. nasociliaris
i	N. oculomotorius
j	N. abducens
k	as d
l	V. (Vv.) ophthalmica (ae)
m	meningeal vessels
n	N. lacrimalis
o	N. frontalis
p	N. trochlearis
q	Anulus tendineus (Zinni)
(r)	M. levator palpebrae sup.
(s)	M. rectus sup.
(t)	M. rectus nasalis
(u)	M. rectus inf.
(v)	M. rectus temp.
()	projections
w	Processus clinoideus ant.
x	radix of Processus clinoideus ant.

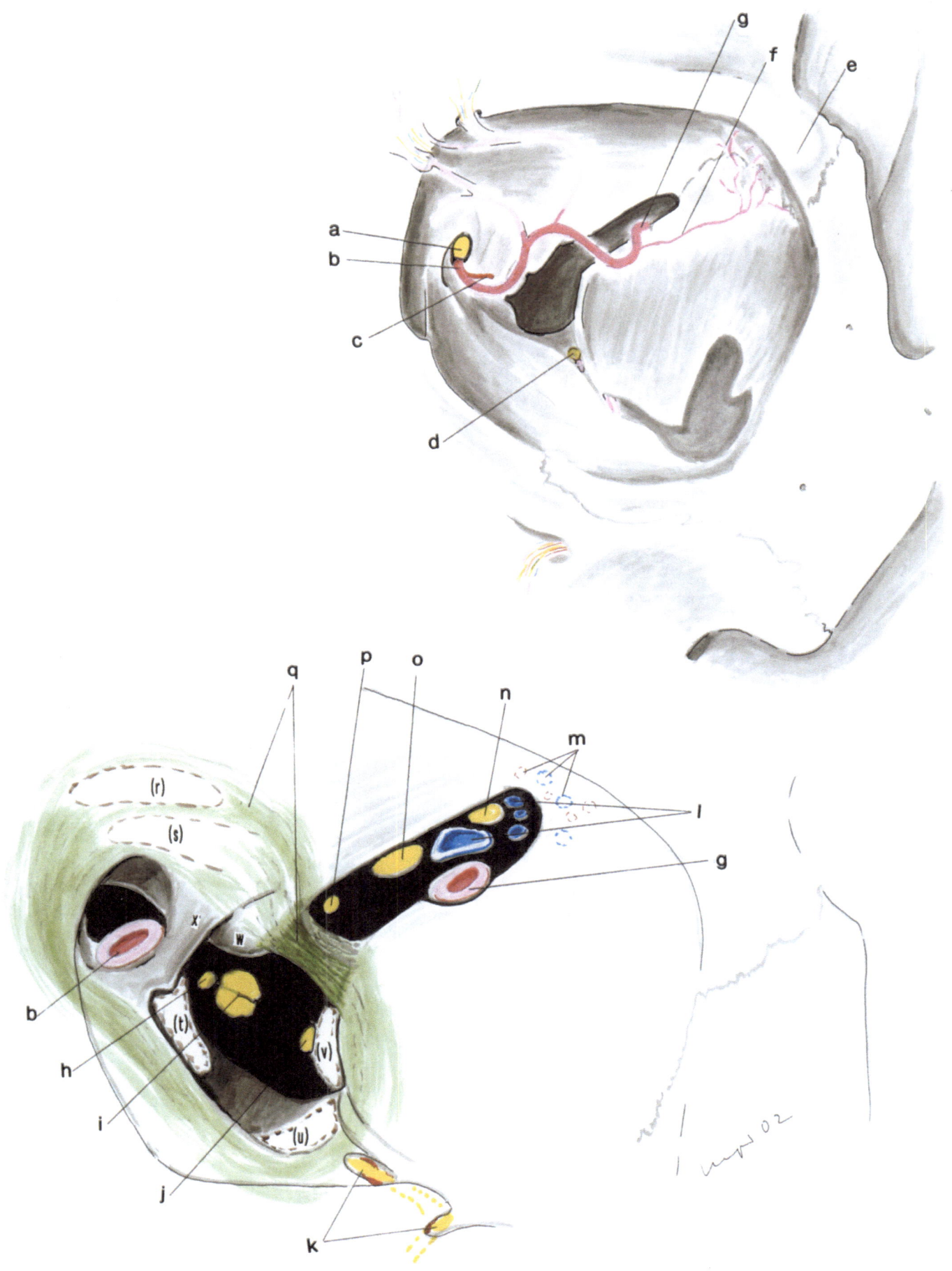

Fig. 58.

Clinical aspects:
Rare variants are often combined with other rare variants. For literature see Lang (1979, 1981)

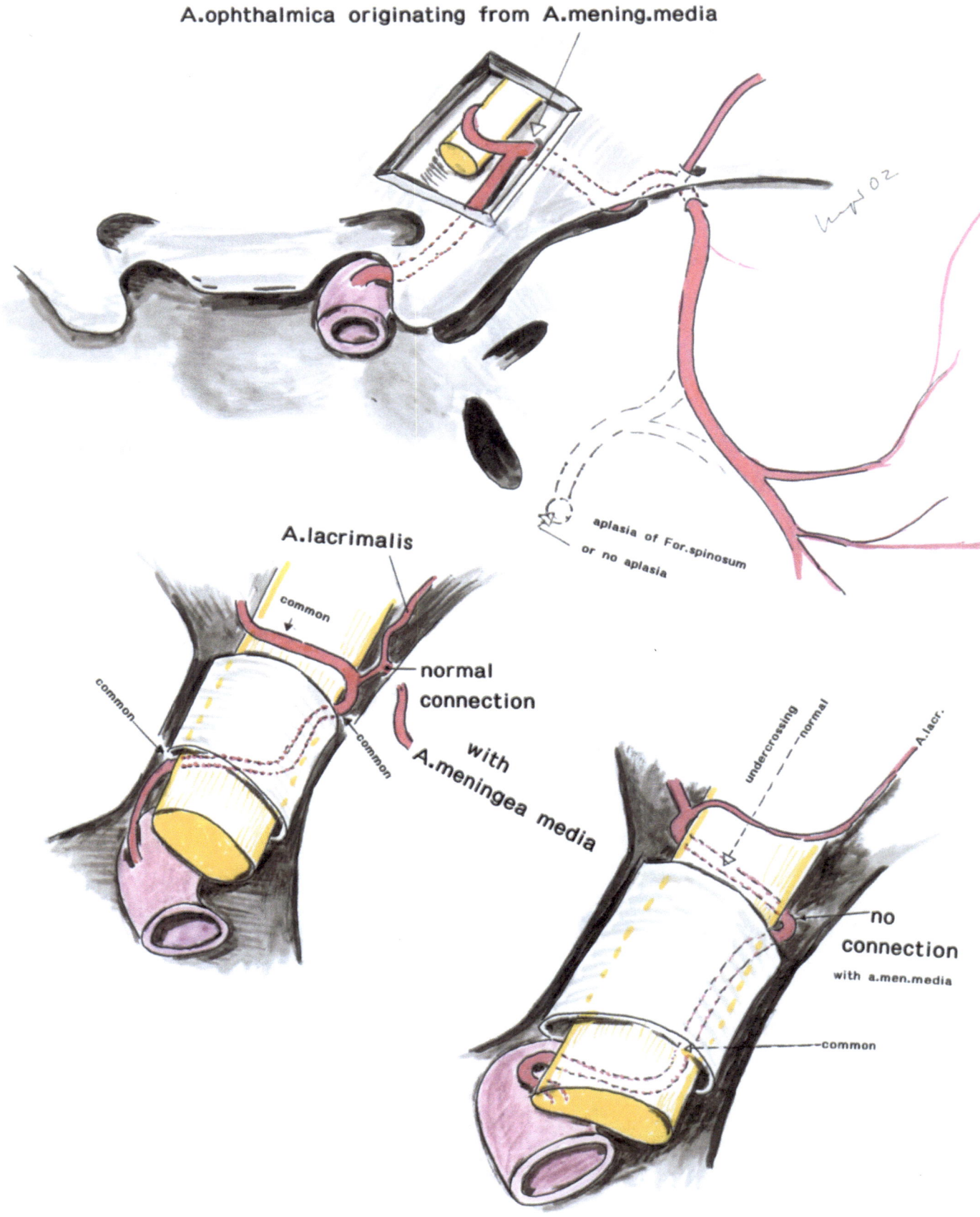

A.ophthalmica originating from A.mening.media

aplasia of For.spinosum
or no aplasia

A.lacrimalis

common

normal
connection

common

common

with
A.meningea media

undercrossing — normal

A.lacr.

no
connection
with a.men.media

common

Fig. 59. **Siphon of A. carotis int., variants***
(distances a, a', b, b')

**Clinical aspects of surgery of pituitary tumors:
Carotid artery may be located close to the midline – b and b' – if tumors are small.
Distance between artery and Dura – a and a' –
should be considered when using transsphenoidal, transcranial microsurgical, endoscopic,
and stereotactic surgery**

* according to skull dissections from the author's collection

FIG. 59

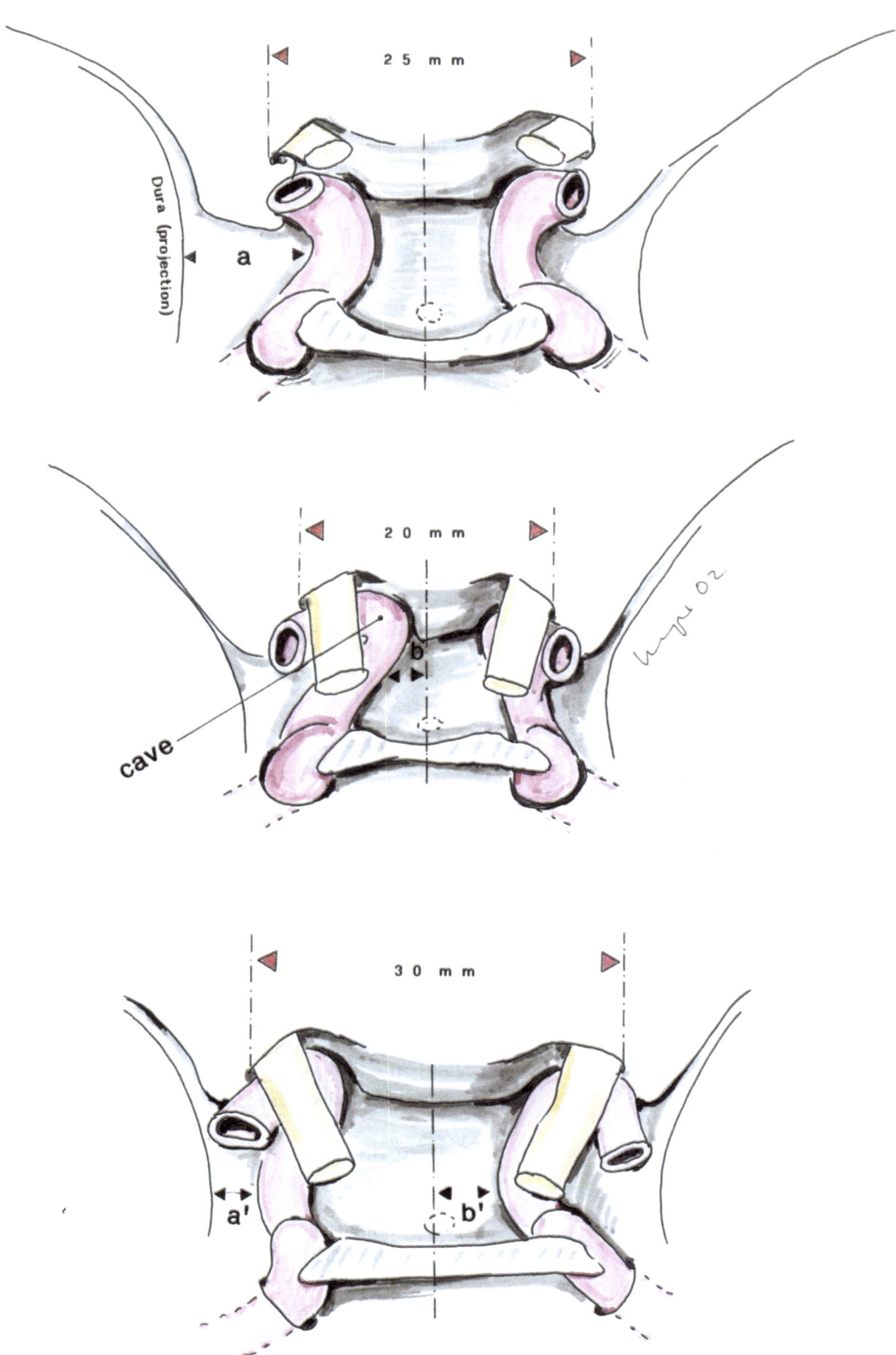

25 mm

Dura (projection)

a

20 mm

cave

b

30 mm

a'

b'

Fig. 60. **Siphon of A. carotis int. and proximal segment of A. ophthalmica**
- normal findings –

Microsurgical aspects:
The distal – epidural located – segment of A. ophthalmica (level S1) is less at risk by surgery of intradural meningiomas, e.g., at the proximal segment (level S2) the artery is more at risk.

Abbreviations
a   Dura mater (see S1) and external sheath of N.II (see S2)
b   bony structures
c   internal sheath of N.II
d   Arachnoidea
e   N. opticus
f   A. ophthalmica

FIG. 60

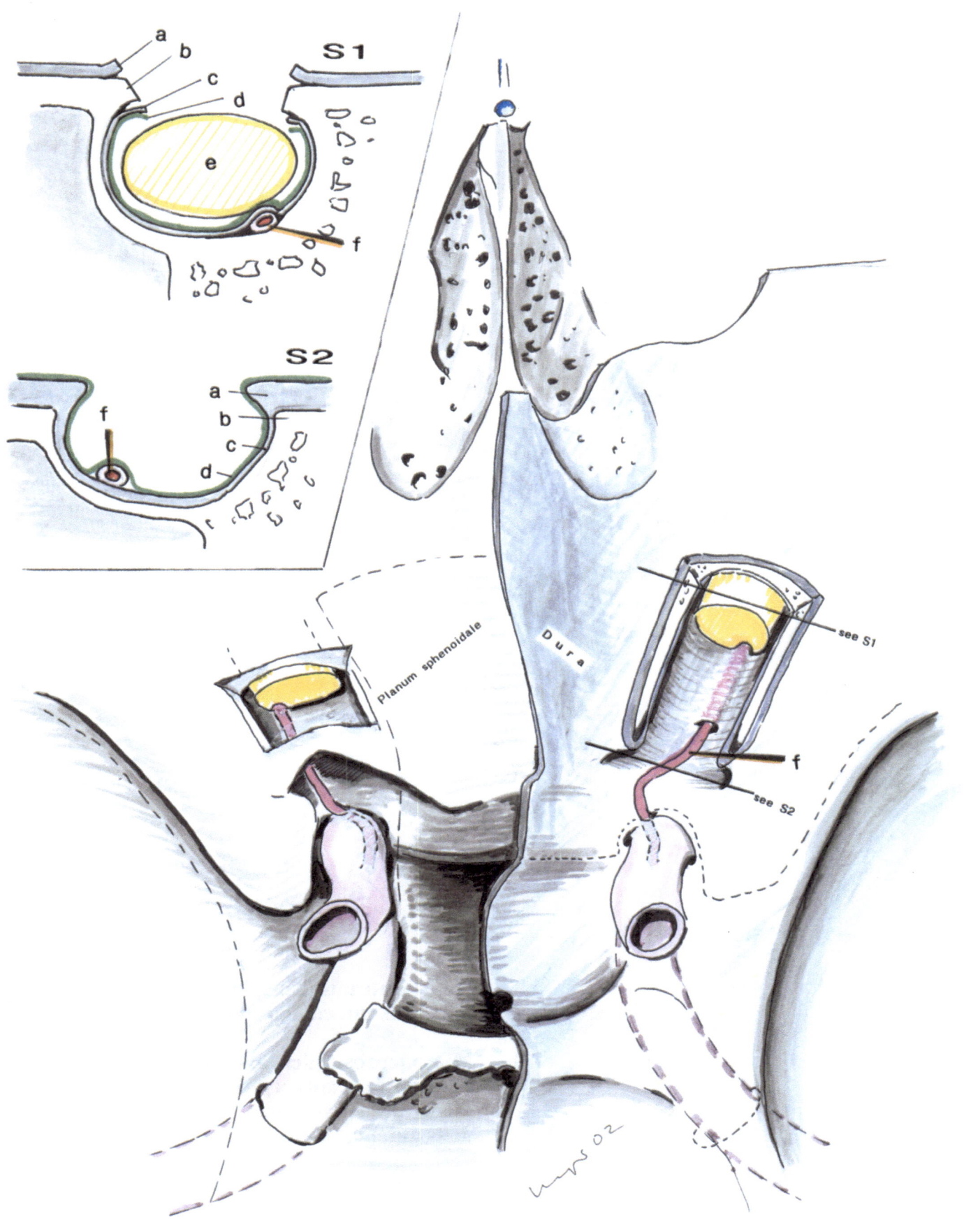

Fig. 61. **Intradural origin of A. ophthalmica from A. carotis int. common variant**

Microneurosurgical aspects:
Favorable for surgery of ophthalmic aneurysms. Here the usual resection of Processus clinoideus ant. may be unnecessary

Abbreviations
a   dural fold
b   bony roof of Canalis opticus
c   Dura

FIG. 61

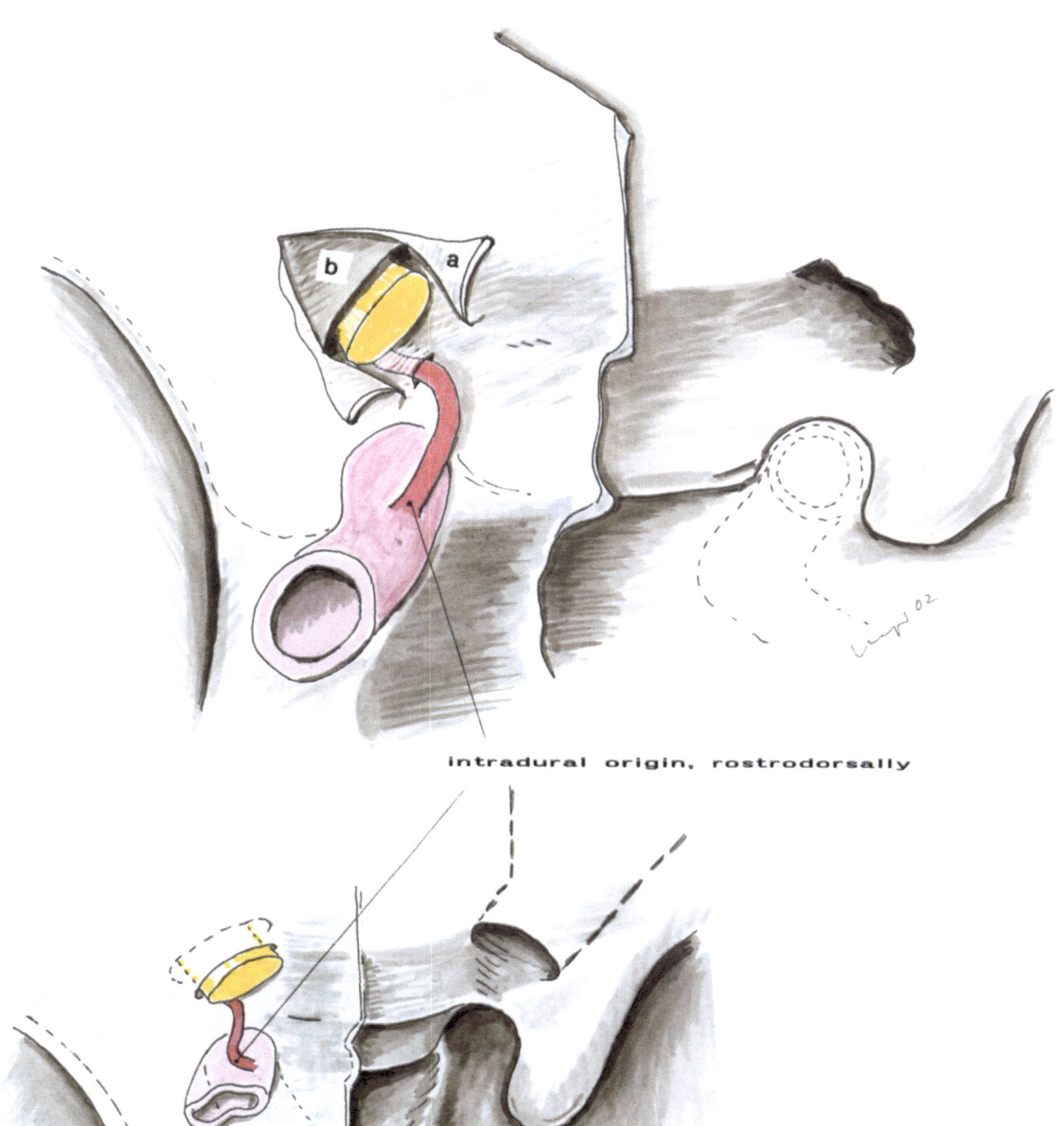

intradural origin, rostrodorsally

Fig. 62. **A. ophthalmica originating from A. meningea media *, rare variants ****

This anatomical variant can be understood by comparing A. cerebri ant. and media in numerous mammals. These vessels are originating from a Rete mirabile of A. maxillaris

Clinical aspects:
Danger for homolateral blindness after embolization of A. carotis ext. or its branches (by reduction of the bloodstream in A. centralis retinae)

---

* A. meningea originating from A. ophthalmica: See Fig. 57
** Lang (1979), p 525 to 533, Lasjaunias and Berenstein (1987) p 33 ff

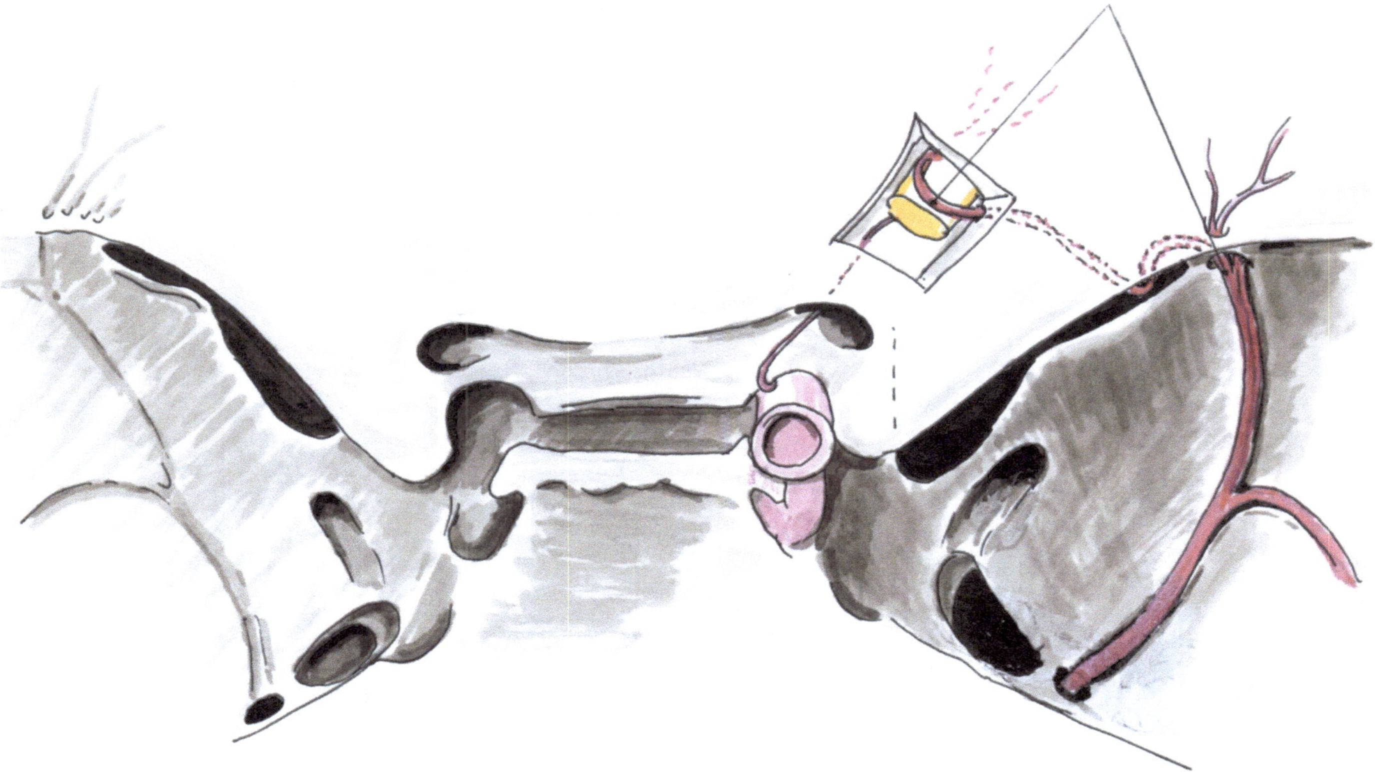

feeding of A.ophthalmica by A.mening.media

from A.maxillaris

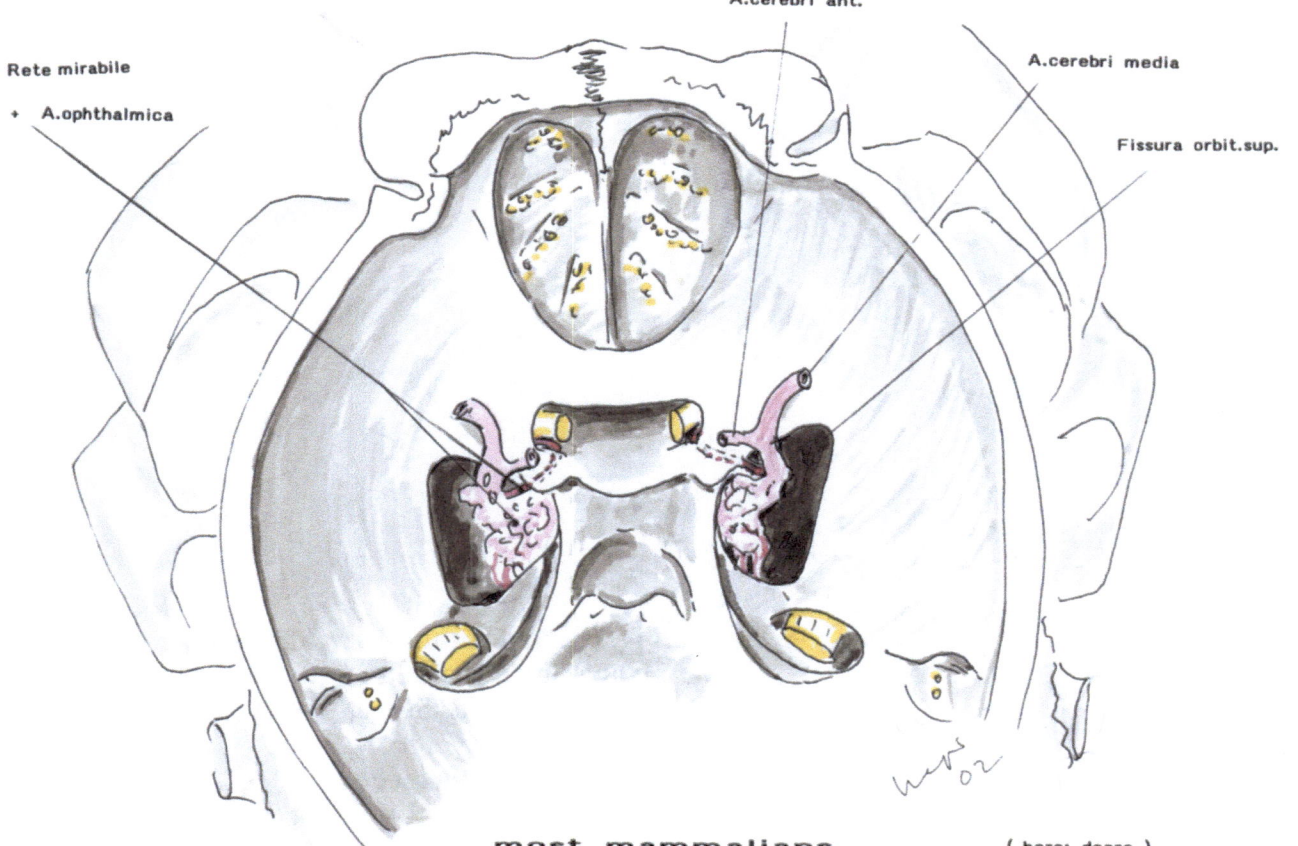

A.cerebri ant.

A.cerebri media

Rete mirabile

+ A.ophthalmica

Fissura orbit.sup.

most mammalians   (here: deere)

Fig. 63. **A. ophthalmica penetrating the radix of Processus clinoideus ant., rare variant.**
(for literature see Seeger, 1983)

Microsurgical aspect:
Drilling of the clinoid process and its root is dangerous (especially during surgery for ophthalmic aneurysms)

FIG. 63

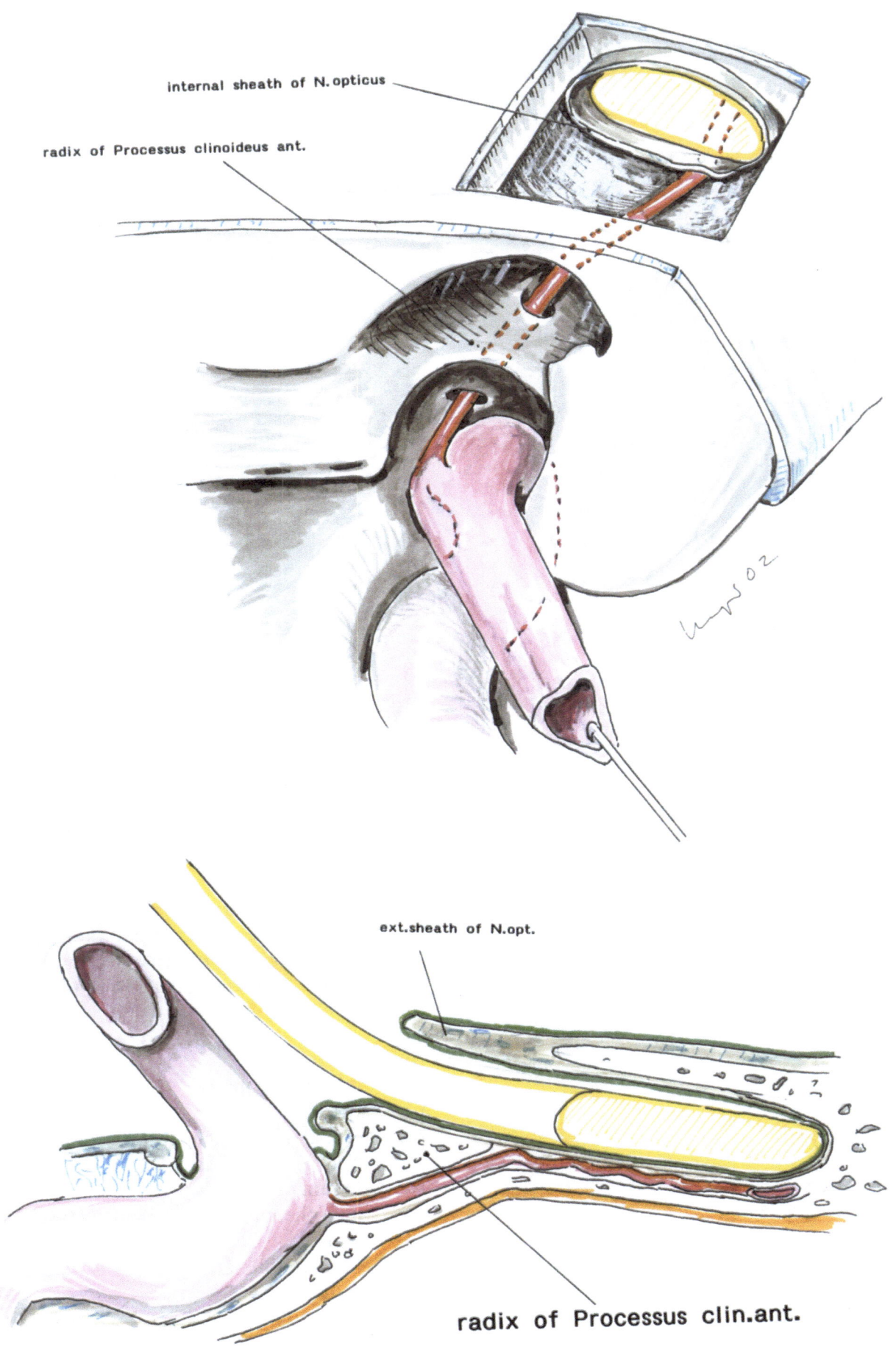

internal sheath of N. opticus

radix of Processus clinoideus ant.

ext. sheath of N. opt.

radix of Processus clin. ant.

Fig. 64 to 68. **Variants of Canalis opticus and surrounding structures**

Fig. 64. **Close relationship of N. opt. and siphon of A. carotis int. are normal findings in humans, but not in primates and other mammals**

Clinical aspects:
In this area the nerve is endangered by even small space-occupying lesions (e.g. tumors, bony stenosis, widening of A. carotis, aneurysms, angiomas, variants of A. ophthalmica)

Abbreviations
a Sulcus chiasmatis
b siphon
c Can. and N. opticus (projection)
d Processus clinoideus ant.
e Fissura orbitalis sup. (dotted: projection)
f Sin. sphenoidalis, thin-walled close to the optic canal
g Rete mirabile and trunk of A. cerebri ant. and media
h Foramen ovale

FIG. 64

homo

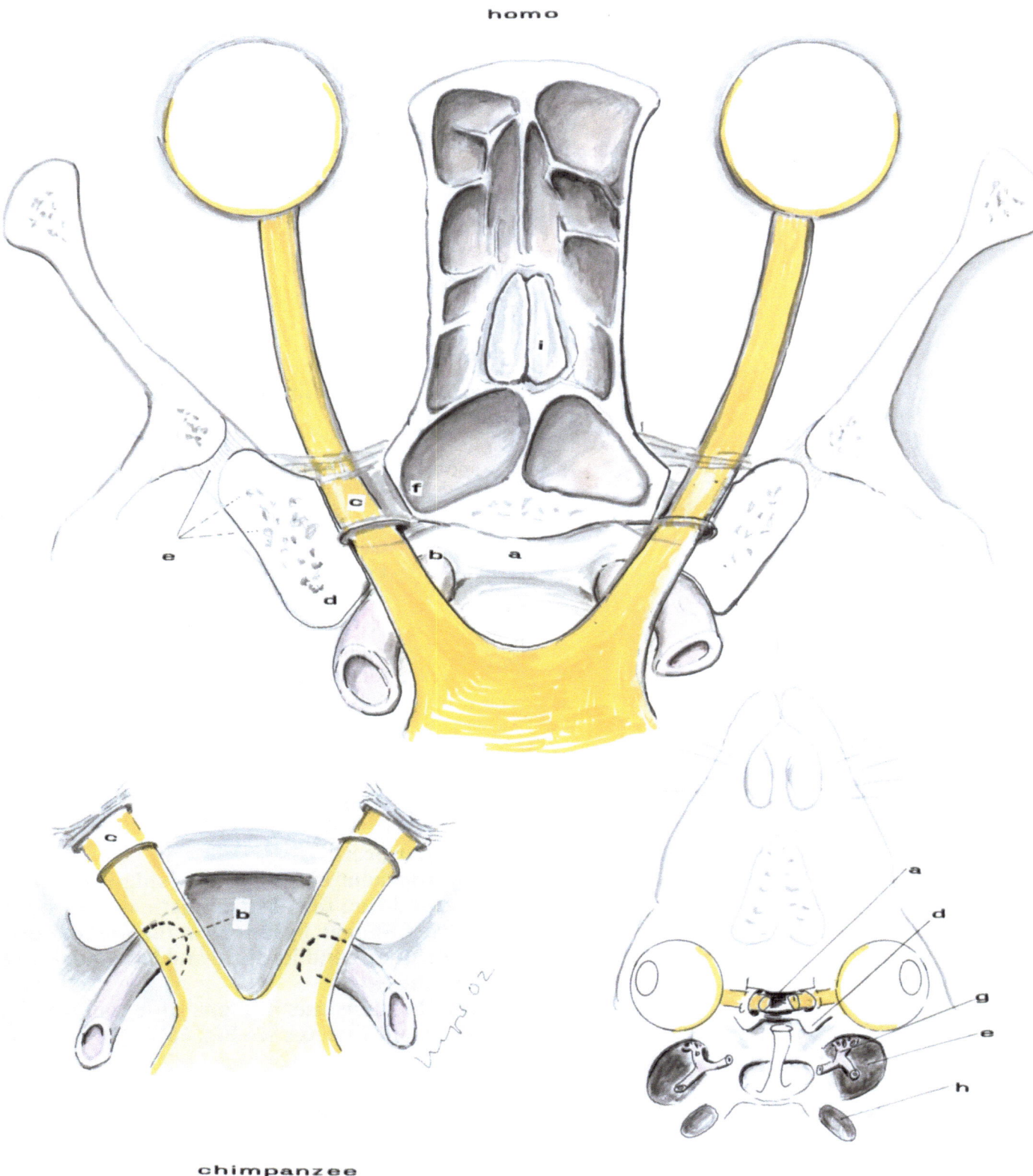

chimpanzee

other mammal

Fig. 65. **Relationships of Canalis opticus and Sinus sphenoidalis**

A and **A'**   Normal findings

B and **B'**   So-called „Pneumosinus" (Hoydt, 1962, Unsöld and Seeger, 1989)
Pneumatization of Processus clinoideus ant., originating from a pneumatized roof of Canalis opticus and/or a pneumatized Radix of Processus clinoideus ant. (anatomical dissection of the author).
Now Canalis opticus is surrounded by the Pneumosinus. Canalis rotundus may be surrounded by a widened Sinus sphenoidalis, too
Arrows: Widening of pneumatization

Clinical aspects:
- **Enlargement of Sinus sphenoidalis may be a precondition for posttraumatic and postoperative CSF leak by lesions of the wall of Canalis rotundus and its dural and leptomeningeal structures**
- **For Pneumosinus dilatans with compression of the optic nerves see Fig. 66**

Abbreviations

a	Processus clinoideus ant.
b	as a, pneumatized
c	pneumatization between Planum sphenoidale and roof of Canalis opt.
d	N. opt.
e	area of siphon
f	medial wall of Canalis opt.
g	medial wall of Canalis rotundus and the sulcus of N. maxillaris before entering its channel

FIG. 65

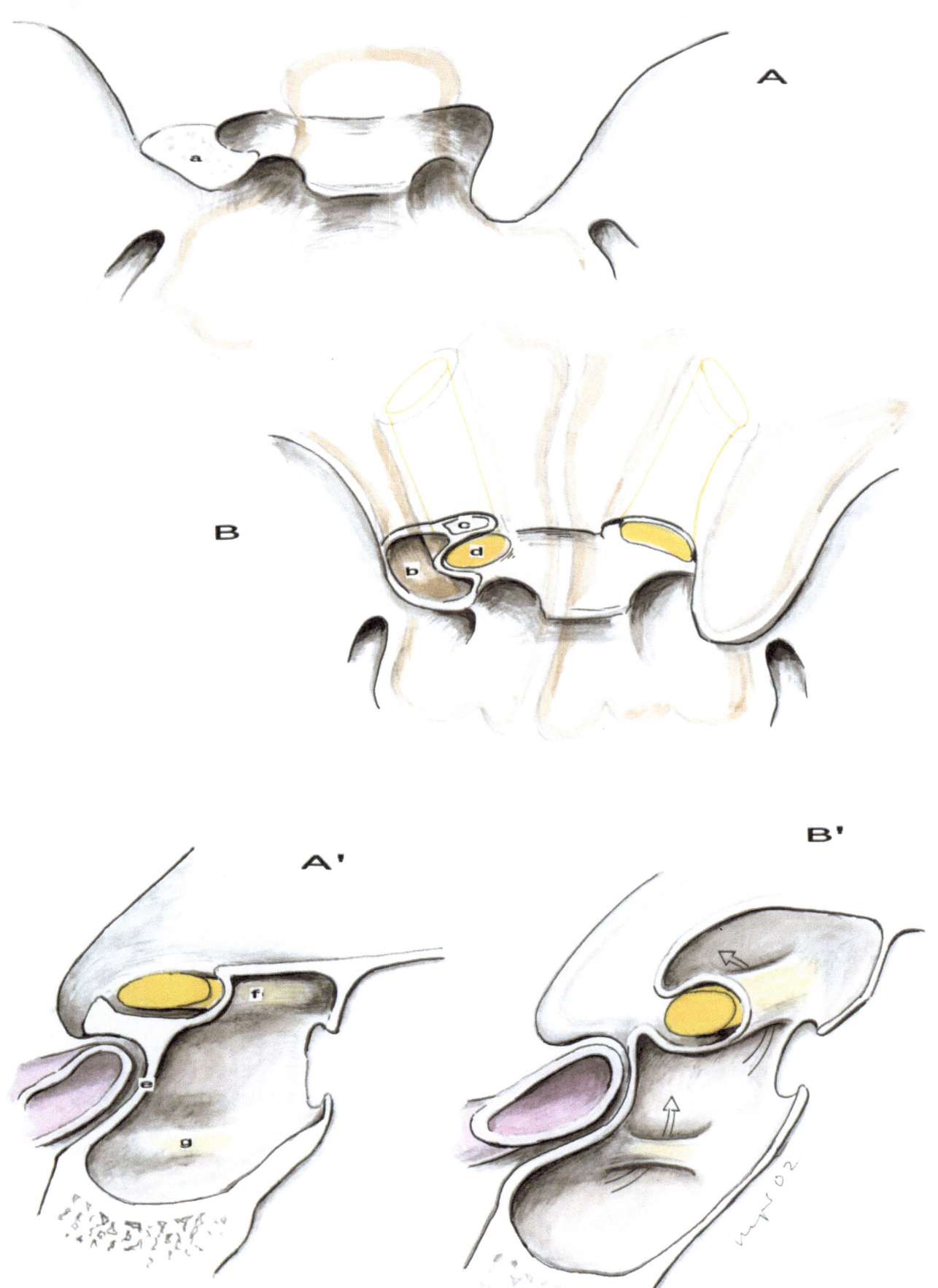

Fig. 66. **Pneumosinus dilatans** (Hoydt, 1962)

= Stenosis of Canalis opticus by a pneumosinus

A   Compression of Can. and N. opticus by a Pneumosinus dilatans (mostly bilateral)

B   X-ray-findings (Rhese). Pneumosinus may be confused with the stenotic Canalis opt.

C   CT

FIG. 66

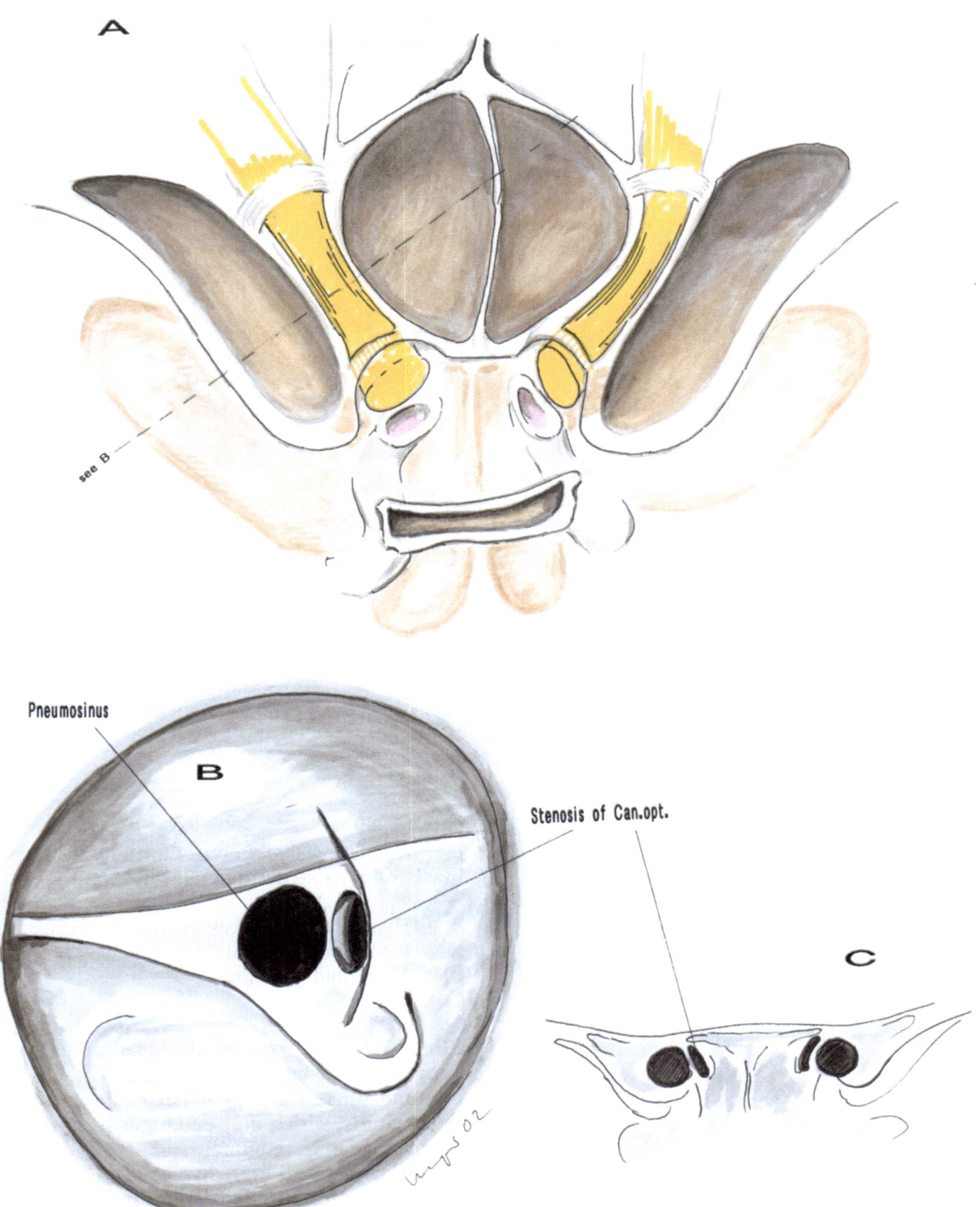

A

see B

Pneumosinus

B

Stenosis of Can.opt.

C

Fig. 67. **Stenosis of Canalis opticus by Hyper-plasia of bony structures** (Hoydt, 1962)

A Hyperplasia of Processus clinoideus ant. and of the lateral wall of Sinus sphenoidalis
B X-ray-findings (Rhese)
C CT

Clinical aspects of Pneumosinus dilatans and of bony hyperplasia:
Pneumosinus and bony hyperplasia are essential variants, which may endanger N. opticus in a late stage of development.
Pneumosinus and bony hyperplasia may occur even in combination with tumors, e.g. meningiomas. Pneumosinus and hyperplasia do not exist prior to tumor growth

A

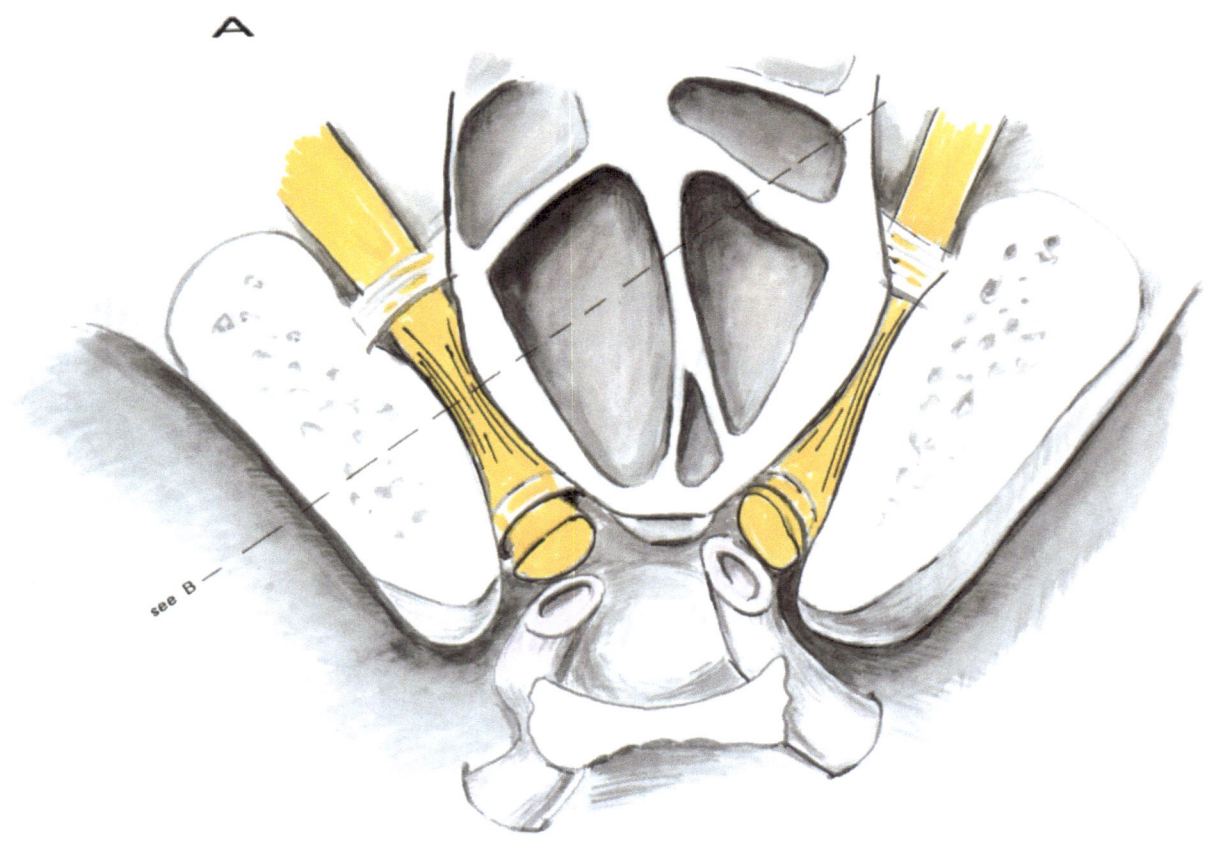

see B

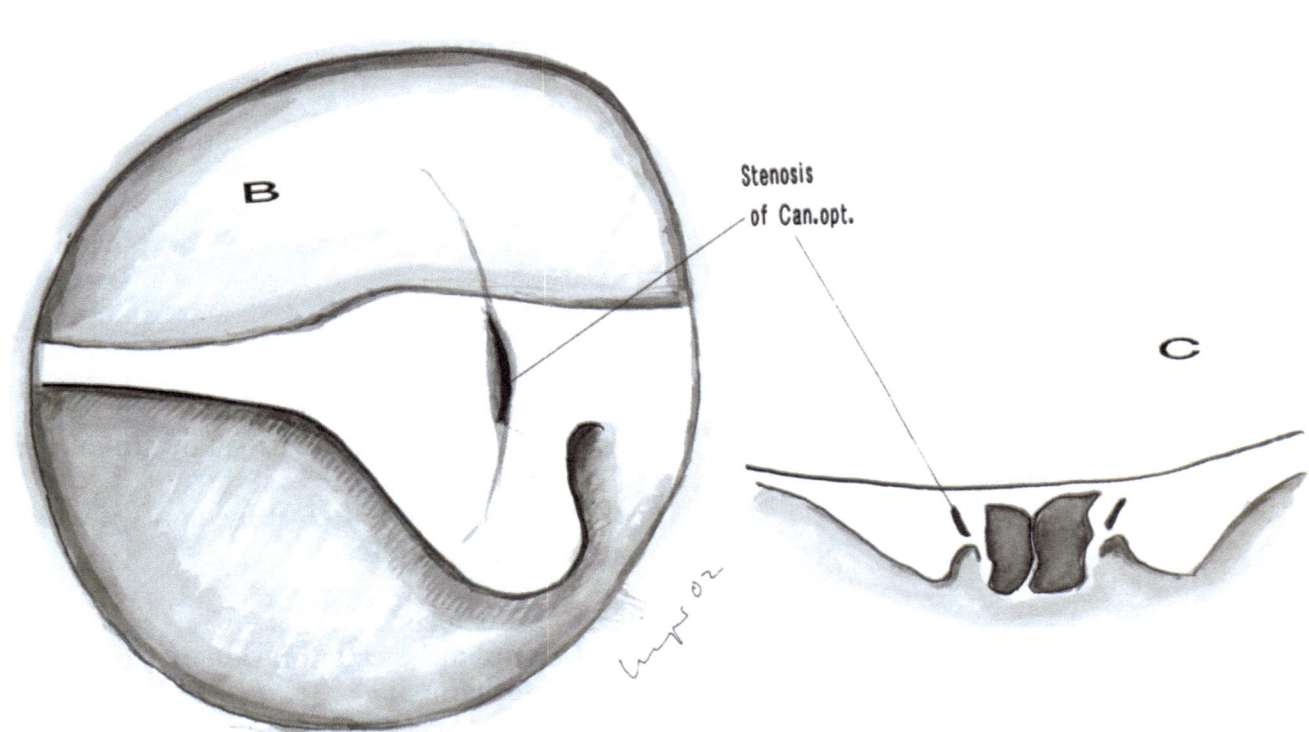

B

Stenosis
of Can.opt.

C

Fig. 68. **Further variants, rare**

A N. II shifted and flattened by **Dolichoectasia of siphon.**

B **Penetration of N. II by an atypical located A.ophthalmica**

Clinical aspects:
Both types or its combination may endanger the visual functions
(Unsöld and Seeger, 1989, according to Hoydt, 1962)

A

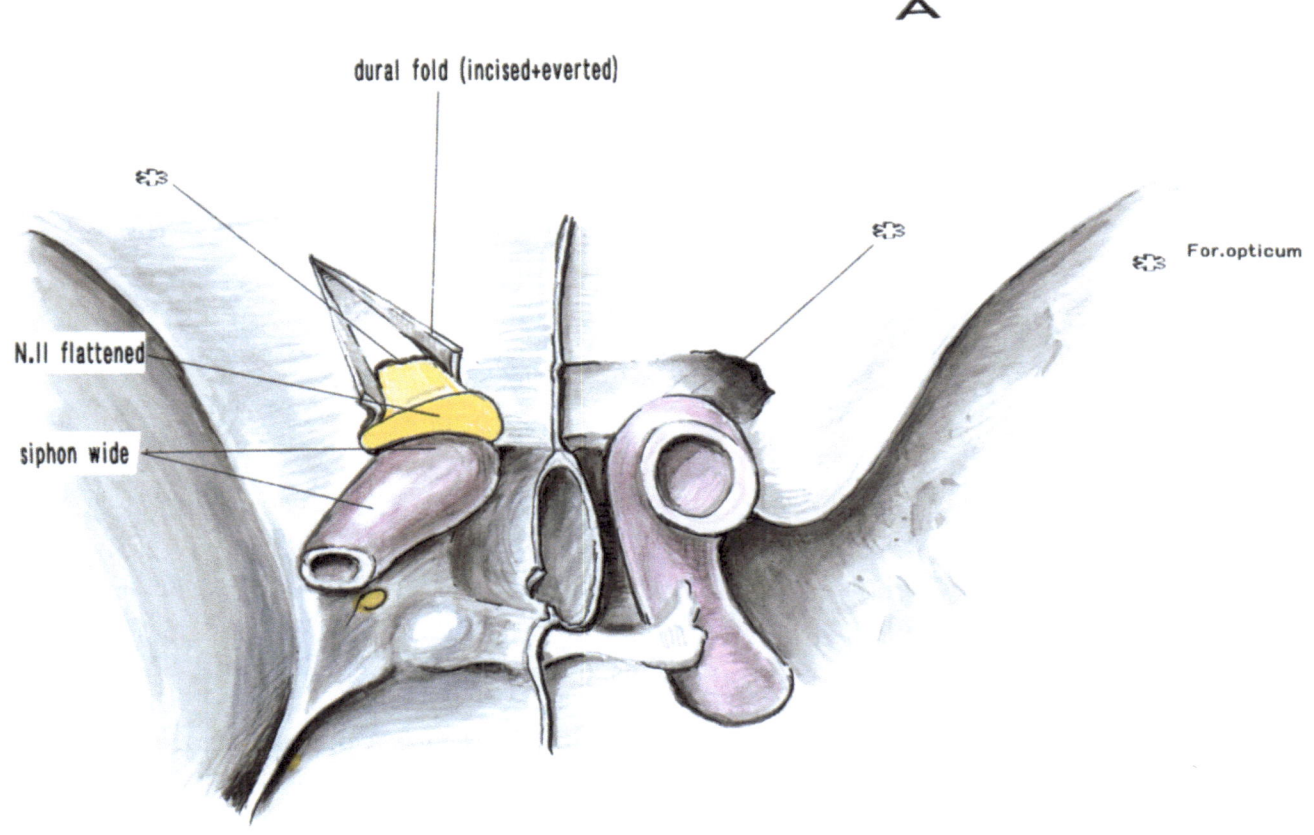

dural fold (incised+everted)

For.opticum

N.II flattened

siphon wide

B

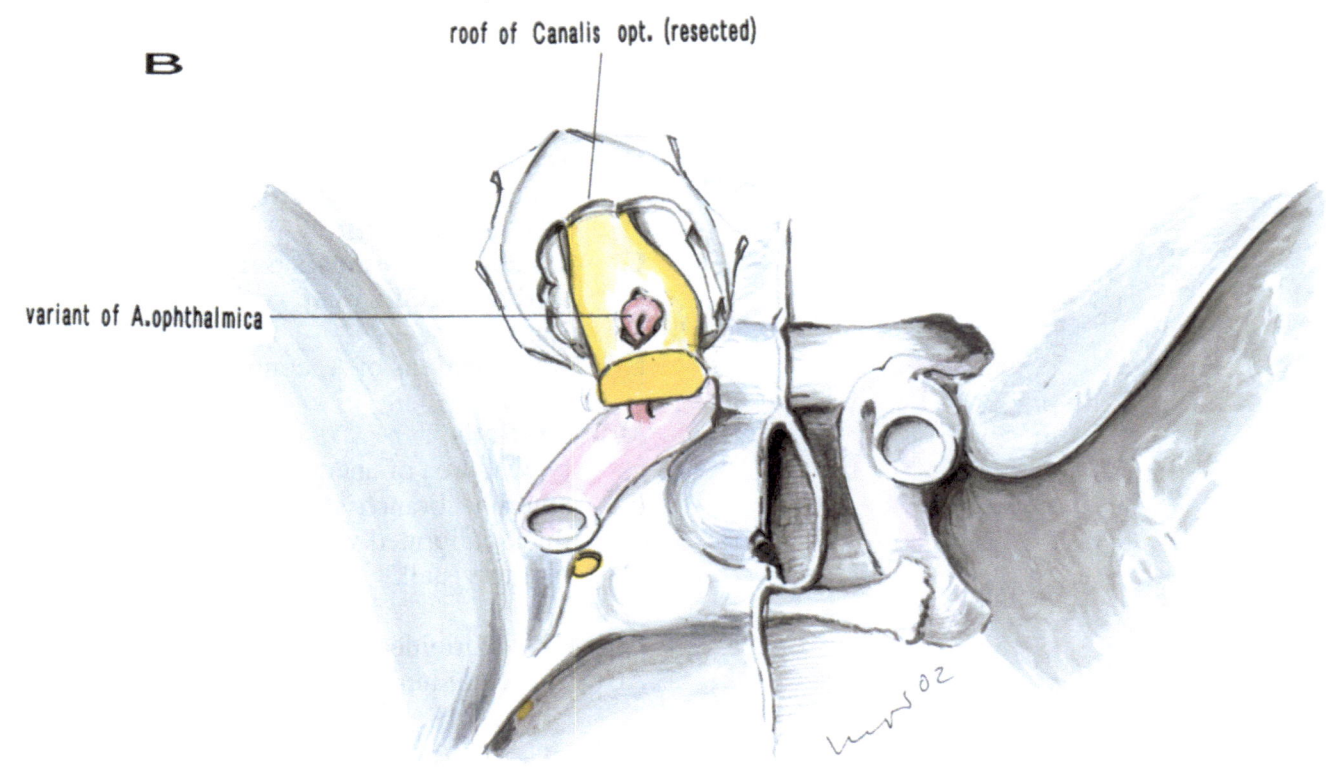

roof of Canalis opt. (resected)

variant of A.ophthalmica

Figs. 69 to 85.

Variants of the supra-parasellar area and of the anterior segment of Circulus arteriosus Willisi

Fig. 69. **Variants of A. communicans post. and A. chorioidea ant.**

Microsurgical and neuroradiologal aspects: Treatments of aneurysms

A   Both arteries and its perforators are located close to each other: **Less favorable for microsurgery**

B   Distal origin of A.chorioidea ant.. **Less favorable for microneurosurgery of aneurysms of A. chorioidea ant.**

C   **Useful anastomosis between A. carotis int. and A. cerebri post.**

D   Rare

E   Ramifications of A. communicans post.

Abbreviations

a   typical area of aneurysms (dotted)

b   anterior branch for hypothalamus (typical thick-calibrated variant – Yasargil*)

c   pituitary stalk

AChA   A. chorioidea ant.
ACoP   A. communicans post.

---

* Personal communication

FIG. 69

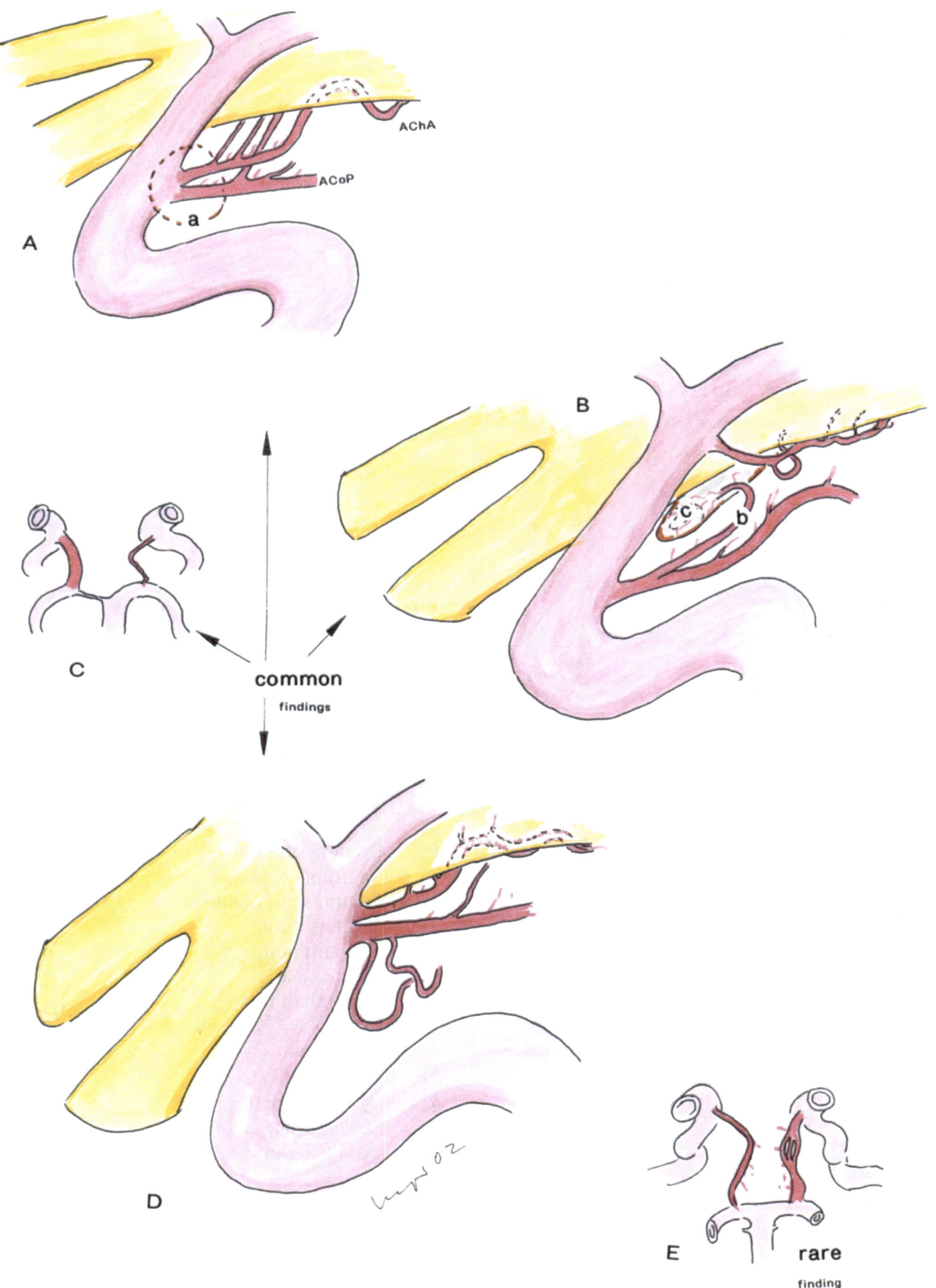

A

AChA

ACoP

a

B

c

b

C

common

findings

D

E   rare

finding

Fig. 70. **Suprasellar area, survey**

Abbreviations

a	Subst.perforata ant.
b	A. carotis int.
c	perforating arteries
d	Recessus opt.
e	Area subcallosa
f	Sulcus longitudinalis
g	A. corporis callosi mediana
h	Commissura ant.
i	Rostrum corporis callosi
j	Cavum septi pellucidi
k	pituitary stalk
l	N. oculomotorius
m	N. trochlearis
n	N. ophthalmicus
o	Fasciculi of N. abducens in the Adventitia of A. carotis int.*
p	Dorsum sellae
q	Gyrus uncinatus
r	Corpus amygdaloides
s	basal ganglia
H	A. recurrens (Heubneri)

---

* Rhoton, personal communication

FIG. 70

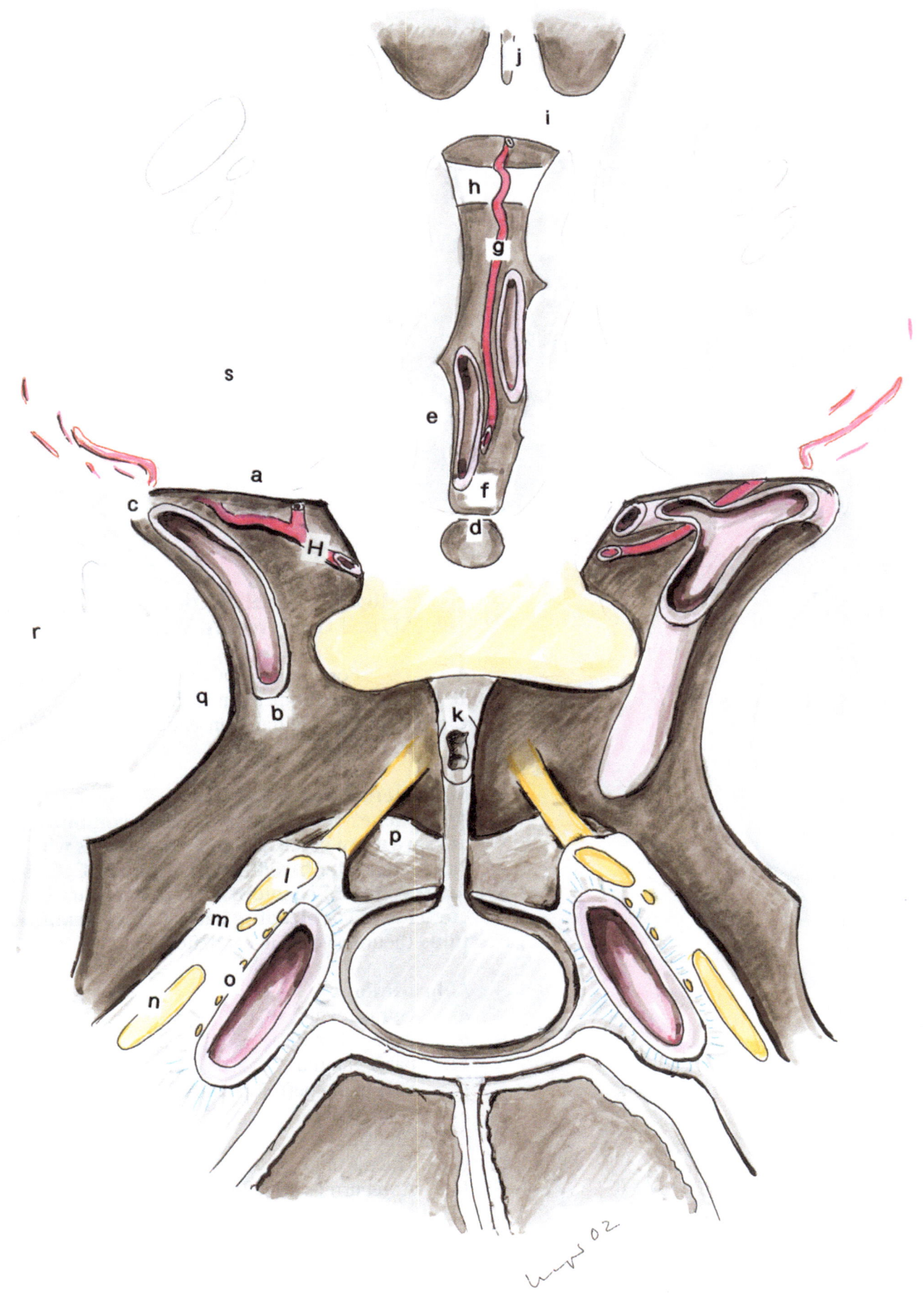

Fig. 71. **Topographical relationship of A. cerebri ant. (A1) and Gyrus rectus. Variants**

A   For comparison with B to D, and Fig. 70
B   For comparison with Fig. 70
C and D   Parasagittal transectional planes, enlarged

Microneurosurgical aspects:
Approaches for aneurysms of A. communicans ant. are more constrained by adhesions with Gyrus rectus after bleeding in D, than in C. Danger for the limbic segment of Gyrus rectus by the coincidence of bleeding with surgical manipulations (neuropsychological deficits*)

Abbreviations
a   Gyrus rectus
b   Area subcallosa
c   basal ganglia
d   Substantia perforata ant. /Cisterna valleculae
e   N. opticus
f   Chiasma
g   Tractus opt.
H   A. recurrens (Heubneri)

---

* according to Hütter (2000)

FIG. 71

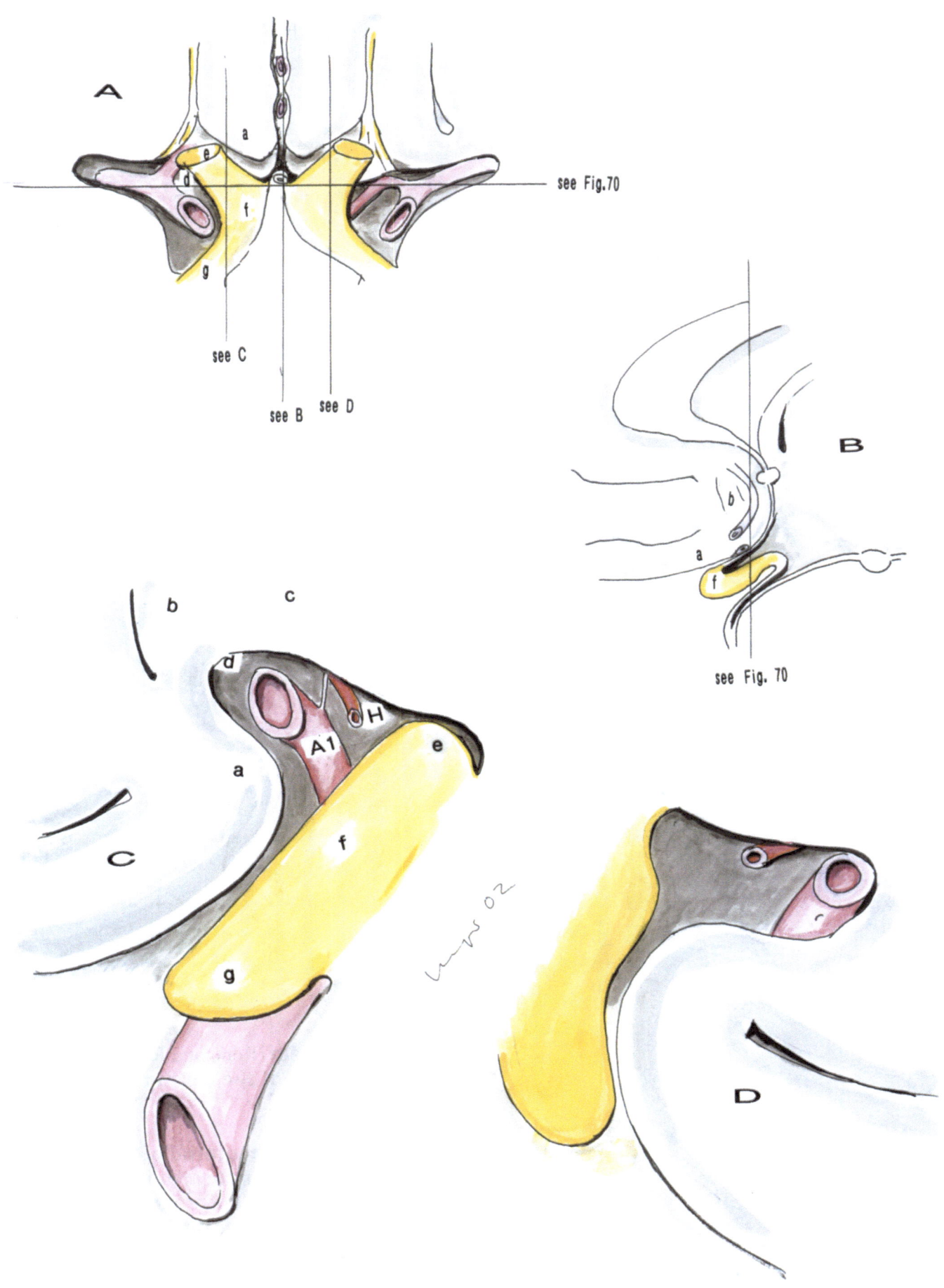

A

see Fig.70

see C

see B   see D

B

see Fig. 70

b   c

d

a

C

A1

H

e

f

g

D

Fig. 72.

- **Prefixed chiasm** (Renn and Rhoton, 1975) and
- **Some variants of the carotid bifurcation and A1**

Microsurgical and neuroradiological aspects: Prefixed chiasm complicates interoptic microsurgical approach. Elongated A1 and short M1 are less problematic for microsurgery than it is in normal conditions.

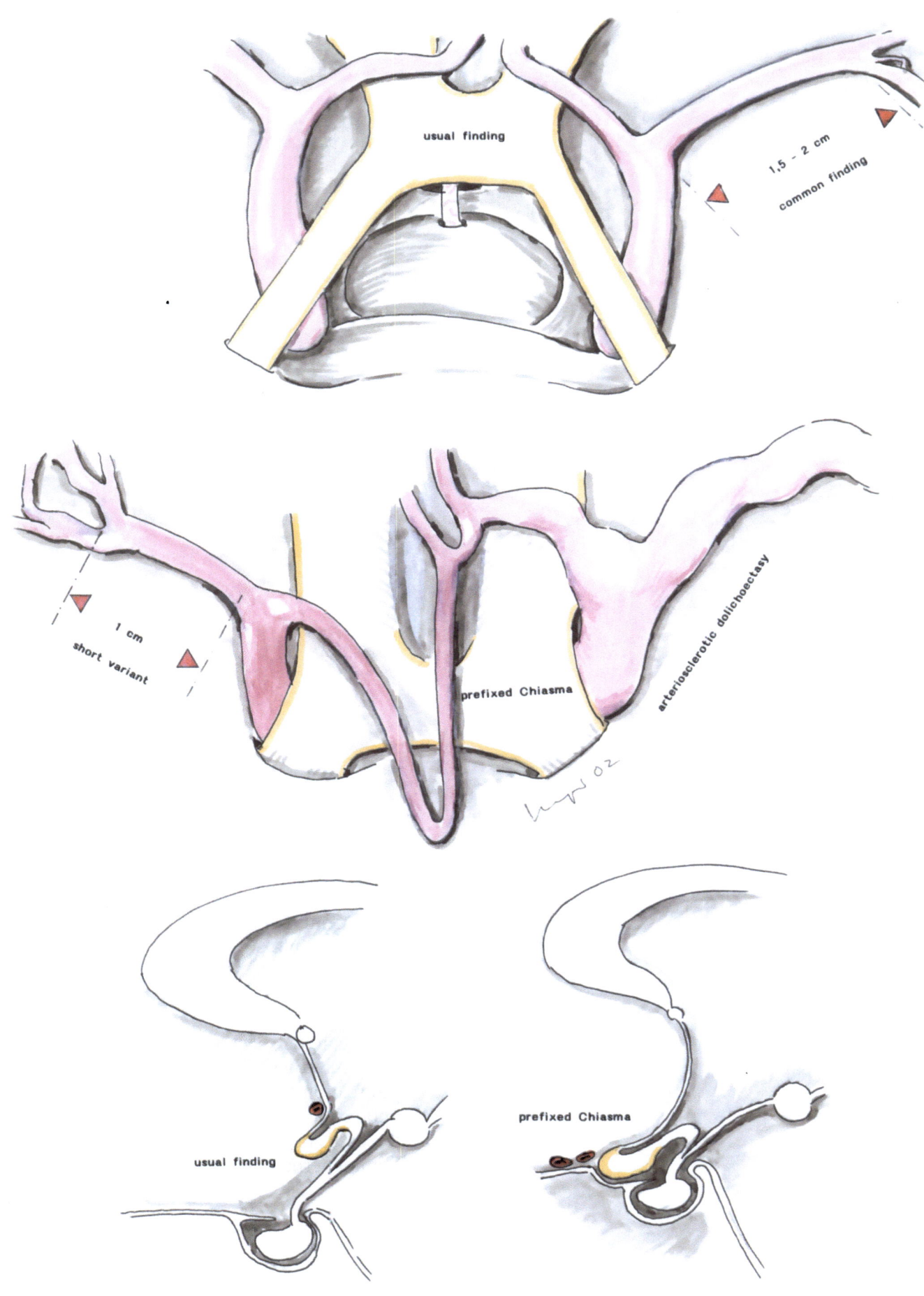

usual finding

1,5 – 2 cm

common finding

1 cm

short variant

prefixed Chiasma

arteriosclerotic dolichoectasy

usual finding

prefixed Chiasma

Fig. 73. **A. recurrens (Heubneri), variants**

A   Common finding
B to **E**   Variants

Abbreviations
a   Subst. perforata ant.
b   basal ganglia
c   Cisterna valleculae and perforators
d   bifurcation of A. carotis int.
e   optic chiasm
f   Recessus opticus
g   Lamina terminalis
h   both A2
i   Tractus opt.
j   Area subcallosa
H   A. recurrens (Heubneri)

FIG. 73

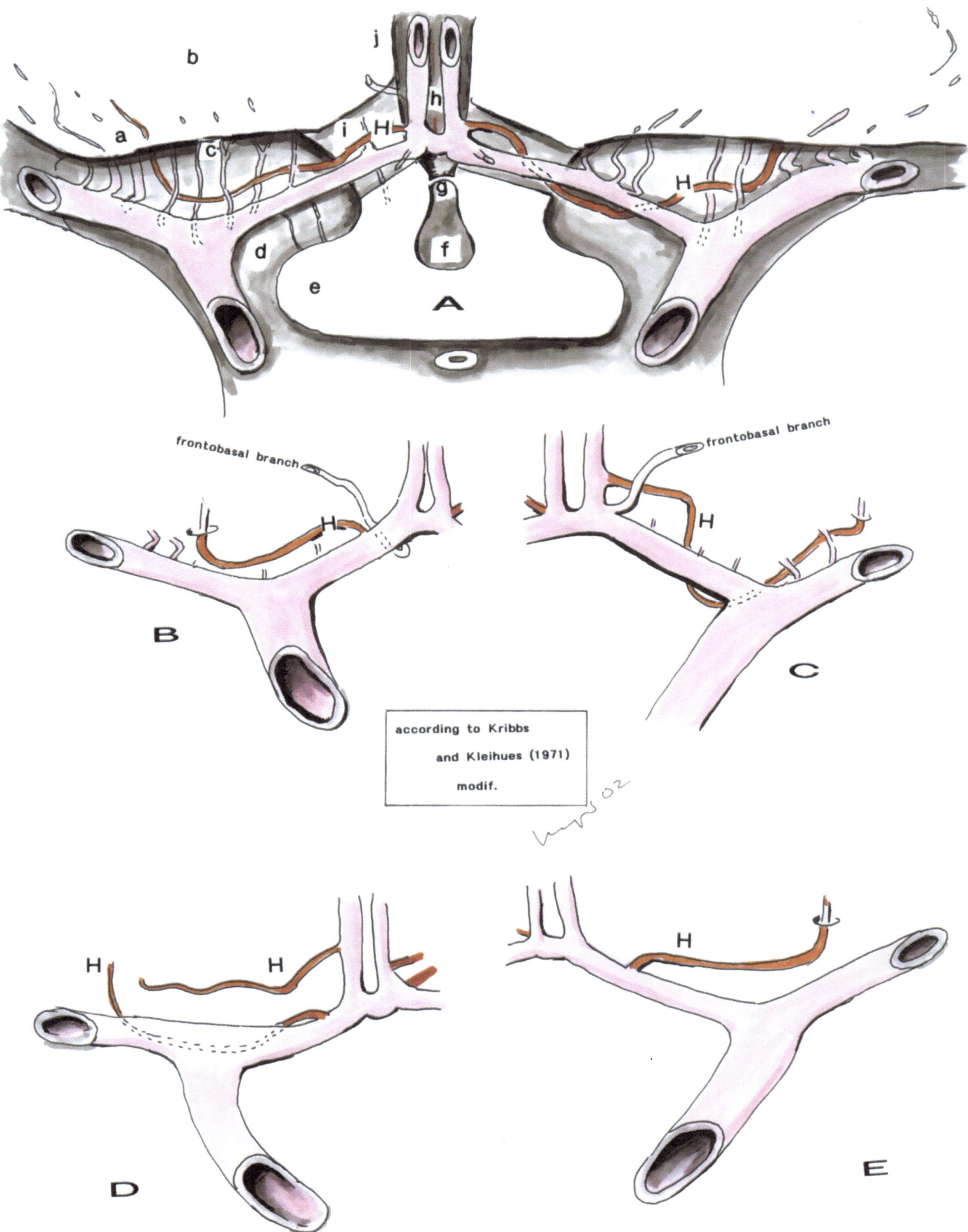

frontobasal branch

frontobasal branch

according to Kribbs
and Kleihues (1971)
modif.

A

B

C

D

E

Fig. 74. Continuation of Fig. 73

Inferior drawing: As upper drawing; senile widening and elongations (dolichoectasy) omitted

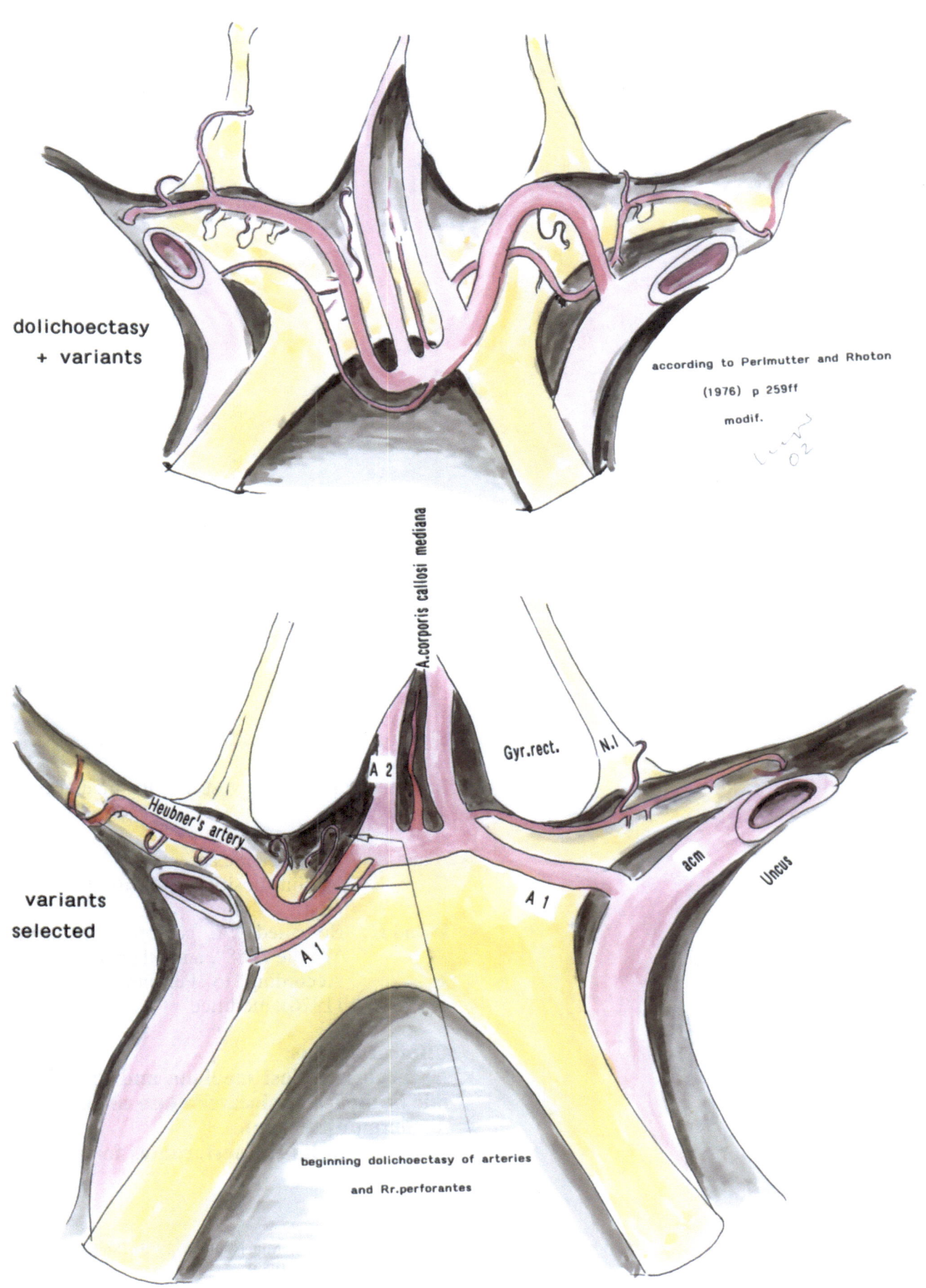

dolichoectasy
+ variants

according to Perlmutter and Rhoton

(1976) p 259ff

modif.

variants
selected

A.corporis callosi mediana

A 2

Heubner's artery

Gyr.rect.

N.I

A 1

A 1

acm

Uncus

beginning dolichoectasy of arteries
and Rr.perforantes

Fig. 75. **A. corporis callosi mediana**

In human, this archaic vessel of the medial cortex, is variable*

A and **B**	Phylogenetic aspects
C	Midline position of the trunk and its division
D and **E**	So-called „A. cerebri ant. media" of Windle 1888", and others (Lang, 1979)
F	According to Perlmutter and Rhoton (1976), modified

Clinical aspects:
A. corporis callosi mediana and its hypothalamic branches are located close to aneurysms of A. communicans ant.
- **Beware of its Rr. perforantes for Hypothalamus –**

---

* Kleiss (1941/42), v. Mitterwallner (1955), Stephen and Stilwell (1969), McCormick (1969)

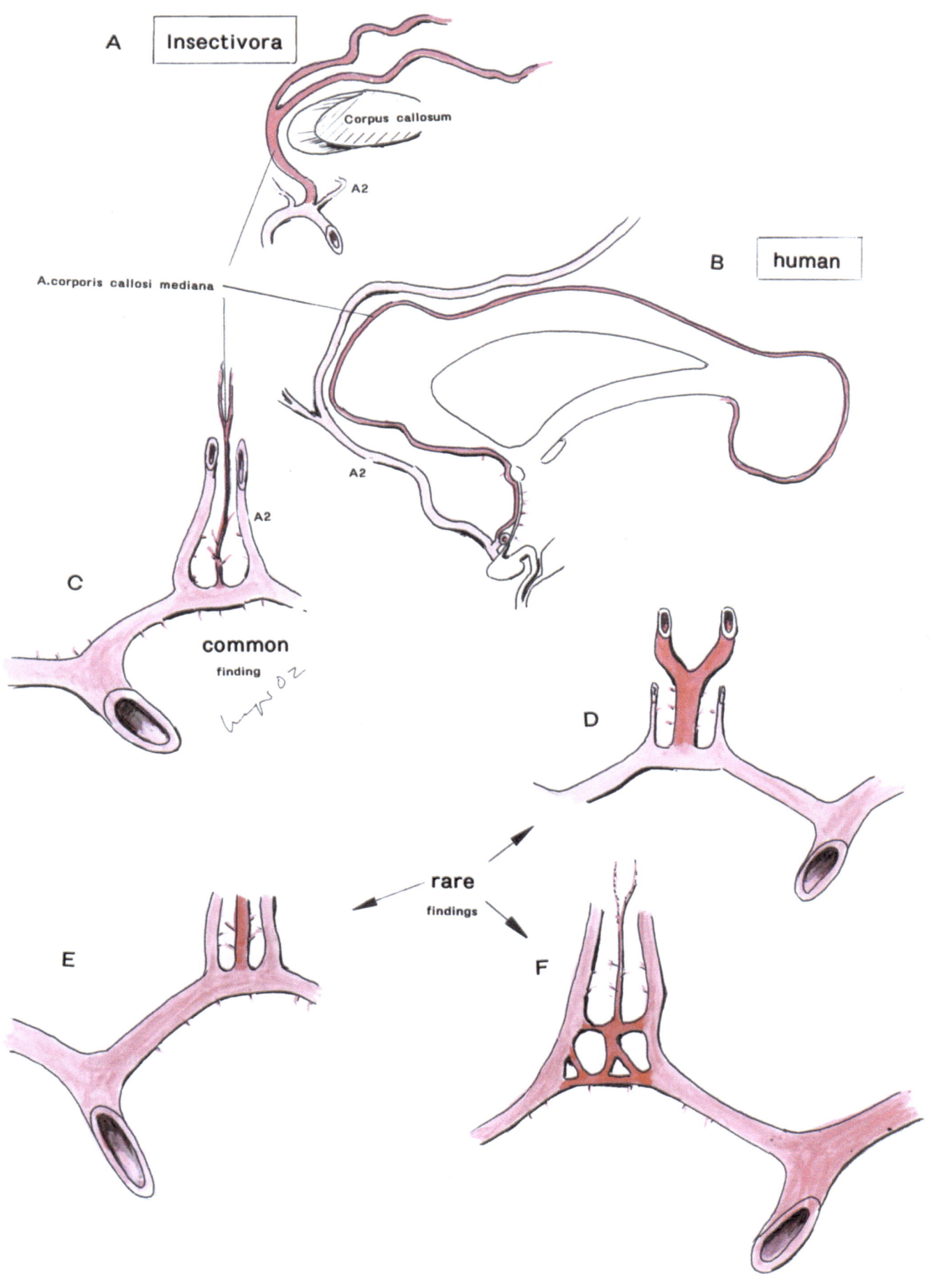

A Insectivora

Corpus callosum

A2

B human

A.corporis callosi mediana

A2

C

A2

common

finding

D

rare

findings

E

F

Fig. 76. **Fenestrations of the anterior segment of Circulus arteriosus Willisi**

A to **C**	A1
D and **E**	A. communicans ant.
F	A. communicans ant. and A1

According to Mitterwallner (1955), and Perlmutter and Rhoton (1976)

Clinical aspects:
All variants are located close to aneurysms of the anterior segment of Circulus arteriosus Willisi

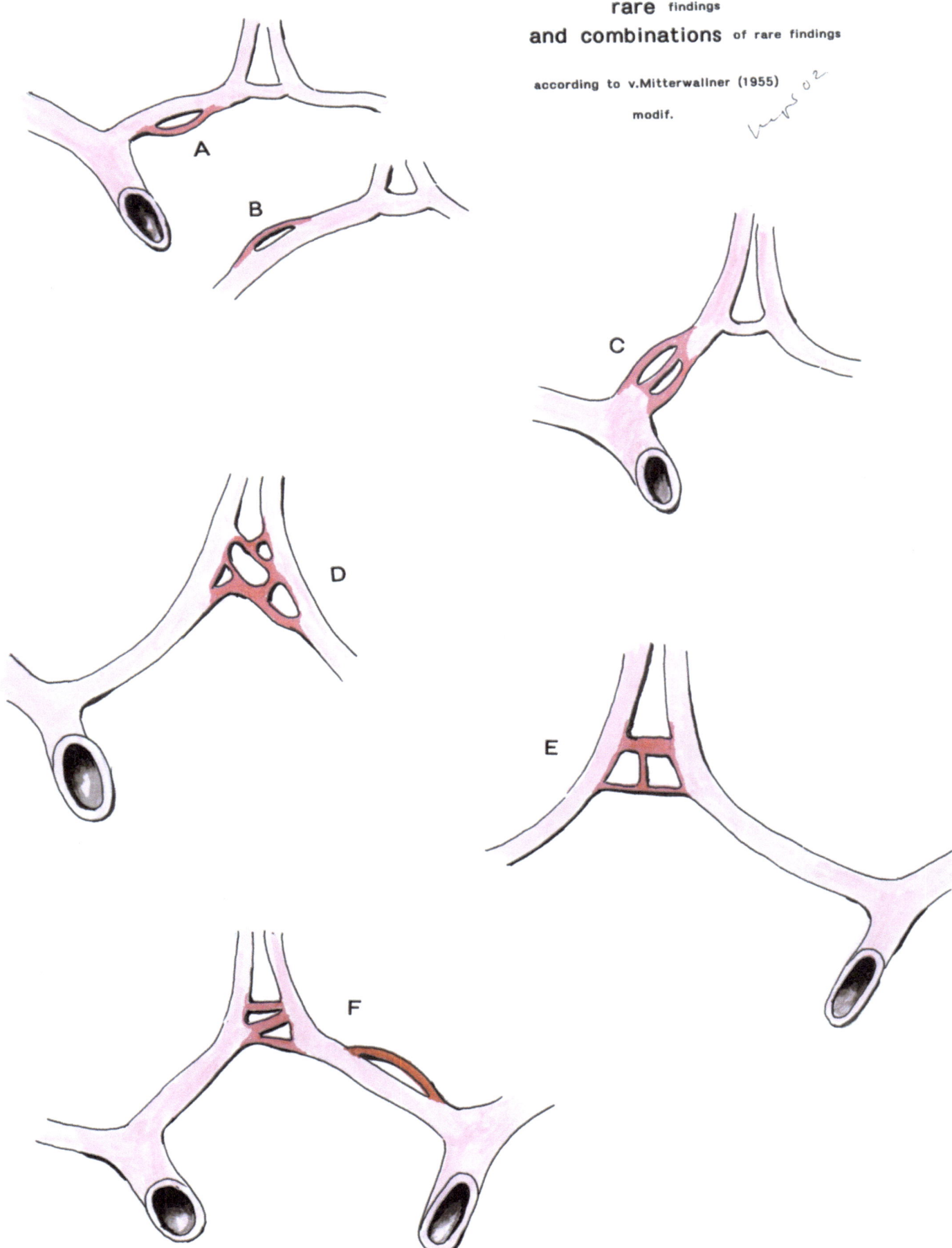

rare findings
and combinations of rare findings

according to v.Mitterwallner (1955)

modif.

Fig. 77. Continuation of Fig. 76

A. recurrens Heubneri (H) and combinations with other arteries

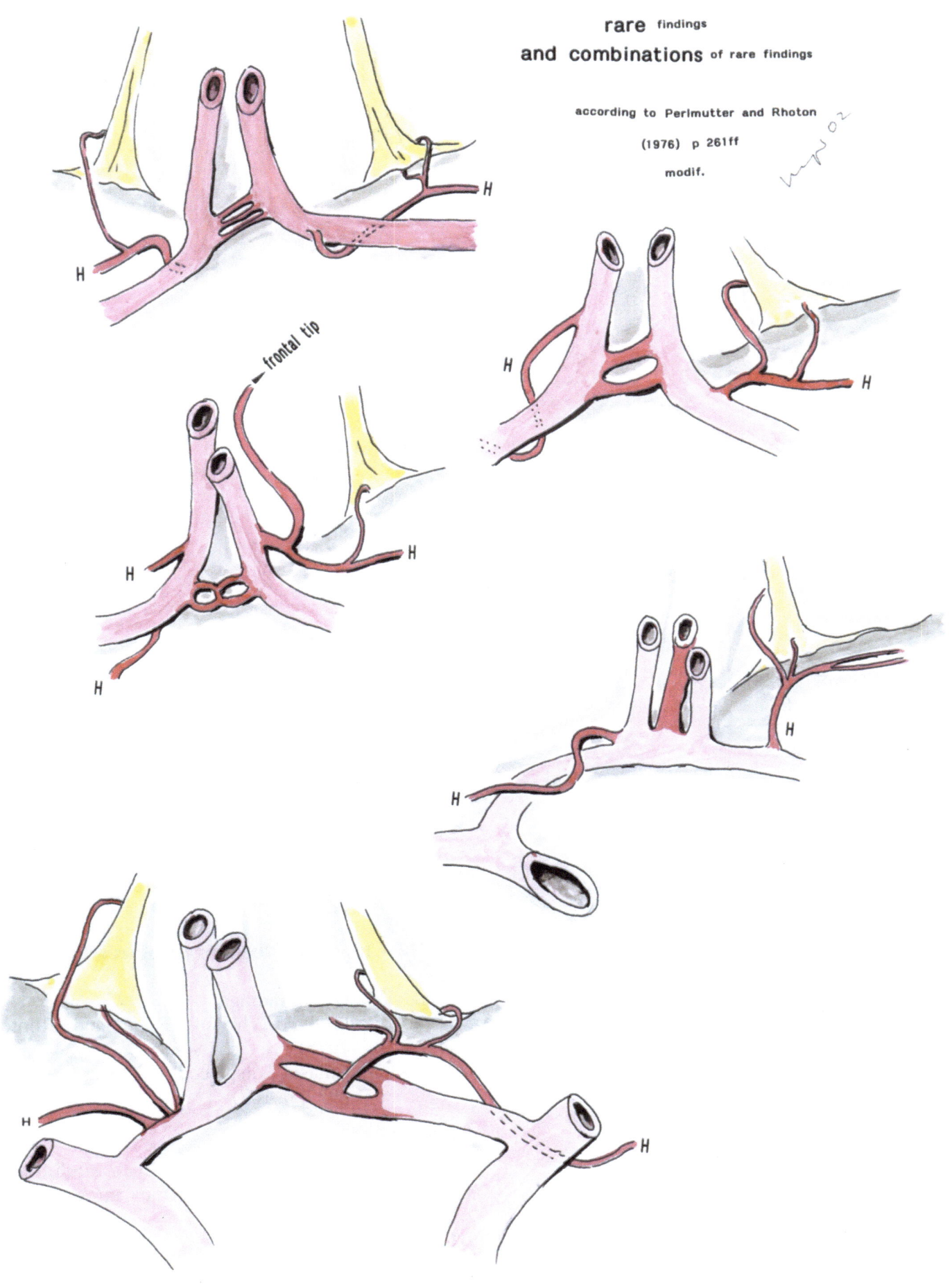

rare findings
and combinations of rare findings

according to Perlmutter and Rhoton
(1976) p 261ff
modif.

frontal tip

Fig.78. **Chiasm prefixed or displaced backwards**

A and **A'**   Usual findings

B and **B'**   Prefixed chiasm (Renn and Rhoton, 1975)

C and **C'**   Chiasm displaced backwards (Lang, 1979, p. 107)

Clinical aspects:

B and B' **unfavorable for the baso-mediane interoptic suprasellar microsurgical approach**

C and C' **favorable**

FIG. 78

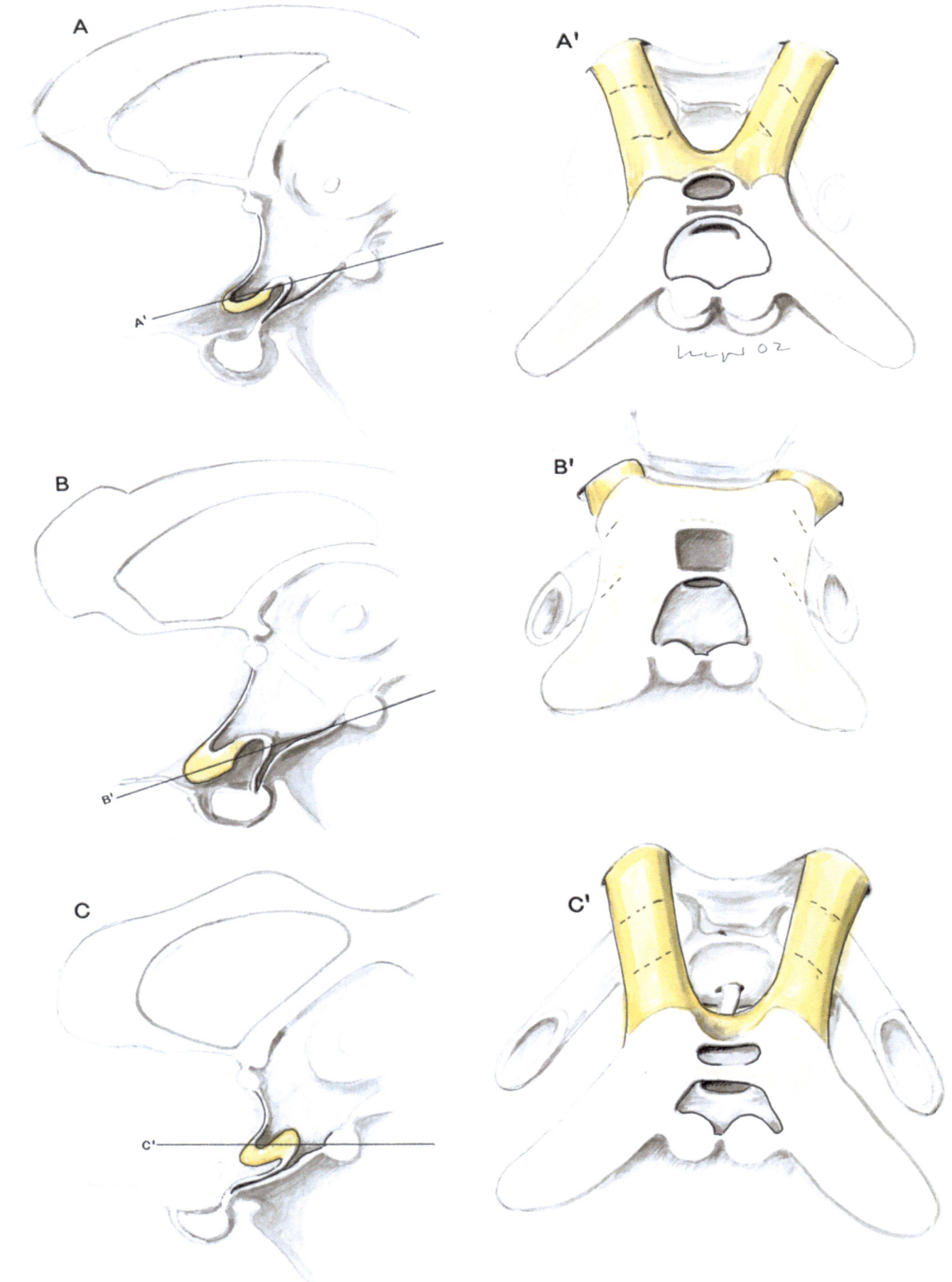

Fig. 79. **Variability of the dural penetration points of the cranial nerves III and IV, and of the carotid artery** (according to Lang, 1979, pp 76 and 77, modified)

A   Usual findings
B   N. III anterior from Dorsum sellae
C   N. III posterior from Dorsum sellae
        N. IV and carotid artery displaced backward
D   normal

Abbreviations
a   Proc. clin. ant.
b   hyophyseal stalk (cut)
c   Sulcus caroticus

Clinical aspects:
These variants must be considered for micro-surgical approaches at the cranial base, especially for basal meningiomas, large pituitary tumors, craniopharyngiomas, and other lesions of the cerebral and cranial base, which are masking the dural penetration points of cranial nerves

FIG. 79

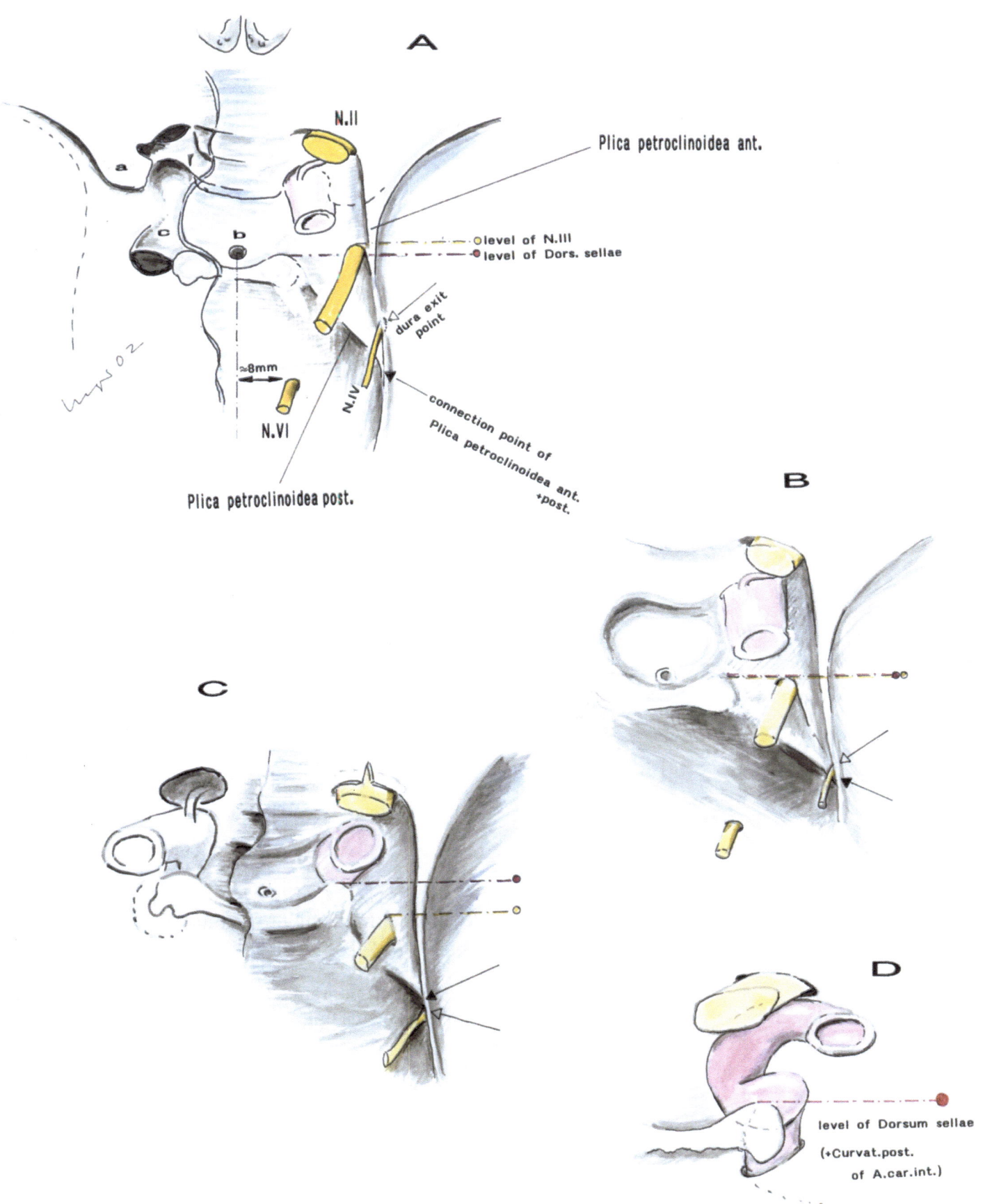

A

N.II

Plica petroclinoidea ant.

○ level of N.III
● level of Dors. sellae

dura exit point

connection point of
Plica petroclinoidea ant.
+post.

≈8mm

N.IV

N.VI

Plica petroclinoidea post.

B

C

D

level of Dorsum sellae

(+Curvat.post.
of A.car.int.)

Figs. 80 to 84. **N. opticus. Errors in MRT-presentations, which may be mistaken for variants or pathological findings**

Fig. 80.* Topogram: B to D not congruent with the axis of Tractus opt.

A   MRT level basal from Corpora mamillaria: error

B   MRT level transsecting the basal segment of Chiasma and the equator of Corpus mamillare: error

---

* Figs. 80 to 82 present sketches for understanding the colored copies of MRT's Figs. 83 and 84

FIG. 80

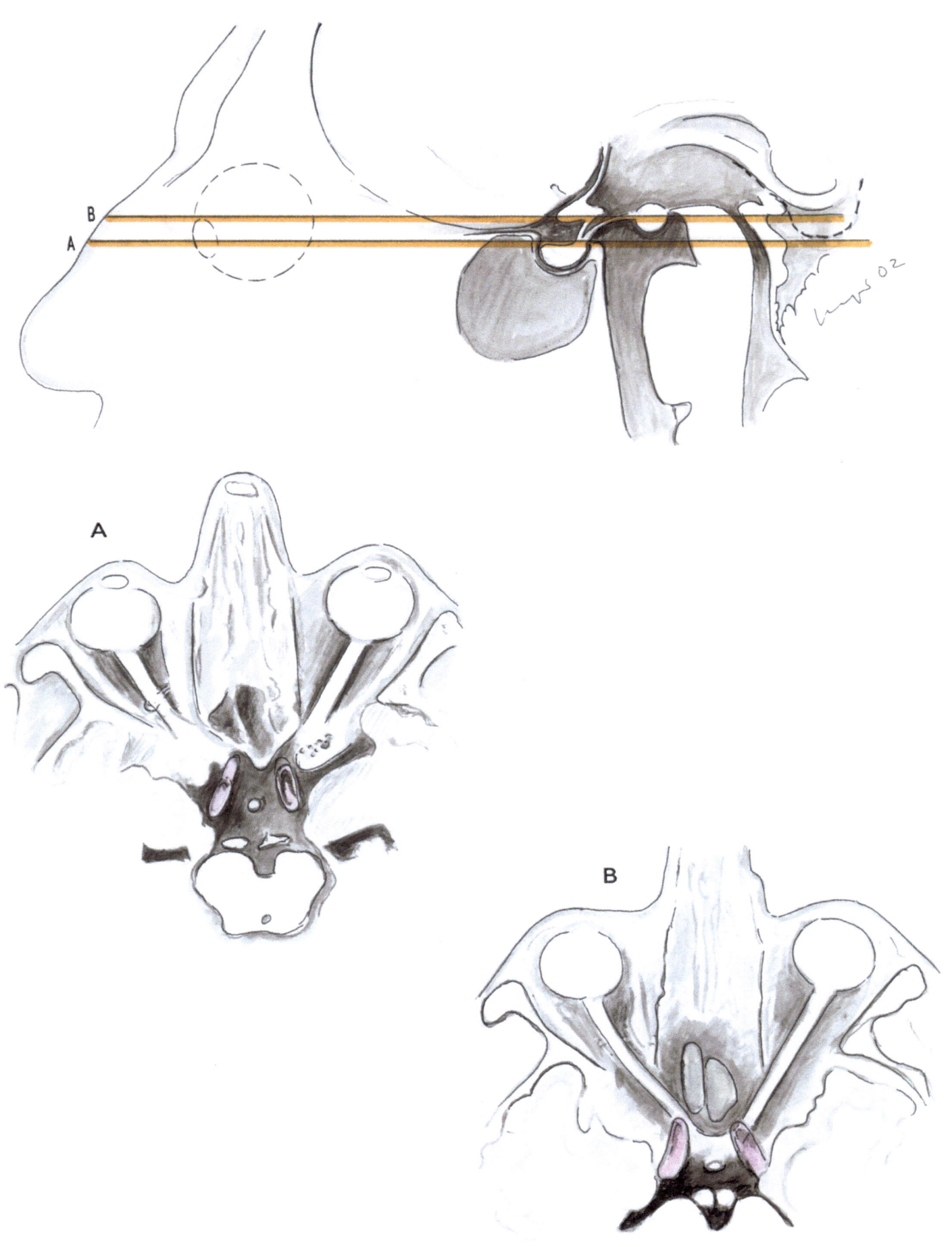

Fig. 81. Topogram: C and D not congruent with the axis of the optic system

C   MRT level rostral too far basally, occipital too far dorsally: error

D   MRT-level leftsided correct

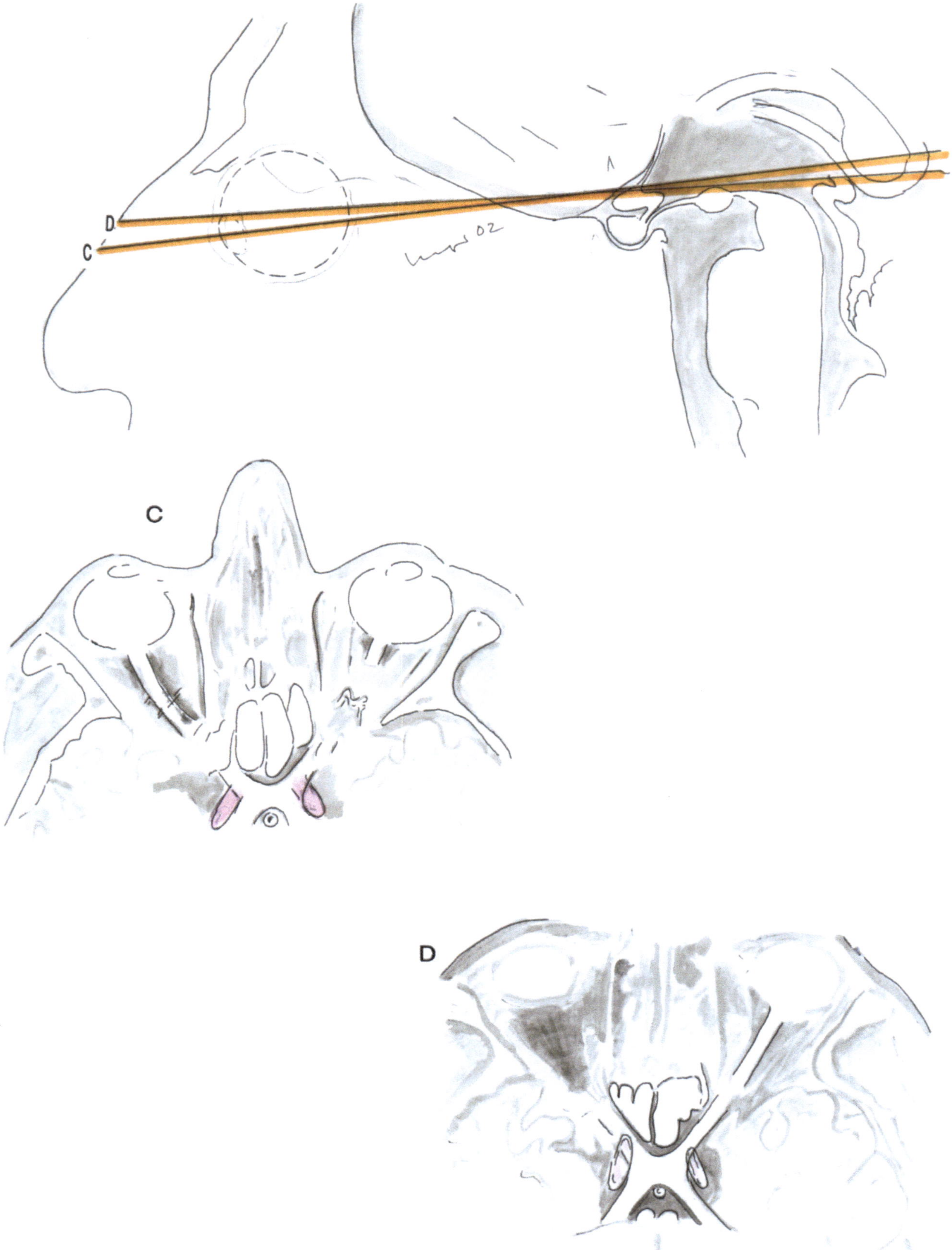

C

D

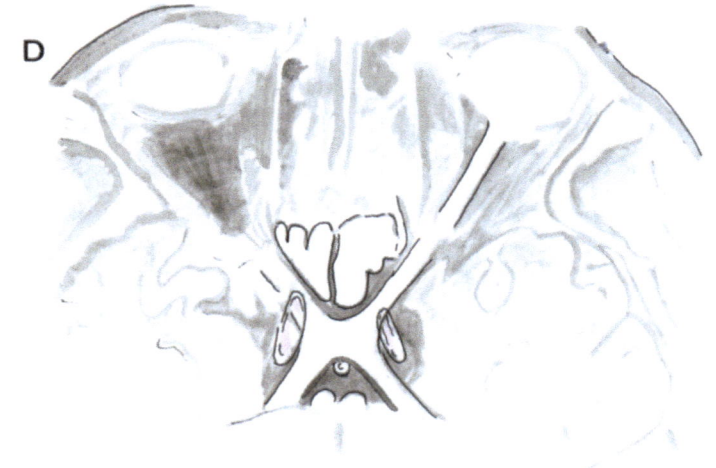

Fig. 82. Topogram: **correct**

E   **Upper segments of Corpora mamillaria and of Infundibulum present**

FIG. 82

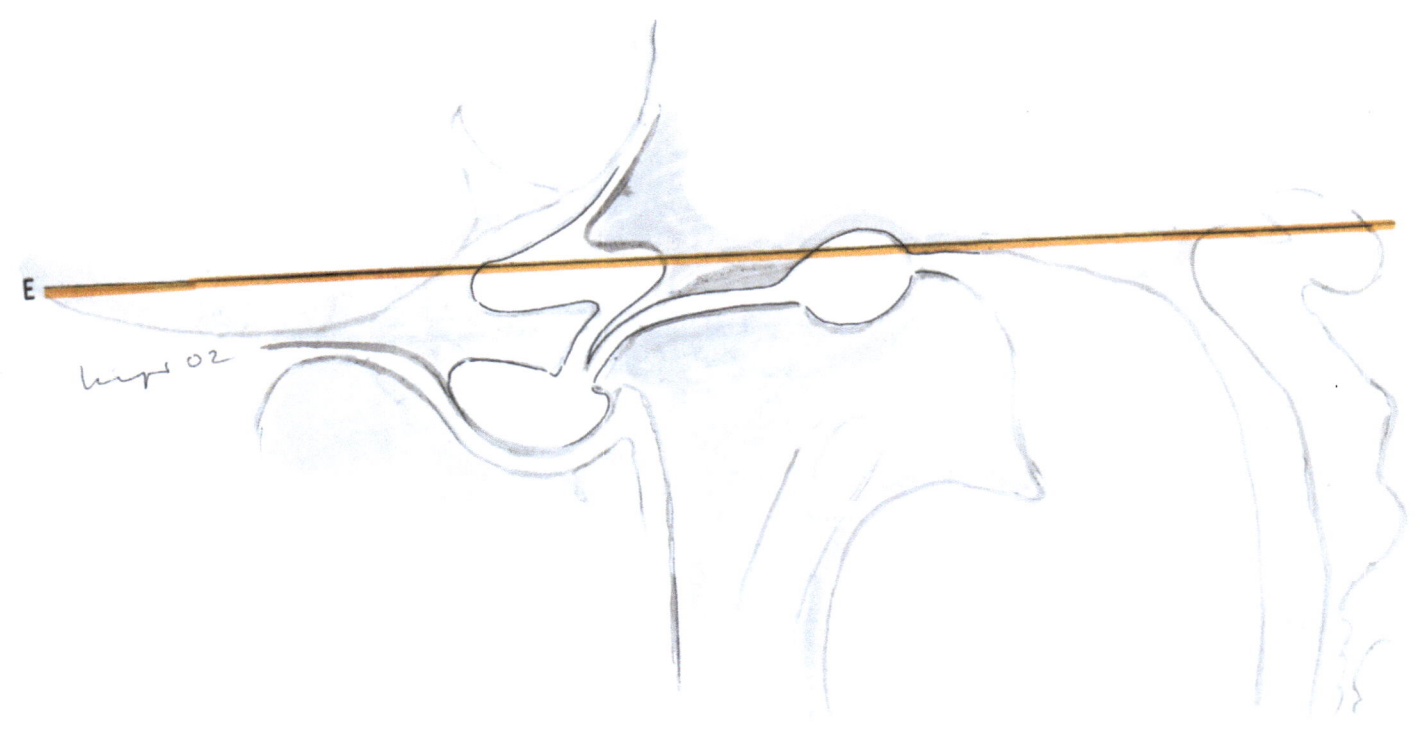

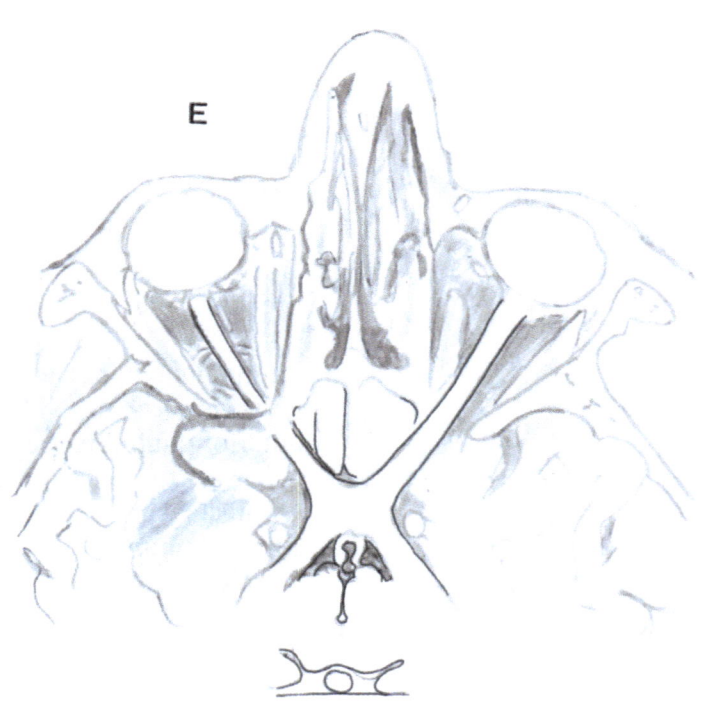

Figs. 83 and 84.

Colored MRT's

A to **E**   According to the sketches Figs. 82 to 82
                     (C is omitted)

Abbreviations

a	N. II
b	A. carotis int.
c	Sinus sphenoidalis
d	Dorsum sellae
e	prepontine cistern
f	P1
g	interpeduncular cistern
h	Crus cerebri
i	anterior clinoid process
j	Gyrus rectus
k	Optic chiasm
l	Infundibulum
m	Corpus mamillare
n	Aquaeductus Sylviii
o	Tectum
p	Colliculus sup.
q	Corpus pineale
r	Splenium corporis callosi
s	Hypothalamus
t	Tractus opt.

a
c
i
b
d
A
e
f
h
g

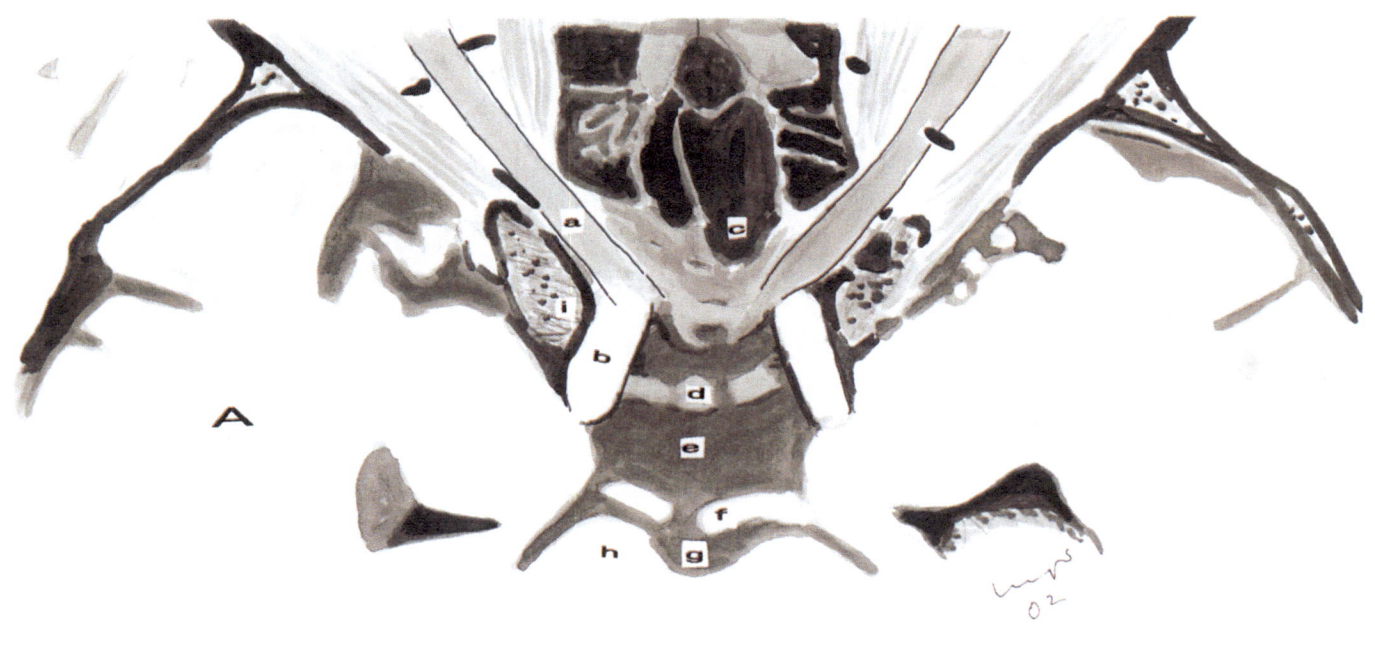

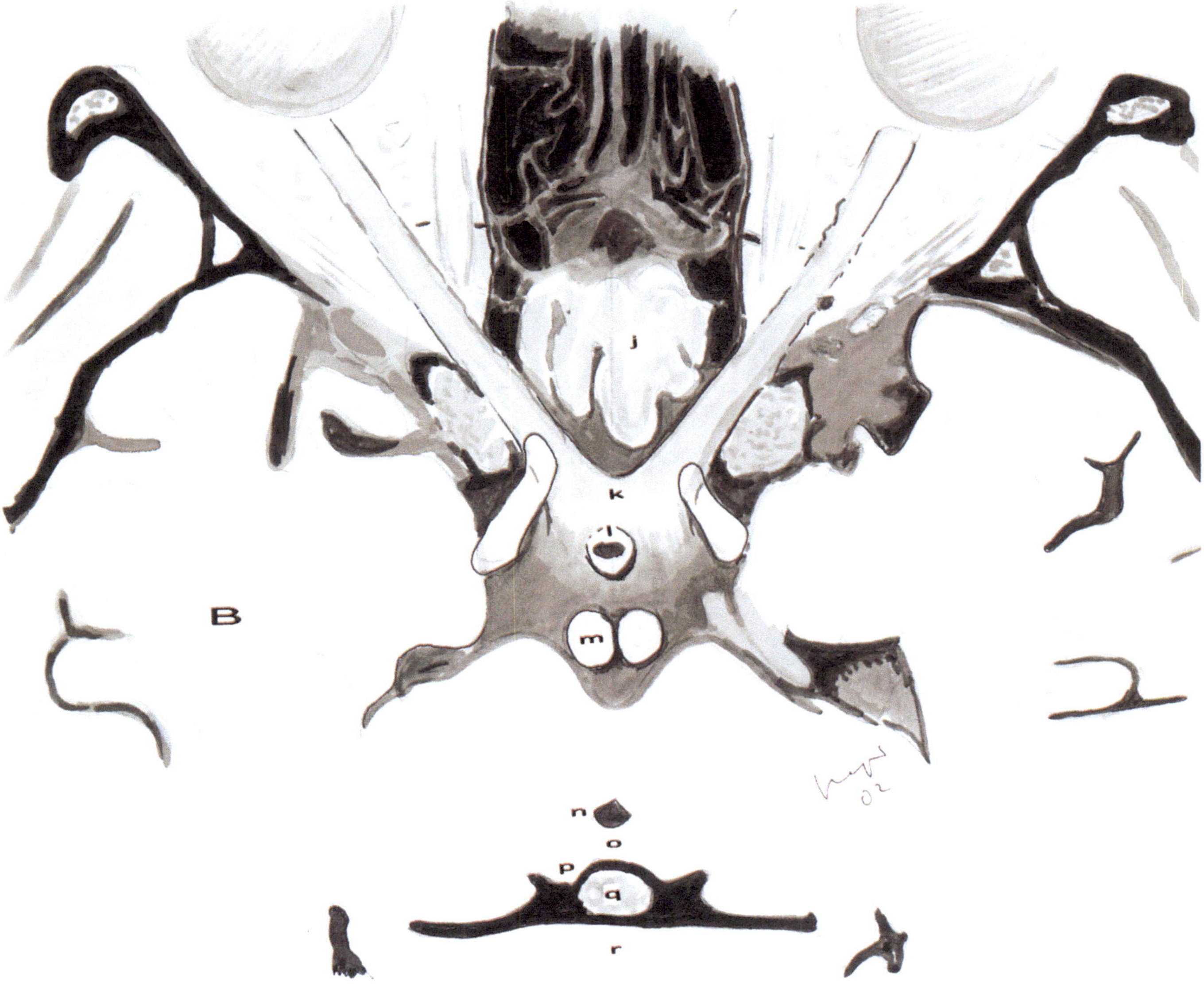
j
k
l
B
m
n
o
p
q
r

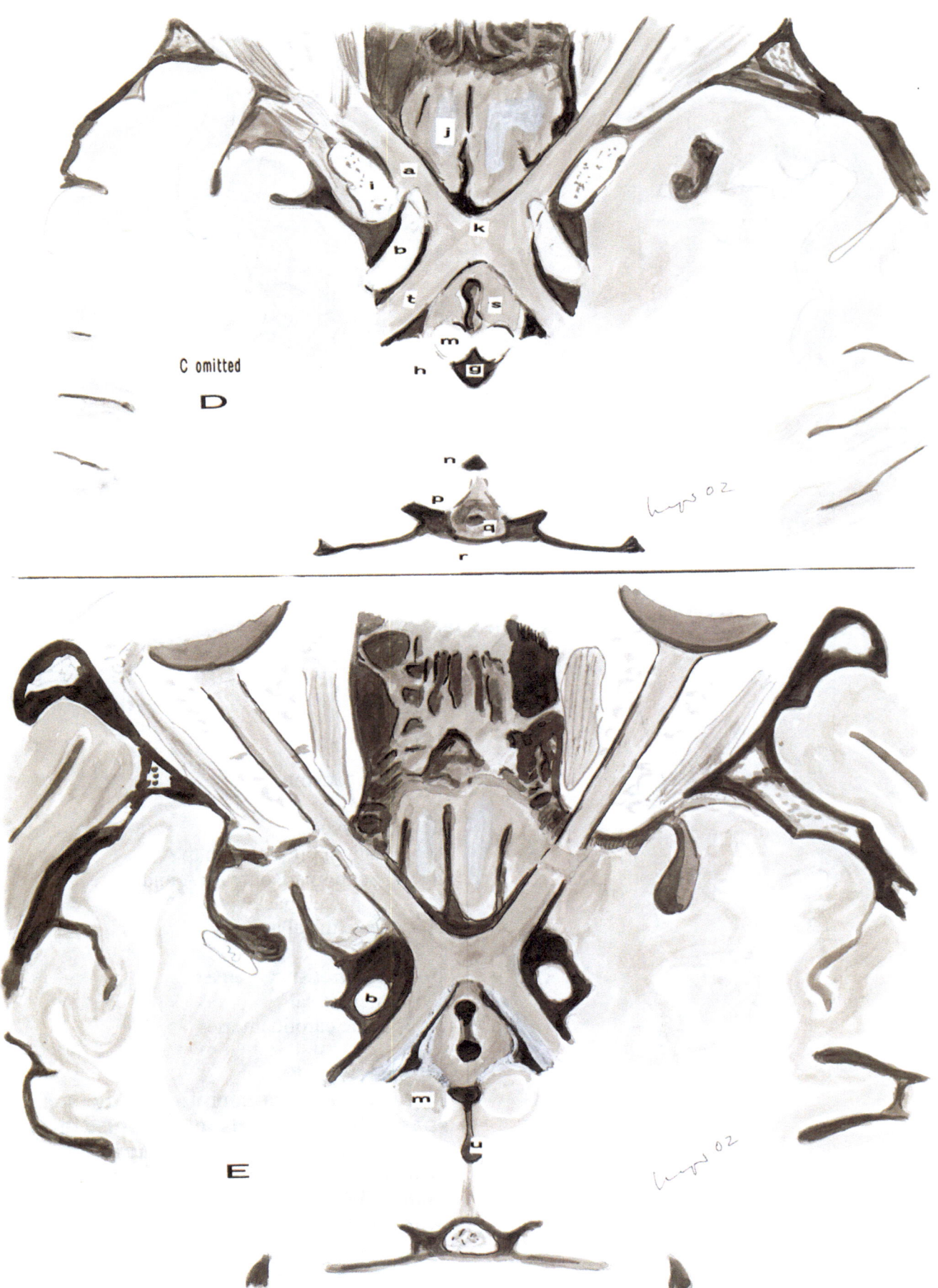

C omitted

D

E

Fig. 85. **Chiasma**

A   **Prefixed**
A' Sagittal presentation to A
B   For comparison with A and C
C   Displaced backwards
D   See clinical aspects

Distance measurements
a   interoptical
b   of the carotid arteries

Clinical aspects:
Prefixed Chiasma complicates the suprasellar approach. All variants may confuse intraoperative orientation, especially during an oblique position of the patients head on the operation table. This must be taken in consideration by untrained neurosurgeons. The optic system may be masked by lesions (see D), **or shifted, combined with variants**

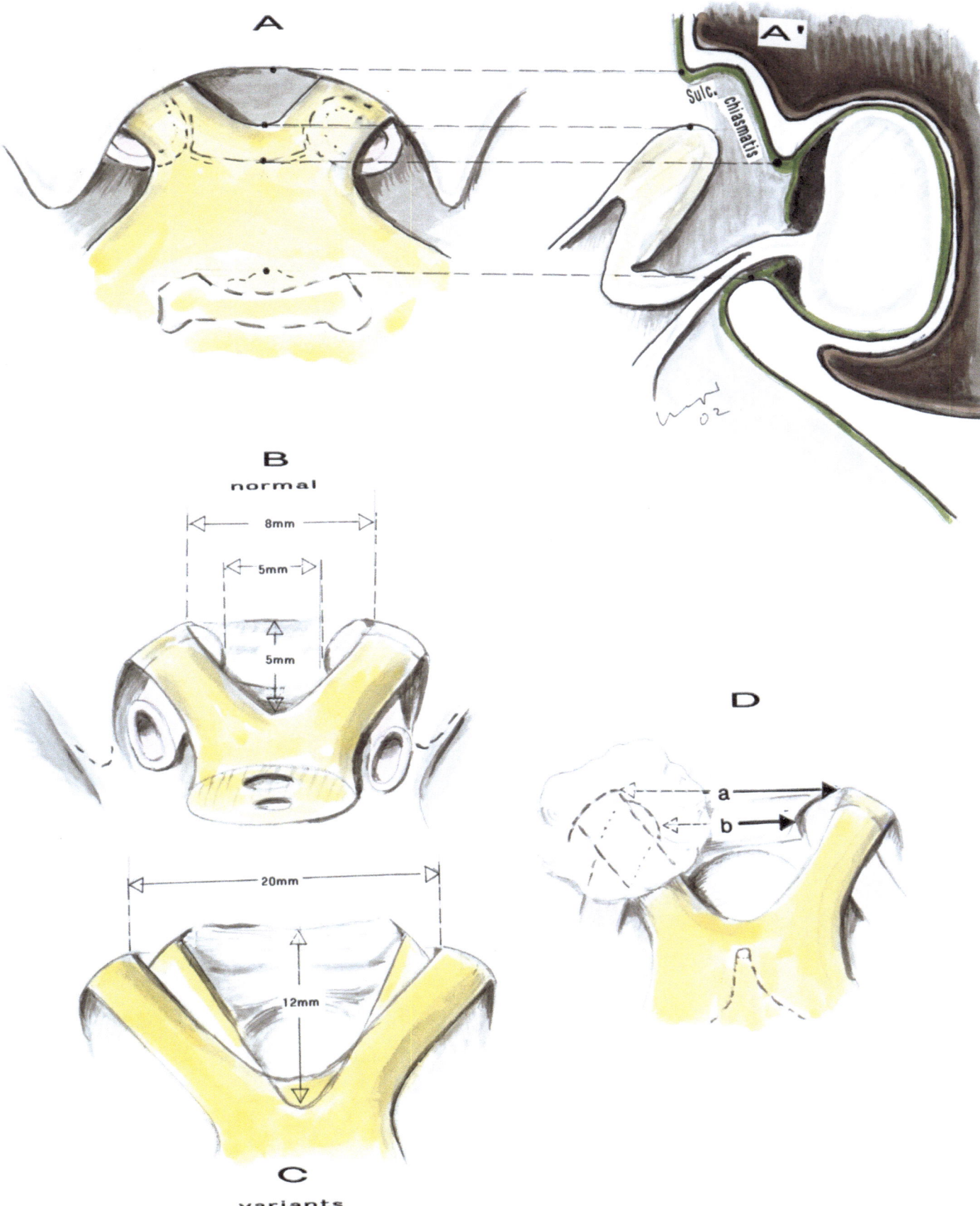

A

A'

Sulc. chiasmatis

B
normal

8mm

5mm

5mm

D

a

b

C
variants

20mm

12mm

A,B,and C according to photographs of Bergland et al.(1968)

considerable modified

Chapter 3
Variants of the temporolateral and temporomedial areas
(Figs. 86 to 100)

Fig. 86. Temporal microsurgical approaches

Schematic presentation

---

* Bony landmarks

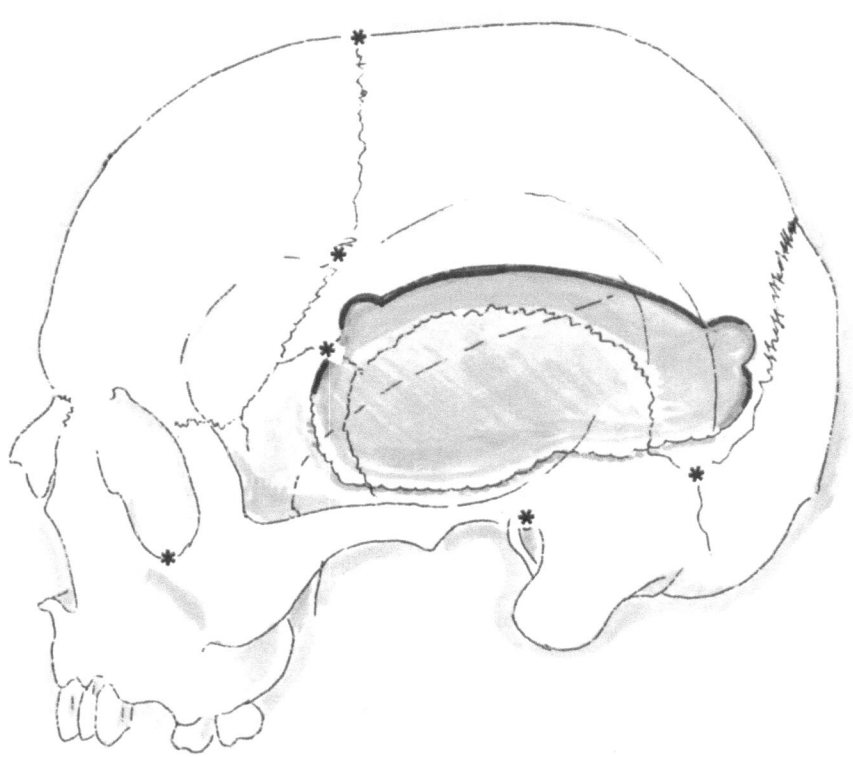

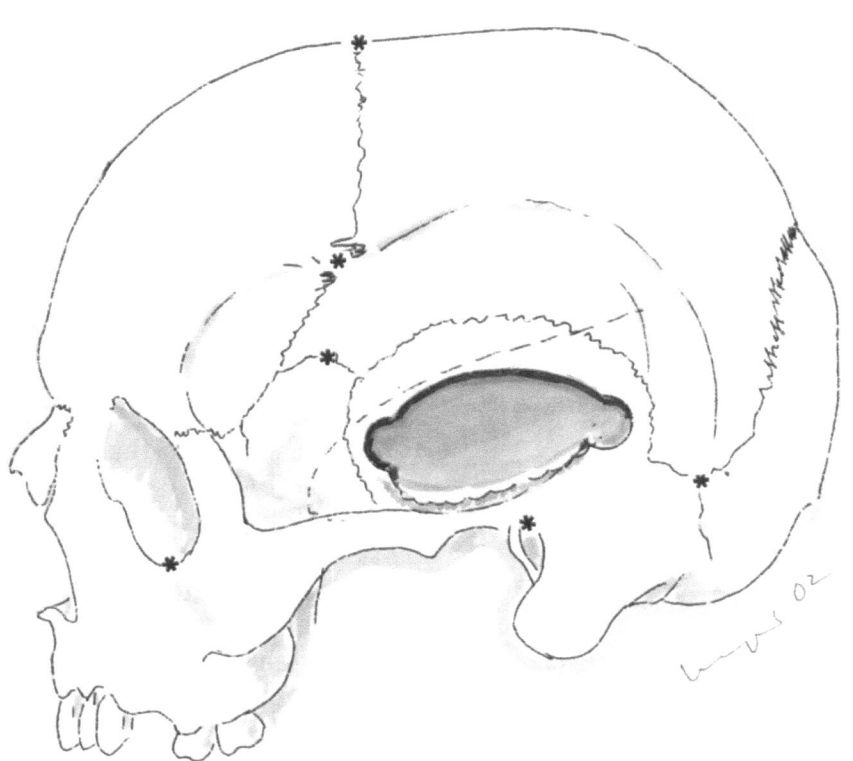

Fig. 87. **Asterion, variants**

3 types of variants from the skull collection of the author

A   Small skull (female, 48 years)
B   Large skull (male, 65 years)
C   Female (?), elderly individual (siphon calcified)

Note Ossa suturalia (B and C)

Clinical aspects:
This anthropological landmark can be useful for defining of the area of Sinus transversus and the knee of the sinus during trepanation. Danger for bleeding from the Sinus sigmoideus during elevation of a traumatic bony impression in the area of Asterion

FIG. 87

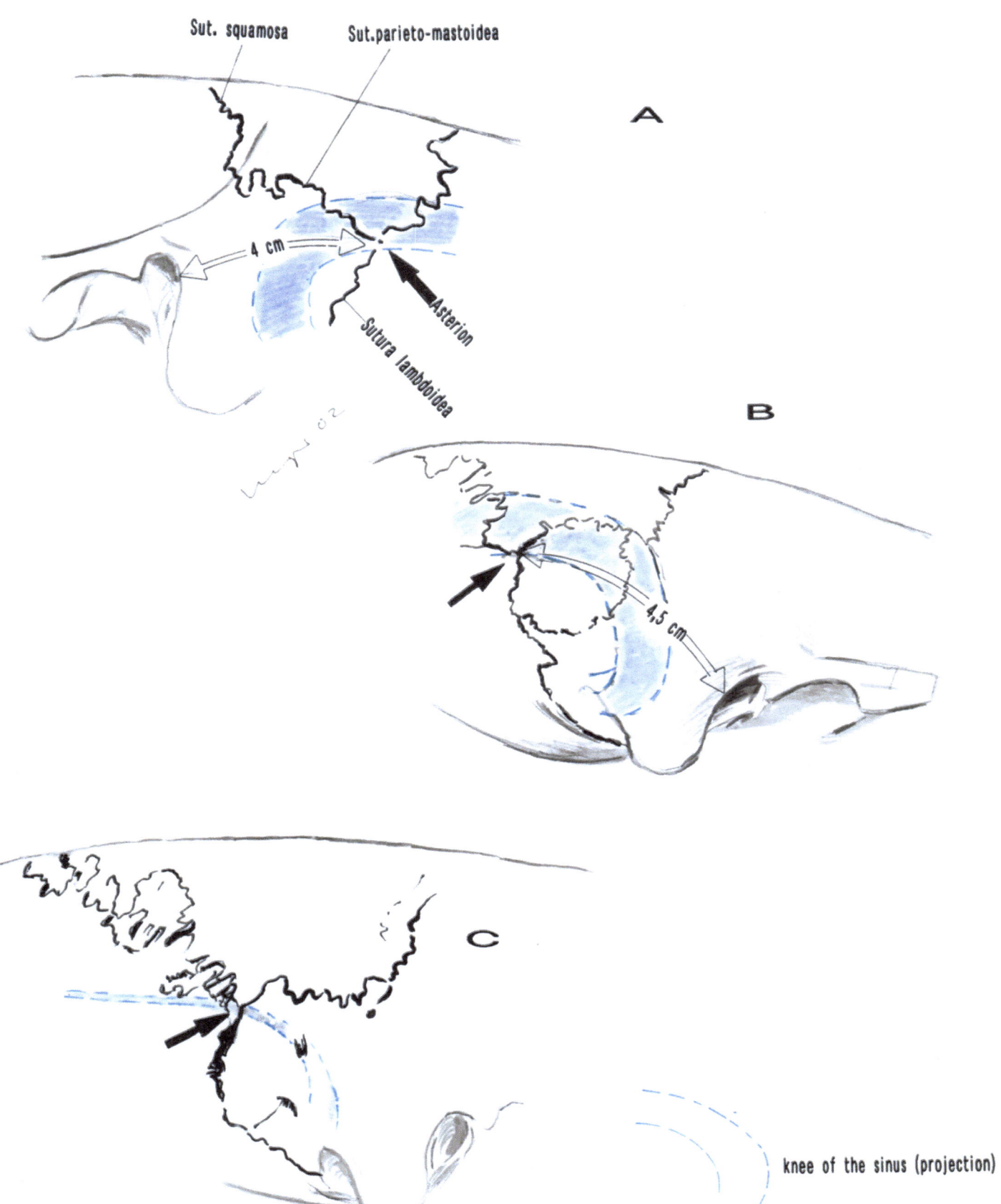

Sut. squamosa

Sut.parieto-mastoidea

A

4 cm

Asterion

Sutura lambdoidea

B

4,5 cm

C

knee of the sinus (projection)

Fig. 88. **Os sphenoidale, survey** (upper drawings)

Aplasia of Foramen spinosum (inferior drawings)

Clinical aspects, if Foramen spinosum is aplastic:
Surgery of traumatic epidural hematoma. A. meningea media – main trunk – will be missed close to the area of Foramen ovale/spinosum
For details see Figs. 55 and 56

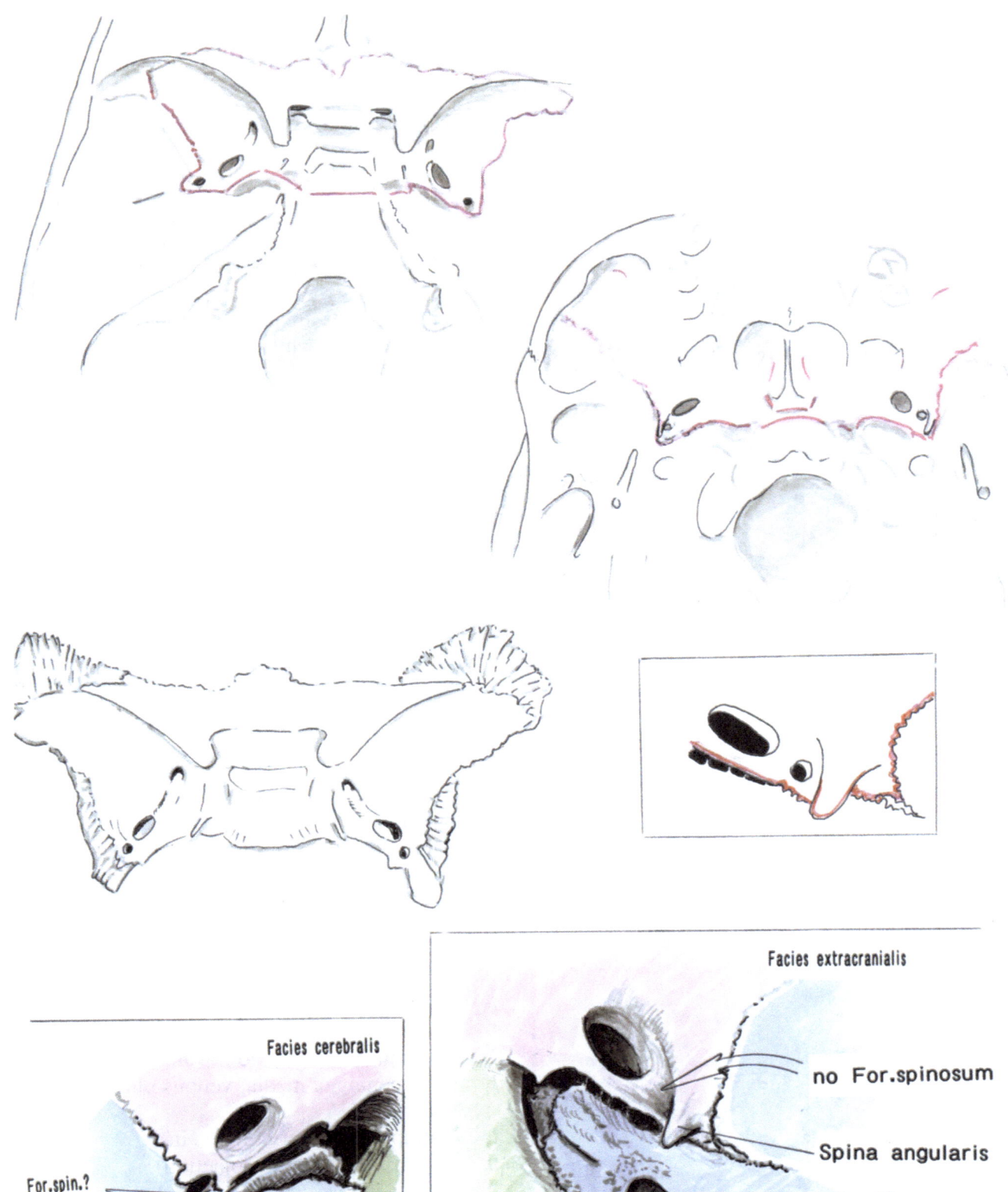

Facies extracranialis

no For.spinosum

Spina angularis

Facies cerebralis

For.spin.?

suture?

Fig. 89. **Dural veins and sinuses temporobasal**

Intracranial veins are extremely variable in size, diameter and course. This drawing presents a collection of typical veins. In one and the same individual all these typical veins never occur simultaneously, only a part of them exist.

Clinical aspects:
Special caution has to be taken to preserve the variable group of temporobasal veins posterior from the petrous bone, e.g. during temporobasal microneurosurgical approaches for elimination of aneurysms of A. basilaris (Drake's method) – danger for temporal brain damage -.
V. Labbé must be preserved, too.

Abbreviations

a	Sinus intercavernosus post.
b	Sinus intercavernosus ant.
c	Uncus vein (variant)
c'	temporopolar bridging veins
d	Venous plexus of For. (Can.) rotund.
e	dural fold enclosing Sinus paracavernosus (typical finding)
f	Ophthalmic vein (projection)
g	V. meningea media
h	Sinus sphenoparietalis (posterior branch of g)
i	Sin. sphenopetrosus, subarachnoid segment (common finding)
j	Venous plexus of For. ovale
k	A. meningea media, venous plexus of For. spinosum
l	temporobasal bridging veins
m	Porus acust. ext. (projection)
n	Eminentia arcuata
o	Sin. petrosus sup.
p	knee of the sinus
q	temporobasal bridging veins
r	V. Labbé and a typical vein, parallel
s	lateral pontine vein (large variant)
t	V. mesencephalica lat.
u	Sinus rectus

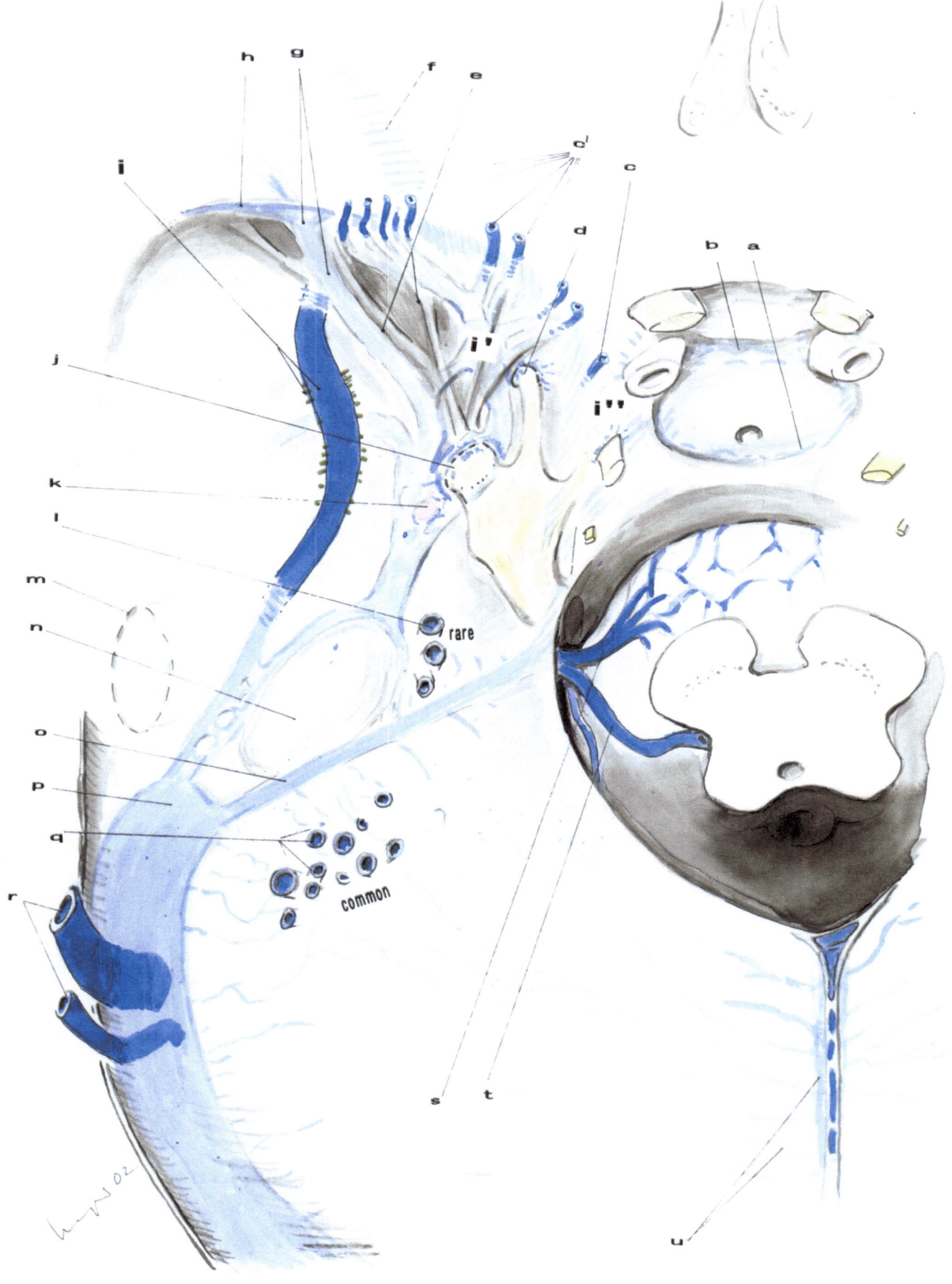

rare

common

Fig. 90. **Temporobasal cerebral veins**

Clinical aspects:
See Fig. 89

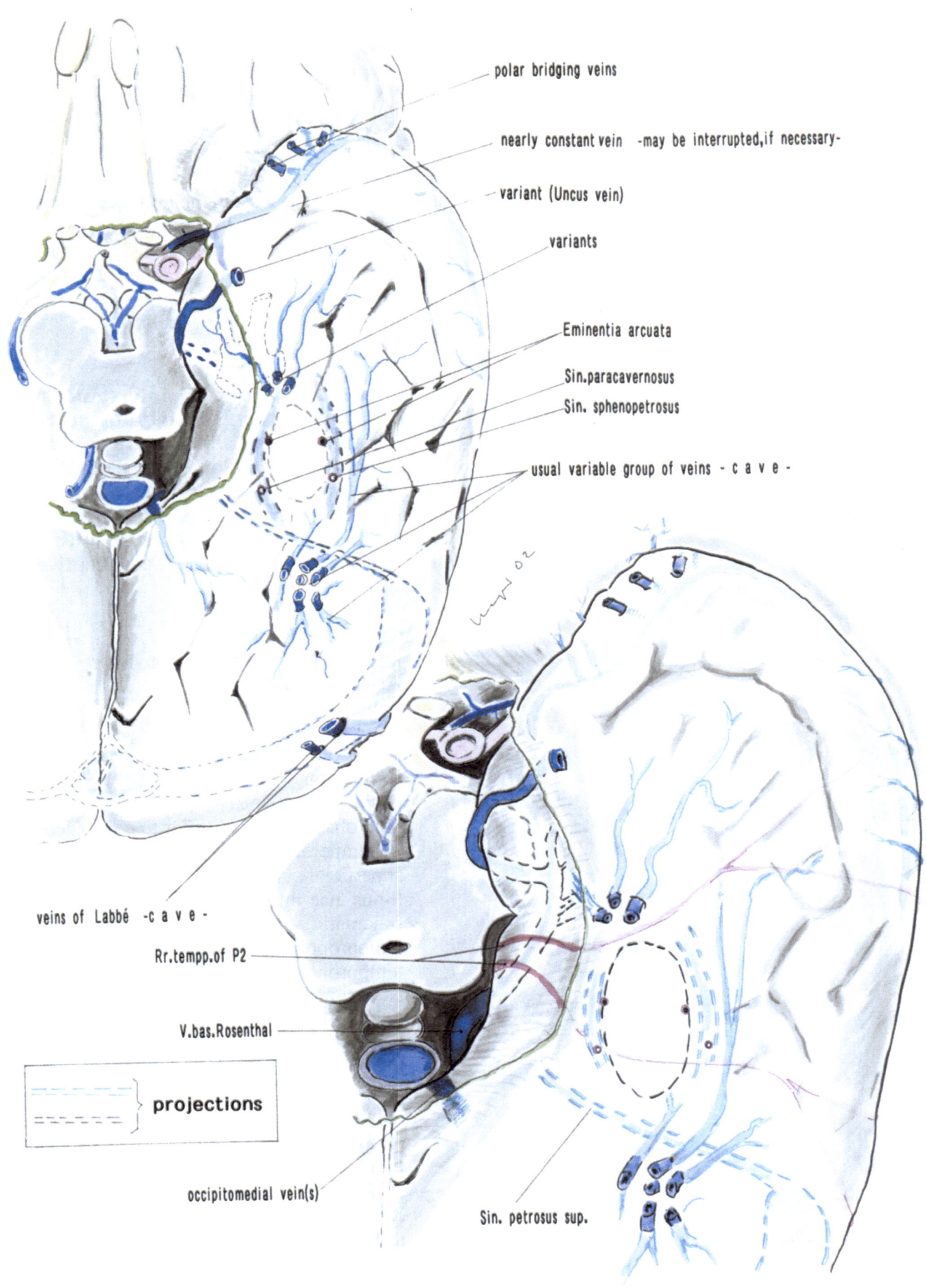

polar bridging veins

nearly constant vein   -may be interrupted,if necessary-

variant (Uncus vein)

variants

Eminentia arcuata

Sin.paracavernosus

Sin. sphenopetrosus

usual variable group of veins - c a v e -

veins of Labbé -c a v e -

Rr.tempp.of P2

V.bas.Rosenthal

projections

occipitomedial vein(s)

Sin. petrosus sup.

Fig. 91 **Temporolateral cerebral veins**
The connections of V. Labbé with sylvian veins are variable. Sylvian veins and their connections with superior and frontopolar bridging veins (with Sinus sphenoparietalis and temporobasal sinuses) are variable, too.

A and **B**   Samples of this variability

Clinical aspects:
These variable veins may be interrupted during operations, but it is unsure, if brain damage will occur or not. Only V. Labbé should be preserved. Testing with application of a temporary clip may sometimes be helpful. If brain volume is increasing after clipping, then this vein should be preserved. If there is no brain swelling after clipping, then it is not sure, if consecutive brain damage will occur or not.

Abbreviations
a   knee of the sinus (projections)
b   Asterion
c   Sutura lambdoidea
d   Sutura parietomastoidea
e   usual variable basal veins, see Fig. temp. 5 (projections)
f   Eminentia arcuata (projection)
g   mediobasal variant similar to e (projection)
h   A. temporalis post.
i   superficial sylvian veins, usual findings
j   Sinus alae minoris (transectional presentation, schematic)
k   temporopolar bridging veins
l   temporopolar artery(ies) originating from A. cerebri media
m   temporobasal arteries originating from A. cerebri media
n   vein of Labbé
o   usual vein running almost parallel to n
p   common variant, testing of stasis by temporary clipping and ultrasonic scan may be helpful (uncertain method, if no brain swelling immediately after temporary clipping)
q   common variant
r   as h, large diameter
s   as n, common variant with large diameter

FIG. 91

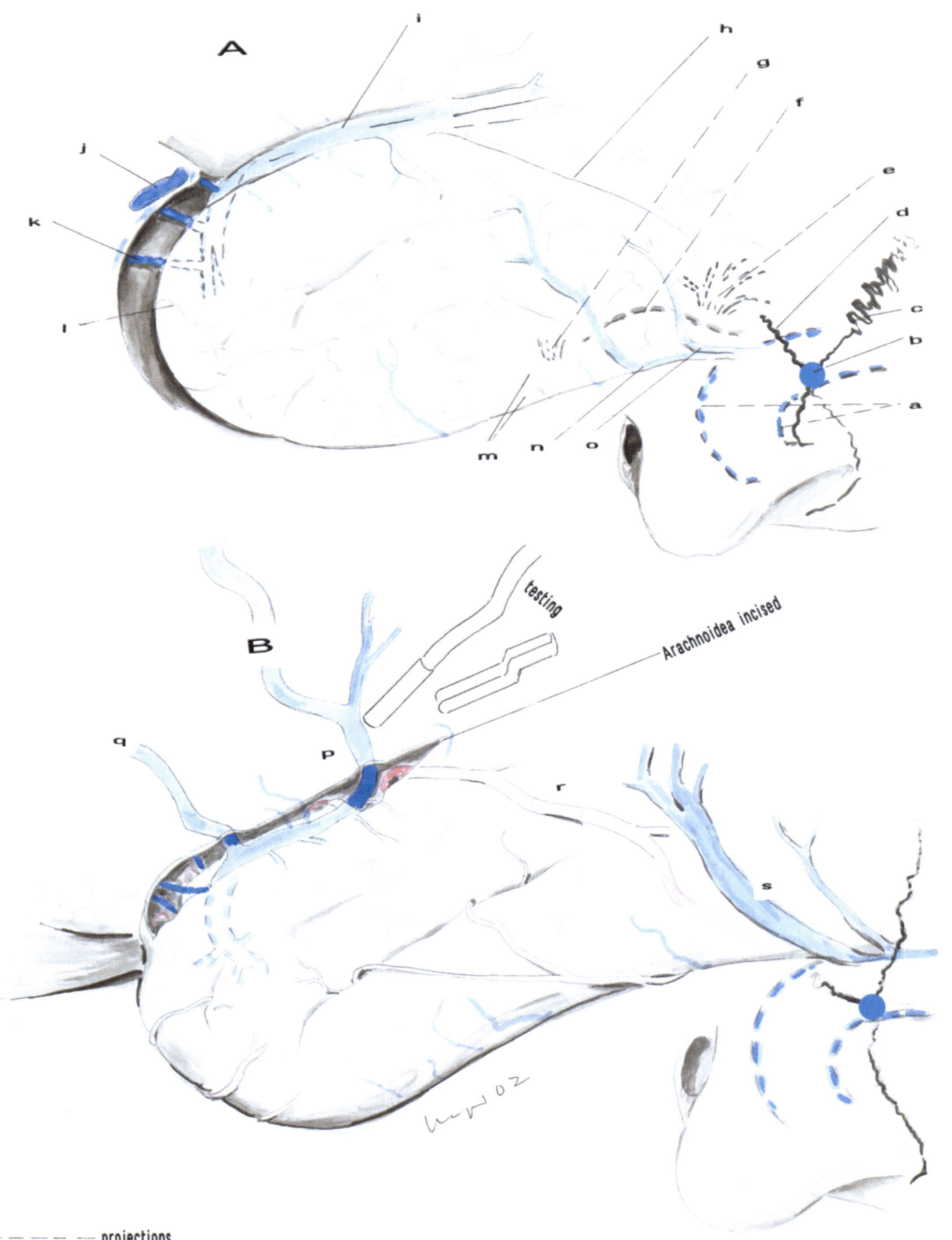

A

i
h
g
f
j
e
k
d
l
c
b
a
m n o

B

testing

Arachnoidea incised

q
p
r
s

projections

Fig. 92. **Deep sylvian veins**
Survey

A   Deep sylvian veins are located close to Insula
with common connections with
  • Vena basalis Rosenthal –1-
  • parietal bridging veins –2-
  • superficial sylvian veins –3- (arrows: blood
    stream)
B   Veins of A, common further connections with
  • veins of the mouth of sylvian fissure, especi-
    ally V. basalis (Rosenthal) 1-
  • Sinus sagittalis (longitudinalis) sup. –2-
  • temporopolar bridging veins and Sinus alae
    minoris –3a-
  • V. Labbé and Sinus transversus –3b-

Clinical aspects:
High variability of sylvian veins and their
connections should be considered in microsur-
gery and interventional neuroradiology
For literature see Seeger, W. (1984)

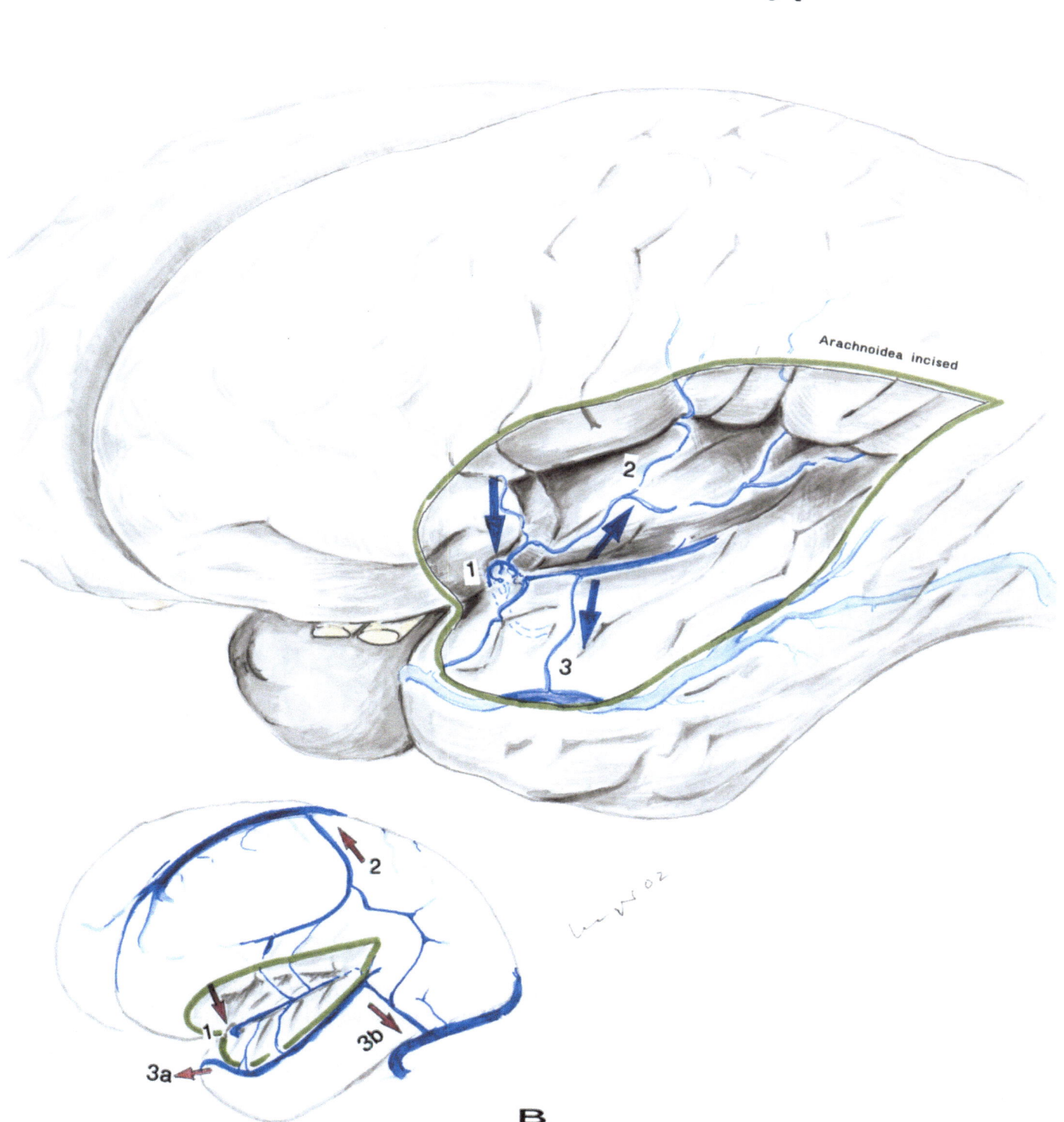

A

Arachnoidea incised

B

Fig. 93. Continuation of Fig. 92

Two groups of connecting veins are variable in size, diameter, and number (common variants), or they may be absent:

- Connections with V. basalis Rosenthal –1-, and with superficial sylvian veins –3-, blue colored

- Connections with the dorsal bridging veins –2-, dark colored

Arrows: typical direction of the blood stream, reversible

A Anatomical drawing, simplified
B A. cerebri media and its branches added. Note the underlying veins

For clinical aspects see Fig. 91

FIG. 93

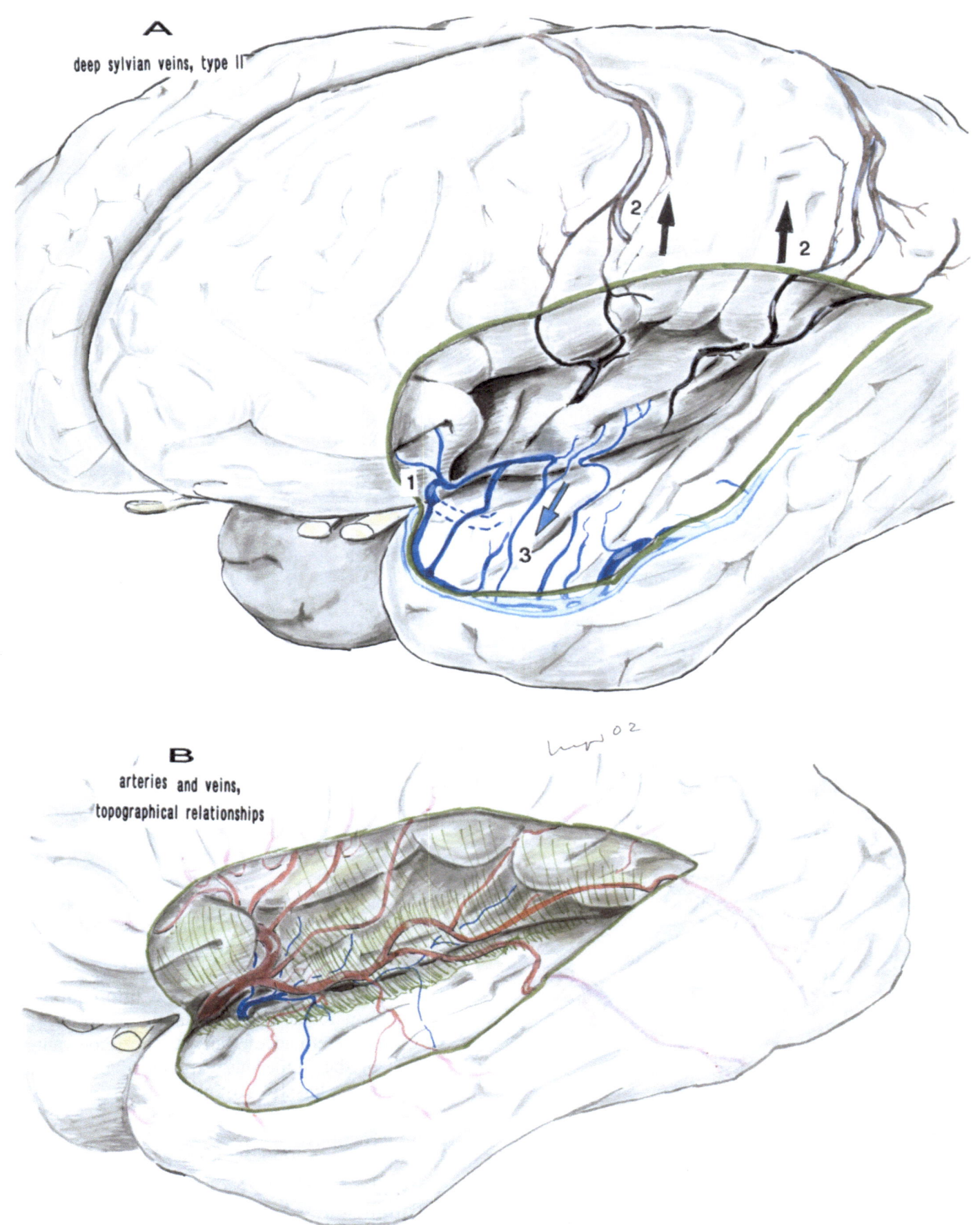

A
deep sylvian veins, type II

2

2

1

3

B
arteries and veins,
topographical relationships

Fig. 94. **Common variants of arteries and veins close to the tip of Cornu inf.**

A   Aa. perforantes of the carotid bifurcation are not only located in Substantia perforata ant. Some of them accompany A. chorioidea ant.
B   One of the numerous variants of the feeding veins of V. basalis Rosenthal

Clinical aspects:

These variants should be considered in Amygdalo-Hippocampectomy and in tumor surgery in this area

FIG. 94

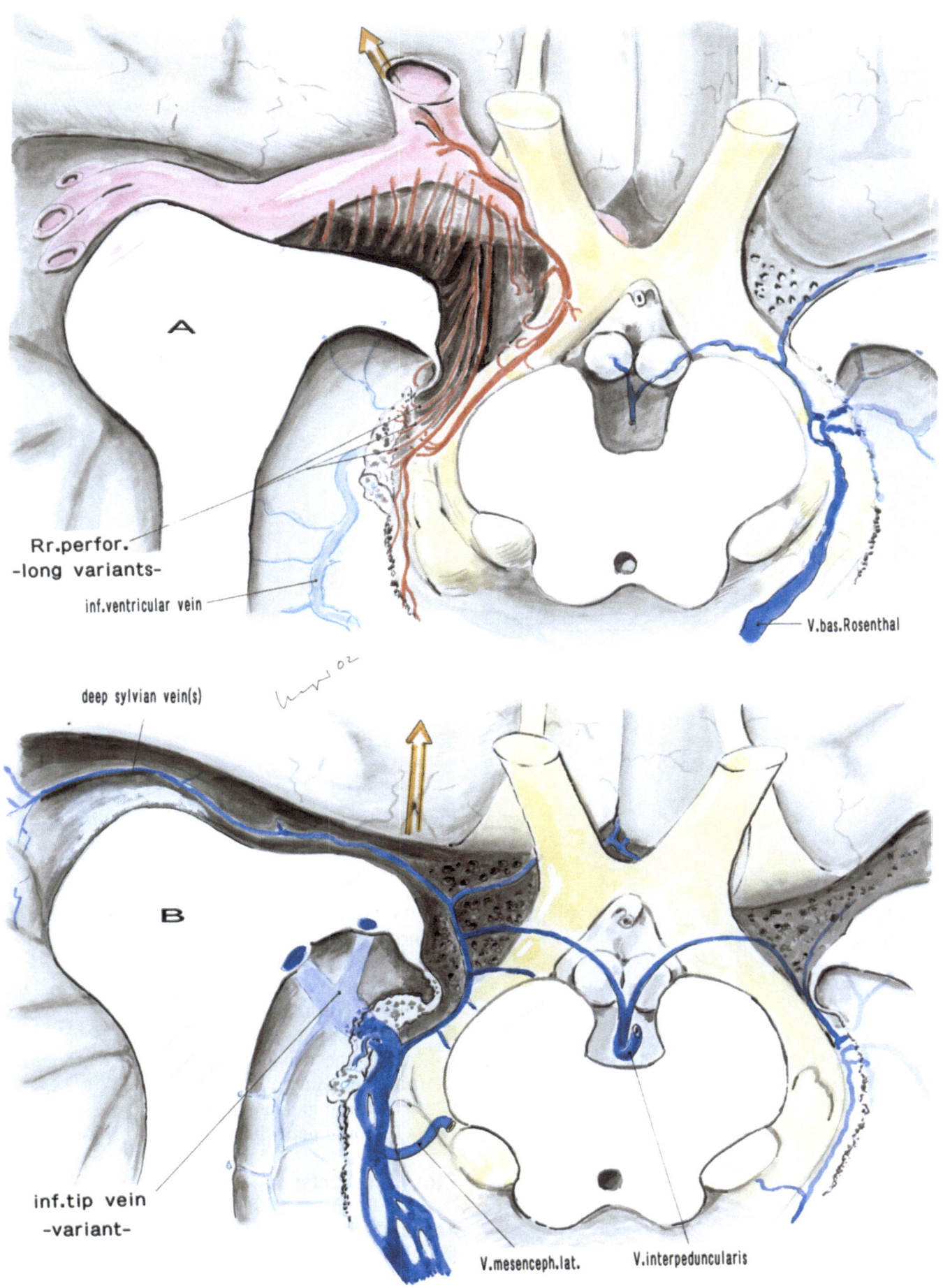

Rr.perfor.
-long variants-

inf.ventricular vein

V.bas.Rosenthal

deep sylvian vein(s)

inf.tip vein
-variant-

V.mesenceph.lat.

V.interpeduncularis

Figs. 95 to 100. **Tentorial gap and surrounding structures**

Fig. 95. **Diameter of the tentorial gap, common variants**

A   Sagittal presentation, used for topogram. Drawing according to MRT's

B   Reconstruction by use of MRT'S (levels a to d). Note the distance between the tentorial edge and the midbrain

C   Tentorial gap narrow

D   Tentorial gap wide

Clinical aspects:

Ad C: With brain-shifting, midbrain compression is earlier to be expected than in B and D

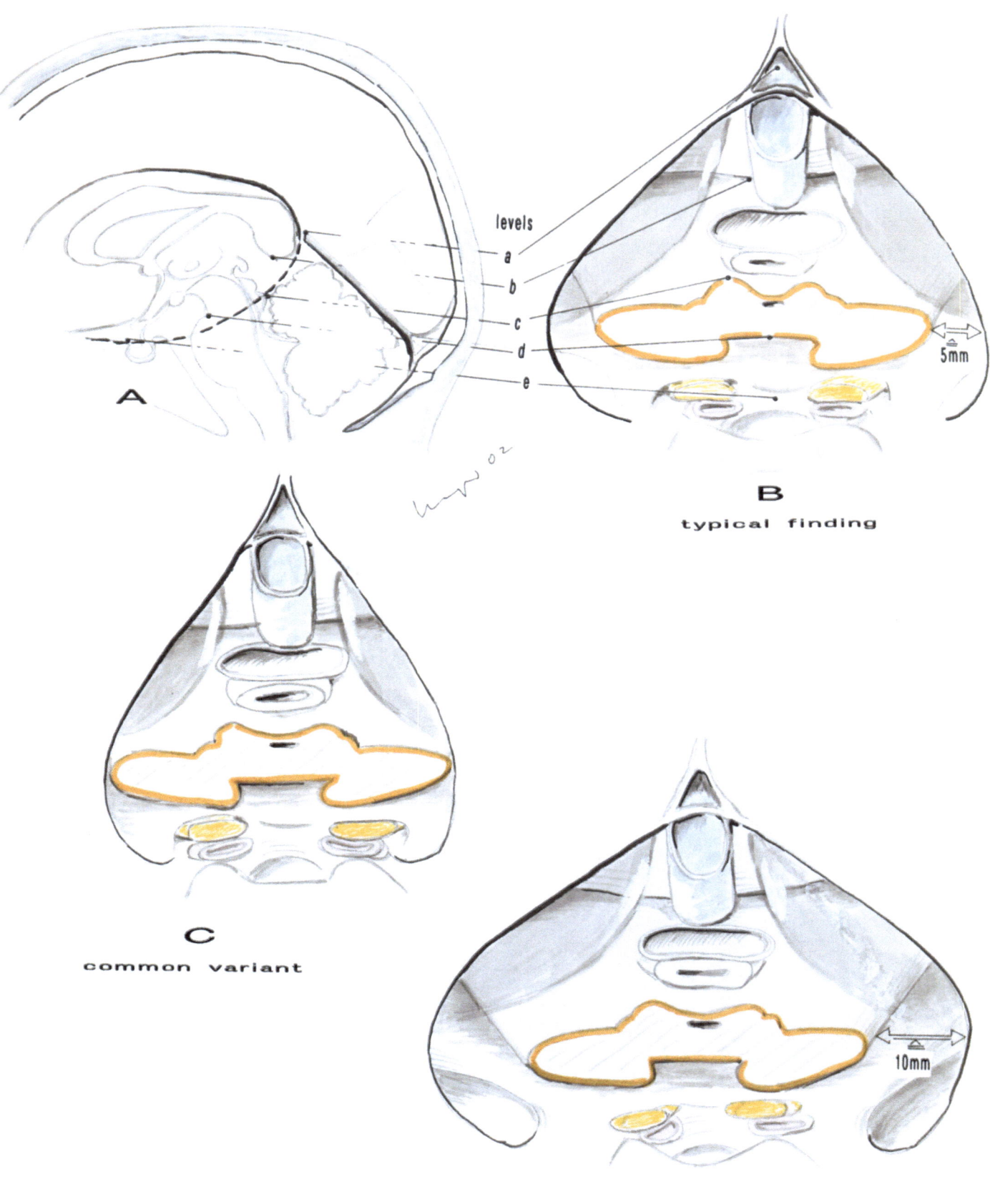

levels
a
b
c
d
e

5mm

A

B

typical finding

C

common variant

10mm

D

rare variant

Fig. 96. Addendum for Fig. 95

A For comparison with B and C
B Narrow tentorial gap and transtentorial herniation
C Wide tentorial gap and transtentorial herniation (leftsided).
Adhesions of Gyrus parahippocampalis and Cerebellum (rightsided)

Clinical aspects:
If the tentorial gap is small, then even a small herniation would produce midbrain damage (see B)
If the tentorial gap is wide, then cerebellar resection may produce damage of Gyrus parahippocampalis with epilepsy

FIG. 96

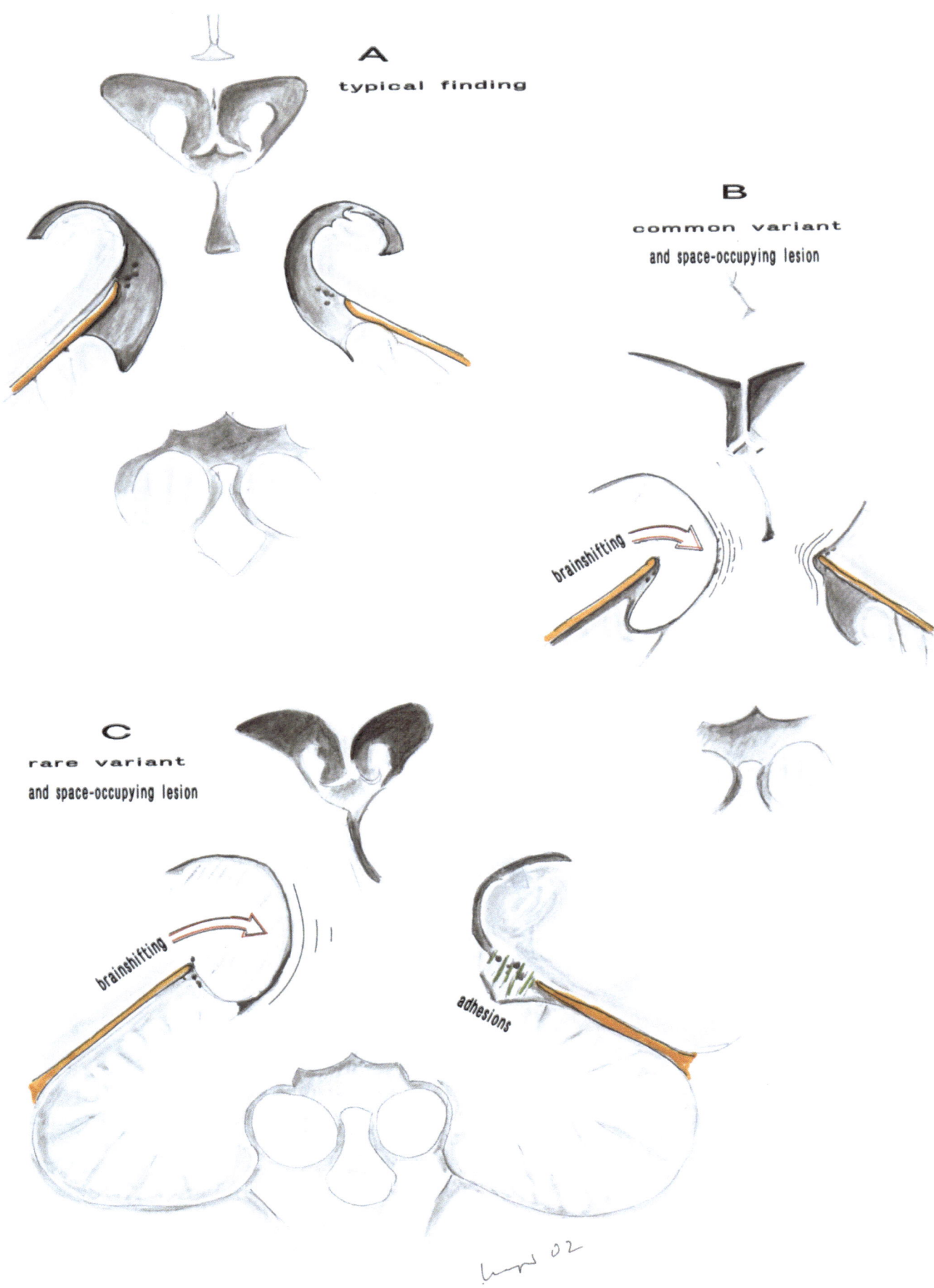

A
typical finding

B
common variant
and space-occupying lesion

brainshifting

C
rare variant
and space-occupying lesion

brainshifting

adhesions

Fig. 97. **Defining of the middle segment of the tentorial gap**
Useful for presentations of transtentorial herniations of the temporal lobe

A	Mapping of intercommissural line
A'	Intercommissural level, not recommendable for definition of the tentorial gap
B	Mapping of a line, vertical to the intercommissural line (coronal)
B'	Coronal level, not recommendable
C and **C'**	Not recommendable

D and D' **MRT-level perpendicular to the axis of Hippocampus**
   • **recommendable for defining of the middle segment of the tentorial gap**

D'' Level vertical to D'. Colored: Level of D' mapped

Clinical aspects:
The axis of Hippocampus is a well-known axis für planning an Amygdalo-Hippocampectomy using MRT. These MRT's can be used for microsurgical procedures close to the tentorial edge, too. The high variability of the level and of the wideness of the tentorial gap must be considered, especially for small lesions (e.g. aneurysms)

FIG. 97

topograms

MRT's

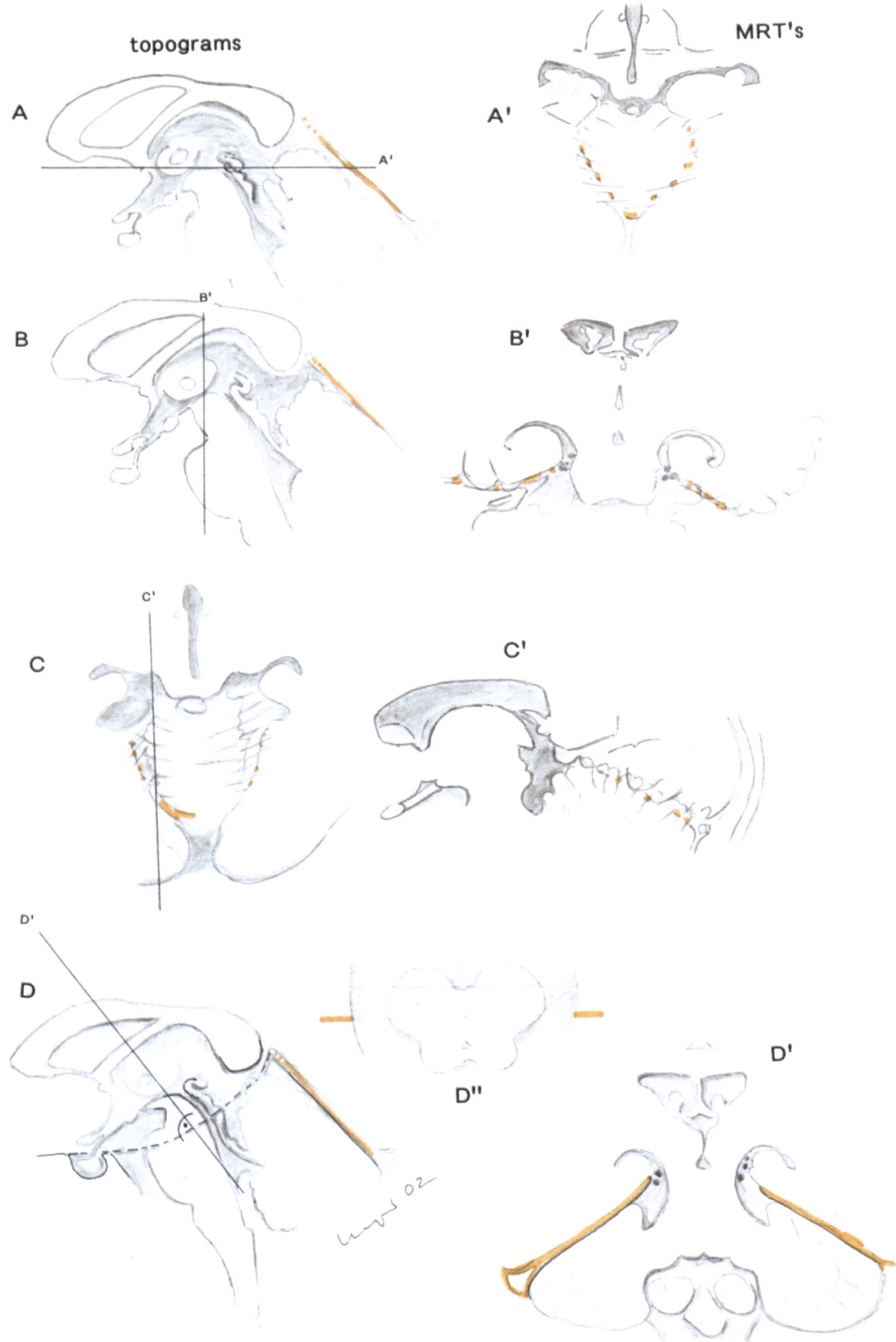

A

A'

B

B'

C

C'

D

D''

D'

Fig. 98. Addendum for A' Fig. 97

Coronal standard projections inaccurate for defining the middle segment of the tentorial gap

Anatomical sketches

- Upper drawing: Midbrain and tentorial gap from a superior view direction. Midbrain transected at level of Isthmus rhombencephali
- Middle drawing: Upper midbrain transected. It overlaps the tentorial gap
  In thick excalibrated horizontal slices of MRT: midbrain diameter seems to be wider than the tentorial gap.
  Vertical projections are running in an oblique direction to the level of the tentorial gap (A to G, for comparison with the upper drawing)
- Inferior drawing: horizontal transectional planes of the upper and inferior (Istmus mesencephali) midbrain added

FIG. 98

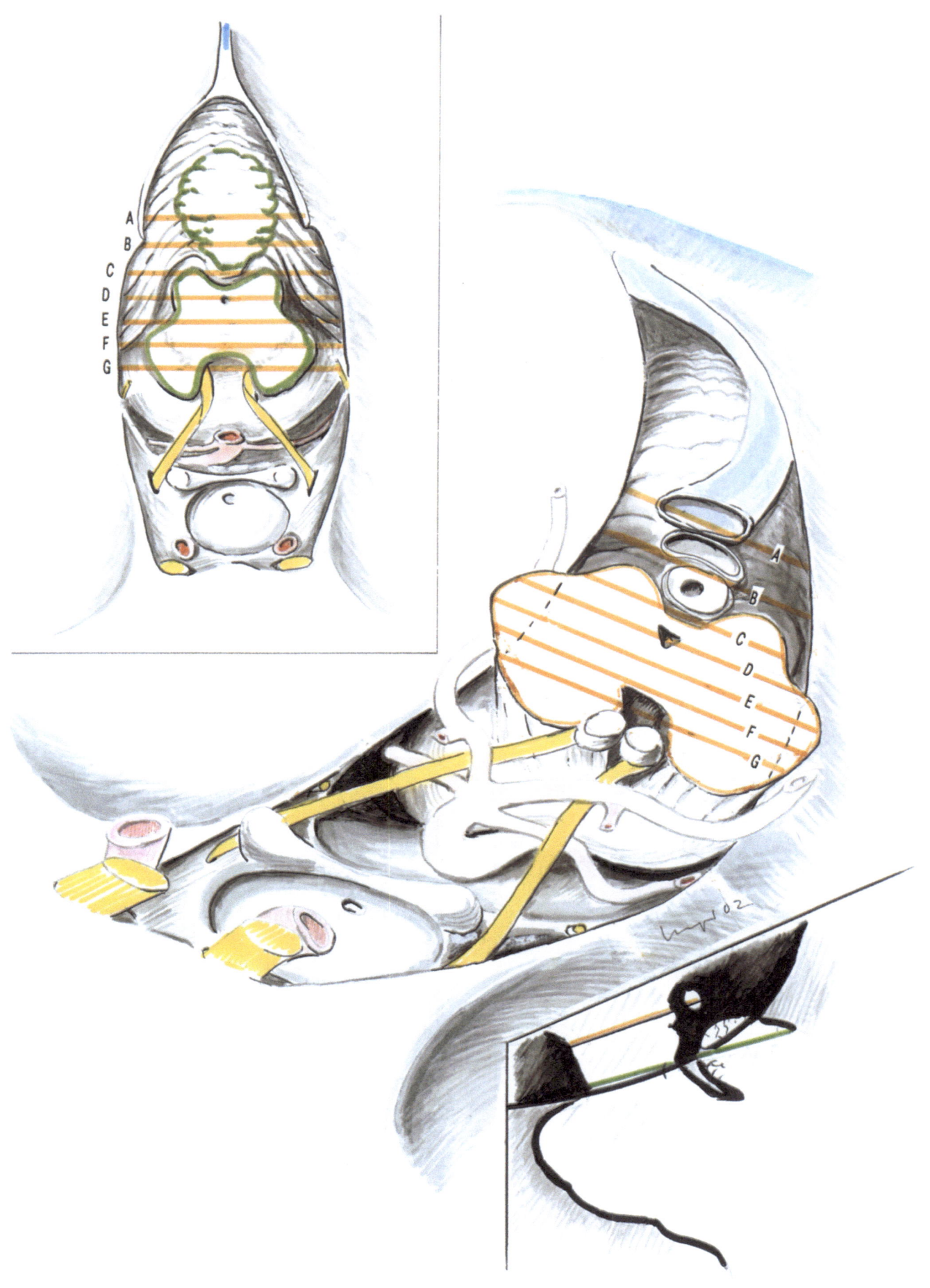

Fig. 99. **MRT's for defining of the posterior segment of the tentorial gap**

MRT-scan perpendicular to the level of the axis of Hippocampus: **The posterior MRT-slices A and B are not positioned correctly. It transcrosses the edges of the tentorial gap (dotted green colored line in the topogram) in an oblique direction. Presention in MRT's unclear**

For definition of the posterior segment of the tentorial gap, **MRT-slices parallel to the Bregma-galenic point-level are** recommendable (samples * and ** in topogram)
The high variability of the galenic point in basal or dorsal direction must be considered

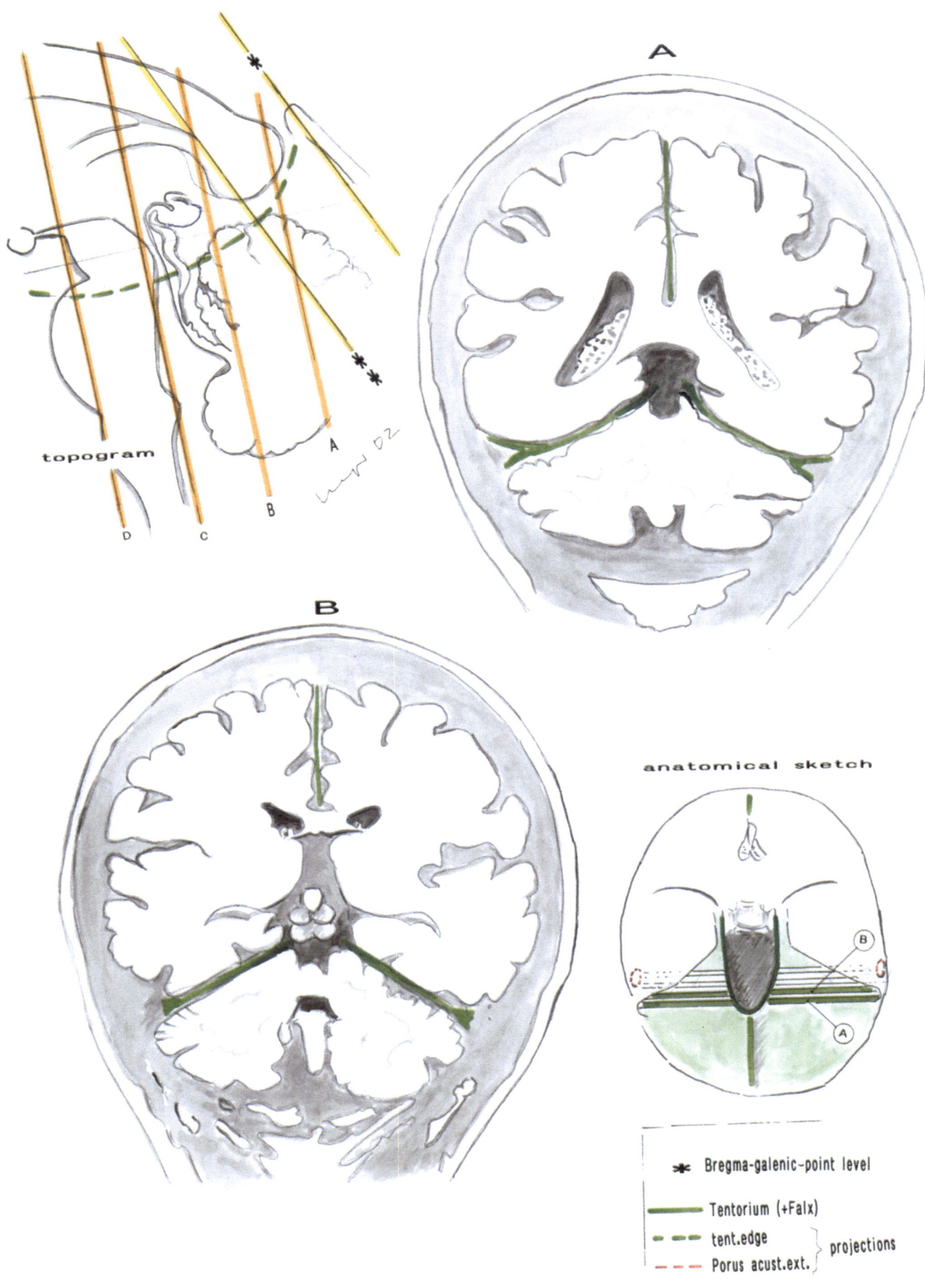

topogram

D   C   B   A

A

B

anatomical sketch

✳ Bregma-galenic-point level

── Tentorium (+Falx)

- - - tent.edge

- - - Porus acust.ext.   } projections

Fig. 100. **Clear presentation of the tentorial edge**

C and **D**   MRT's for defining the middle segment
of the tentorial gap
MRT-slices vertical to the axis of Hippocampus as Figs. 97 D to D"

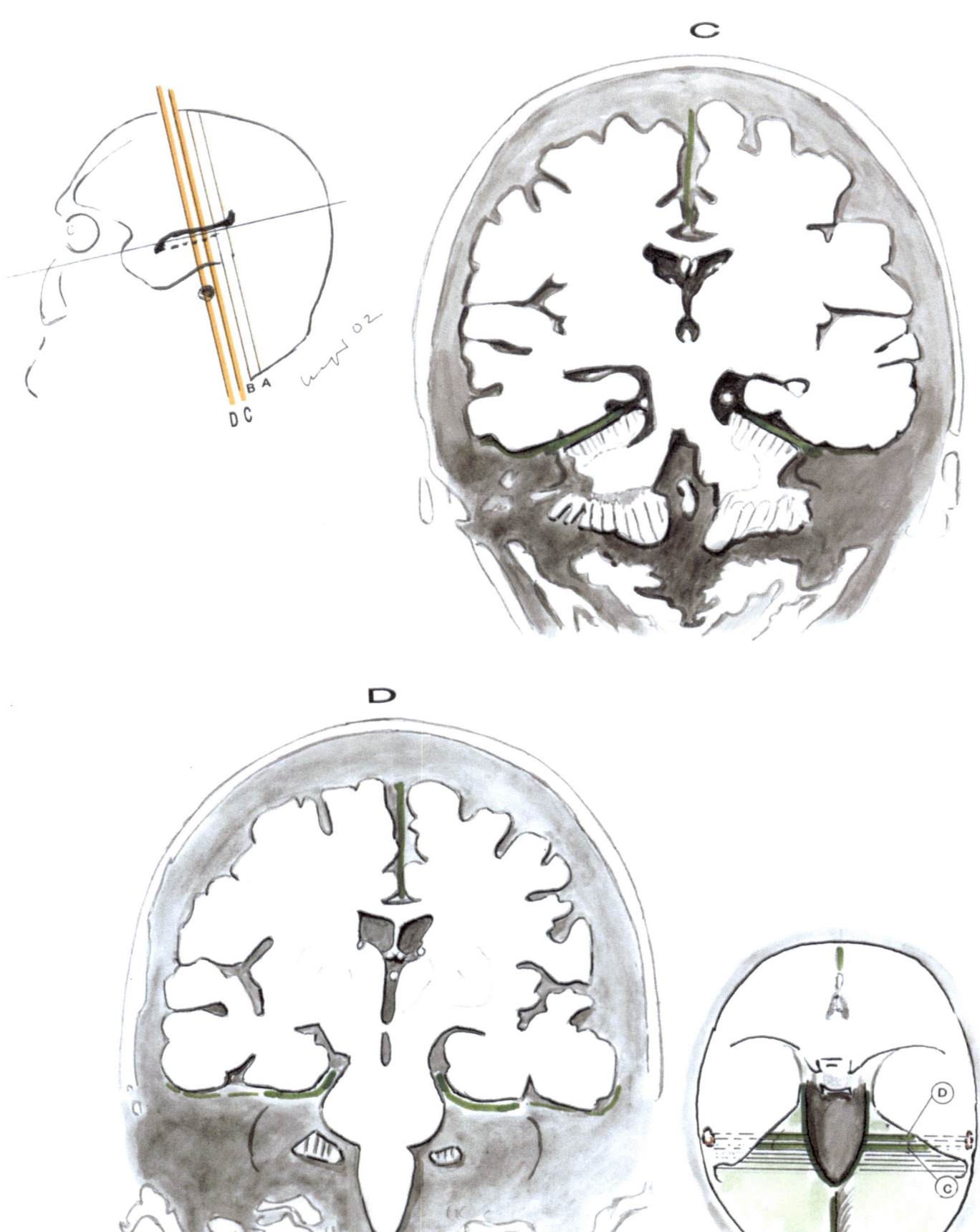

C

D

Chapter 4
Variants of the parietal and occipital areas
(Figs. 101 to 142)

Figs. 101 and 102. Parietal and occipital microsurgical approaches

Schematic presentation

───────────
* Bony landmarks

FIG. 101

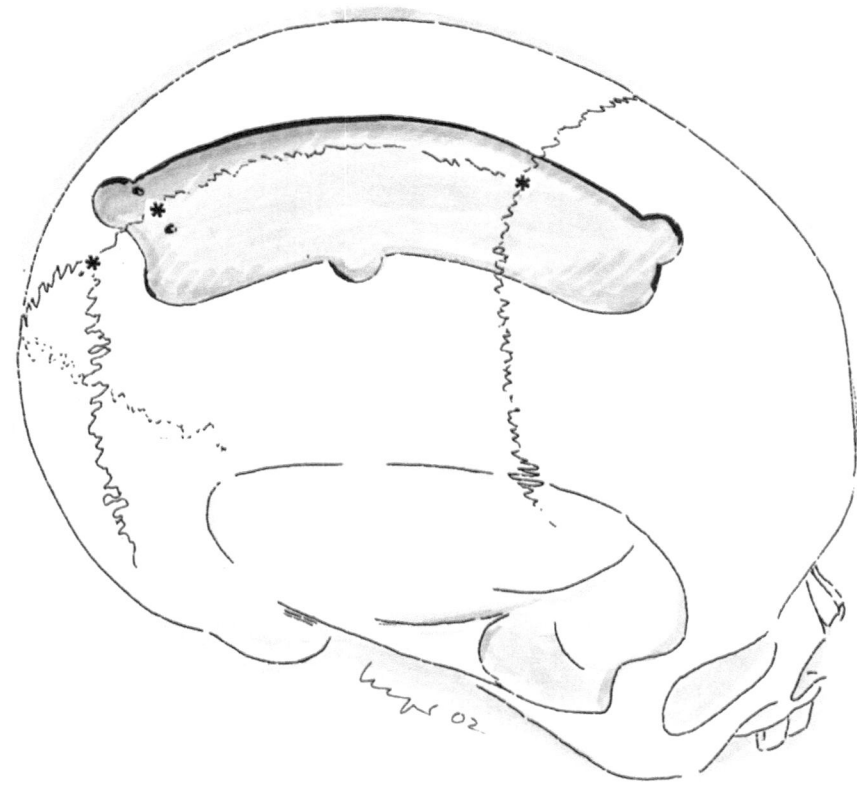

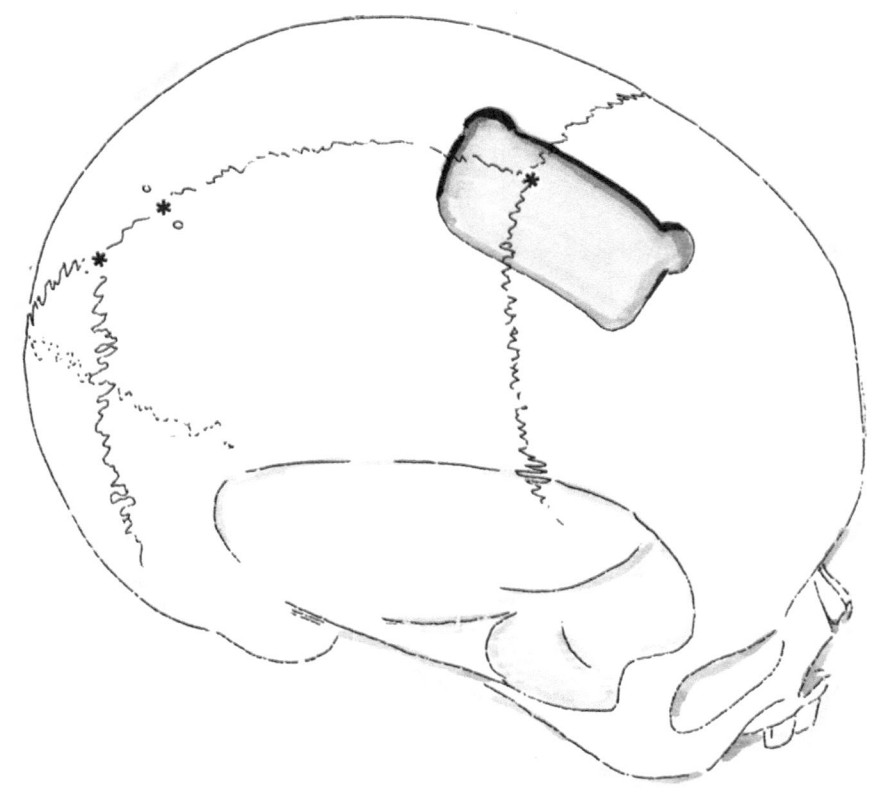

FIG. 102

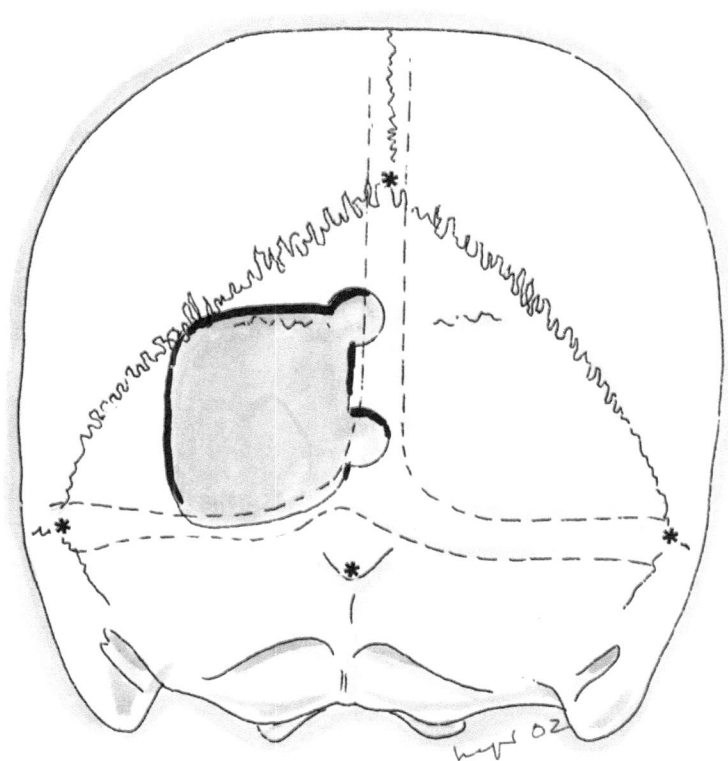

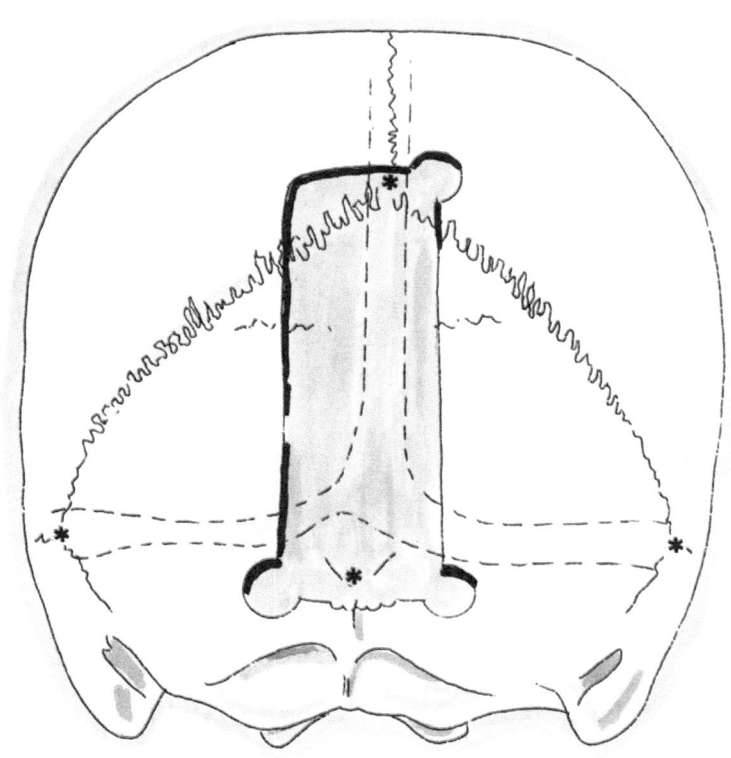

Figs. 103 to 128. Supratentorial cerebral veins

Fig. 103. **Embryological aspects**
For understanding the numerous variants of dural si-
nuses and veins
Simplified

A   Early stage: The dura and its duplications (Falx,
      Tentorium) are containing a large network of
      dilated veins

B   Later on: A great number of the dural veins be-
      comes obliterated, except in Tentorium and
      some segments of Falx cerebri and Falx cere-
      belli. Other portions of the wide dural veins di-
      late. These veins will form the dural sinuses.
      The reduction of the dural venous network
      occurs very slowly, especially in the infantile
      Dura of the dorsal cerebellar area

C   Schematic presentation of dural veins and of a
      sinus (transection) in adults

*Clinical aspects:
Unexpected alarming bleeding during surgery or punctures, es-
pecially in childhood. Neuroradiologic assisted planning strate-
gies are not helpful.

FIG. 103

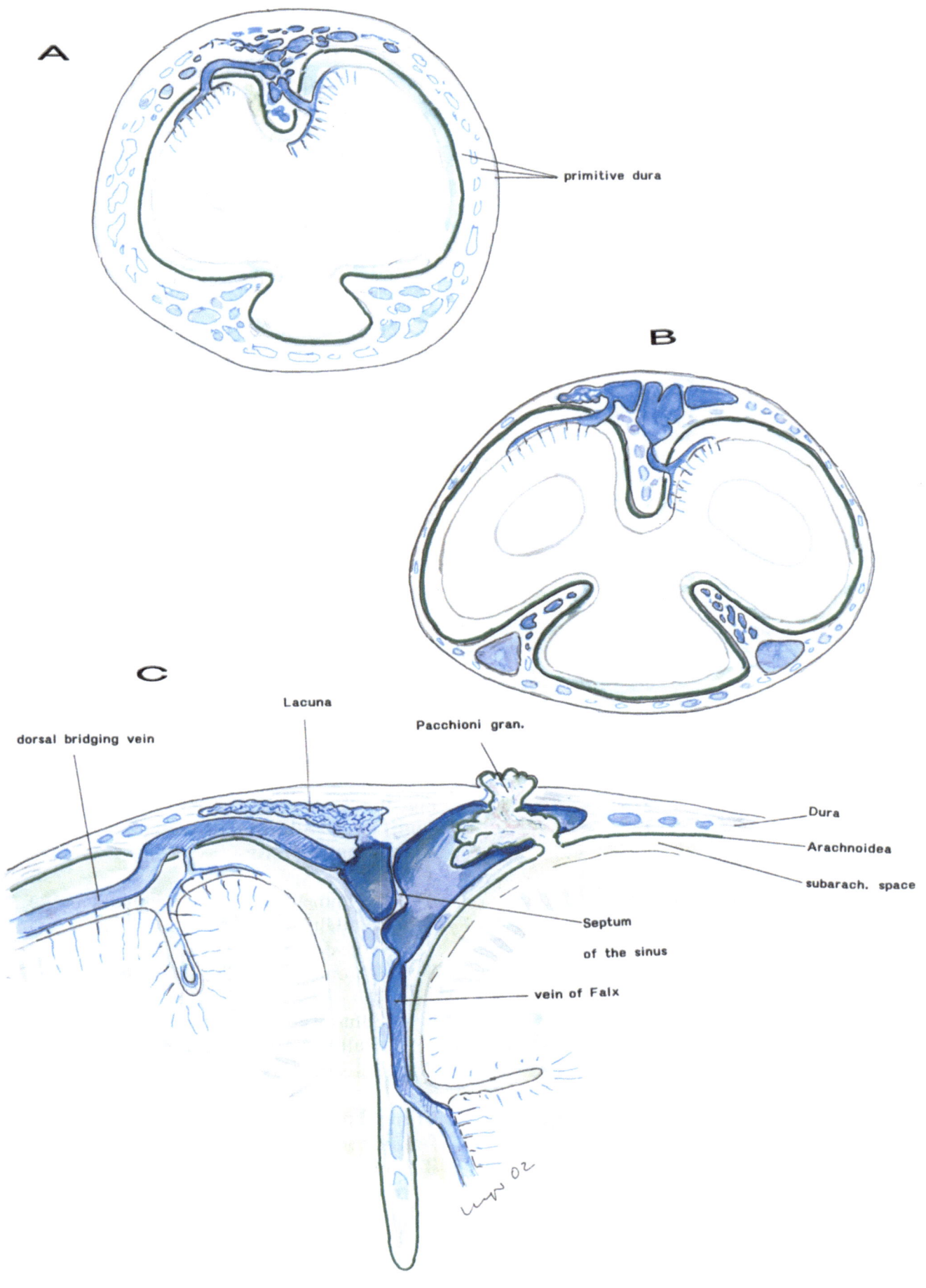

A

primitive dura

B

C

Lacuna

Pacchioni gran.

dorsal bridging vein

Dura

Arachnoidea

subarach. space

Septum

of the sinus

vein of Falx

Fig. 104. **Sulcus (≙ Sinus) sagittalis and Sutura sagittalis (≙ midline) are not always overlapping – common variants –**

A Asymmetry of Sulcus sagittalis and of Sutura sagittalis*
Note deep impressions of Tabula int. by vessels (chronic venous hypertension by tuberculosis* ?)

B Sutura sagittalis present on the outside of the skull, not present on the inside. Asymmetry of Sulcus sagittalis

Clinical aspects:
Danger for microsurgical midline approaches, if the relief of the sinus is not recognizable during operation

---

* age 20, female, + tuberculosis

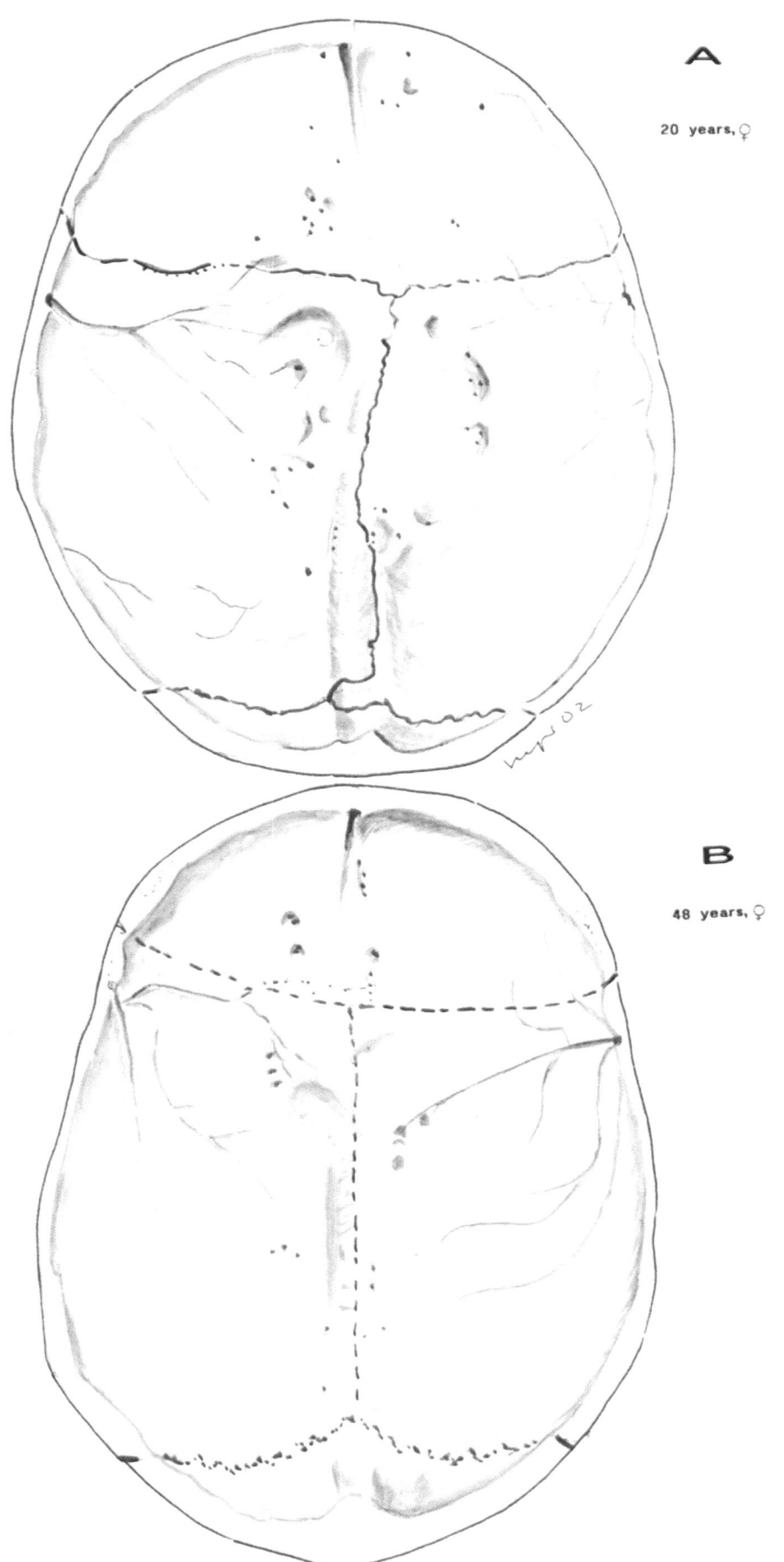

A

20 years, ♀

B

48 years, ♀

Fig. 105. Continuation of Fig. 104

C   Deep impressions of Tabula int. by Sulci meningeales are typically in elderly individuals. Note numerous Emissariae on the left side at the sulci. Suturae on the outside of the skull present

D   The widened Sulcus sagittalis corresponds with a widened sinus. Sutura and sinus are not in accordance

Clinical aspects:
As Fig. 104

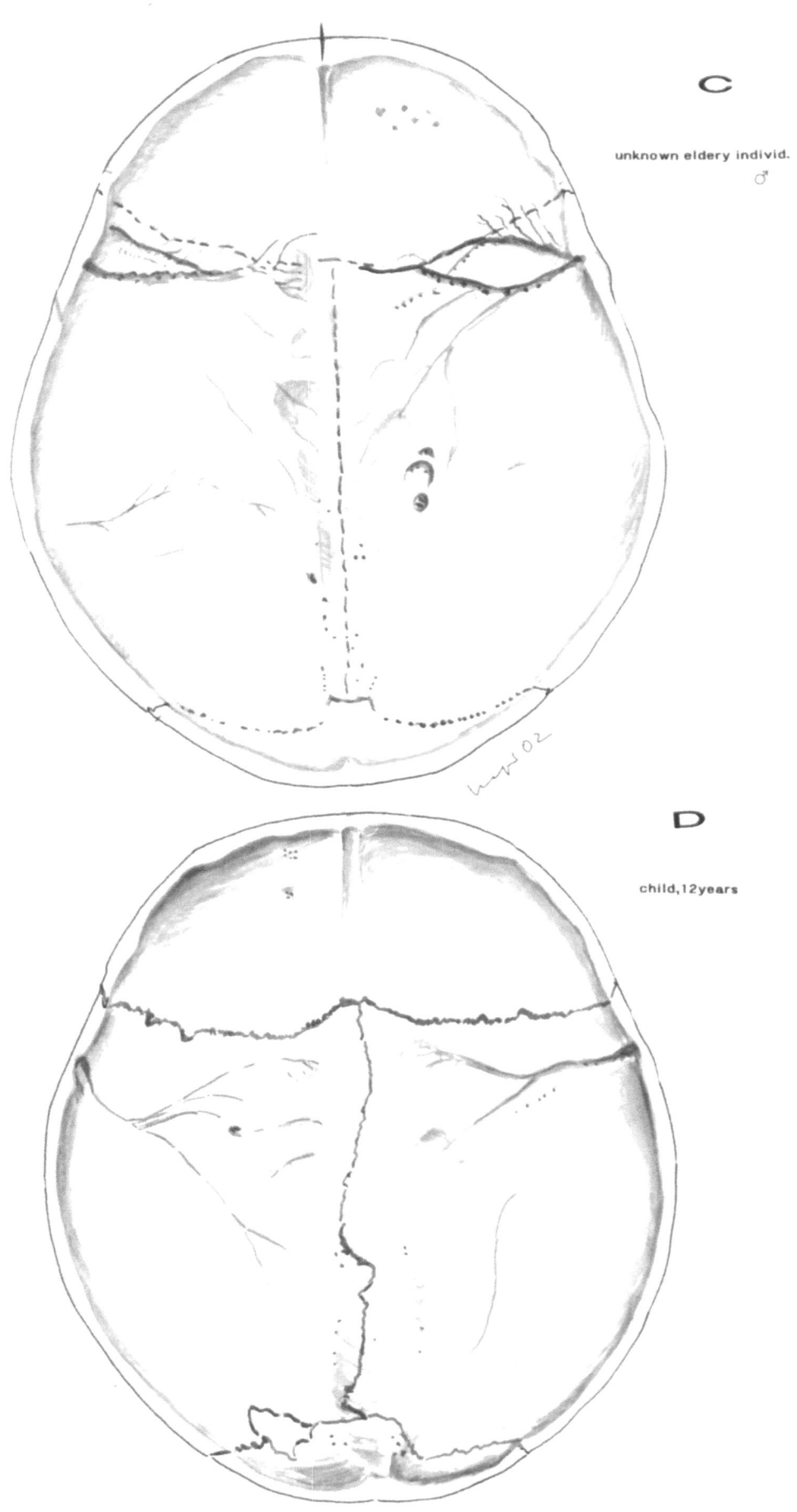

C

unknown eldery individ.
♂

D

child, 12 years

Fig. 106. **Relationship of Sulcus sagittalis and of Sinus sagittalis sup.**

A   Squama frontalis (polar and frontoparietal)
A'   Sinus sagittalis sup. (segments as A)
B   transitional area of Squama parietalis and Squama occipitalis
B'   Sinus sagittalis sup. (segments as B)

Clinical aspects:
Preoperative definition of the lateral edge of the sinus before performing midline approaches. A and B can be defined by CT, A' and B' by MRT, if necessary

Arrows: Lateral edge of the sinus

FIG. 106

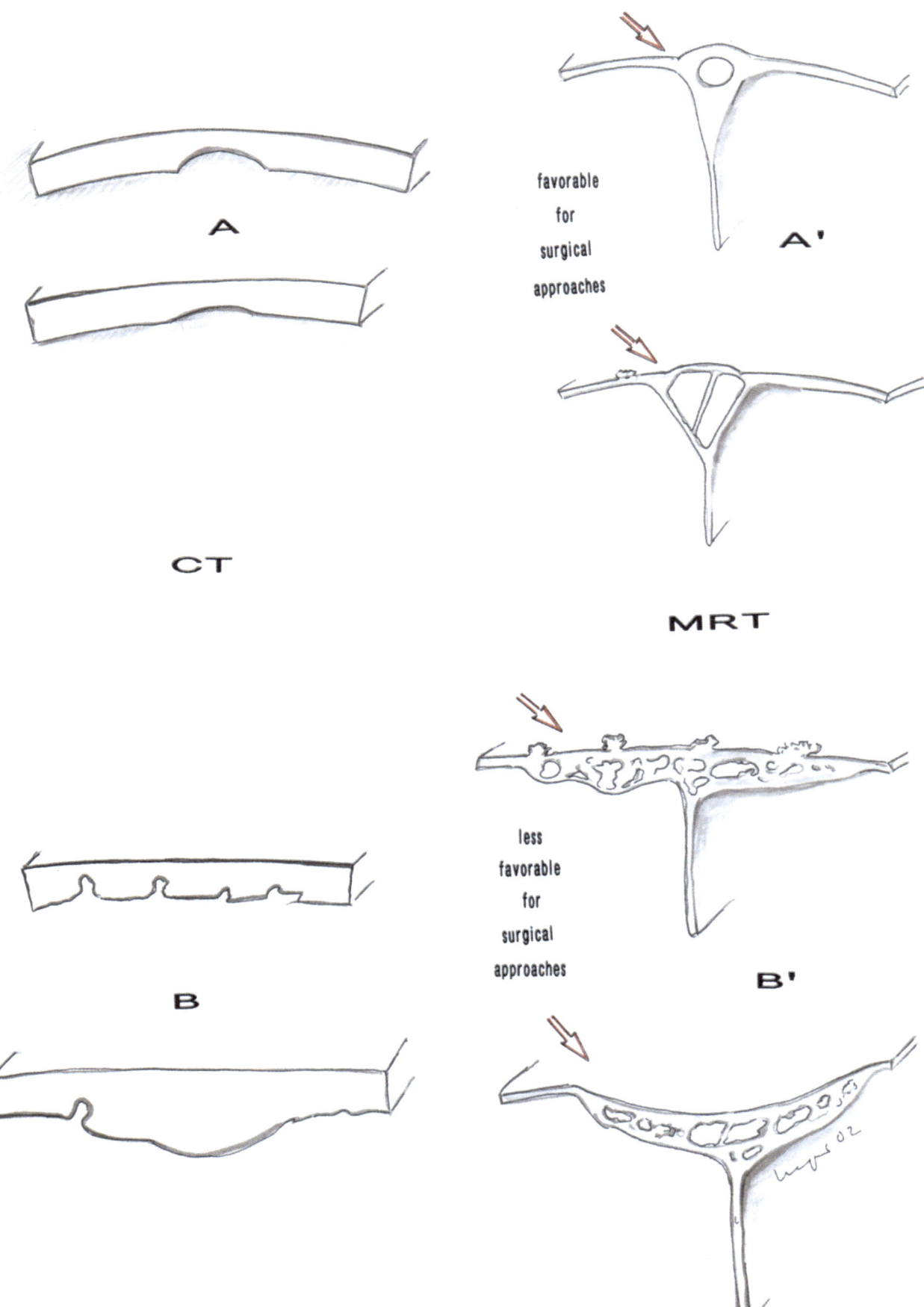

A

CT

B

A'

favorable
for
surgical
approaches

MRT

less
favorable
for
surgical
approaches

B'

Fig. 107. **Dural veins and sinuses, common findings, schematic**
Survey

A	Skull dissection. Foveolae granulares (blue colored)
B	Cerebral veins (blue colored) and longitudinal septa of the sinus added
C	Skull dissection. Sinus, Foveolae granulares (= Granulationes arachn. Pacchioni): some dural and cerebral vessels drawn in (blue colored)
D and **E**	Typical septa of Confluens sinuum (> 60%)

Clinical aspects:
- **Danger for microsurgical midline approaches, if outline of the sinus is not visible during operation**
- **Danger of bleeding during microsurgery, if the Pacchioni-granulations are located atypically (e.g. frontopolar) or if their number and size are variable**
- **The asymmetric course of the rolandic vein must be taken into consideration, especially its dural segment before entering the sinus wall**
- **Variability of septa in all segments of the sinus.**

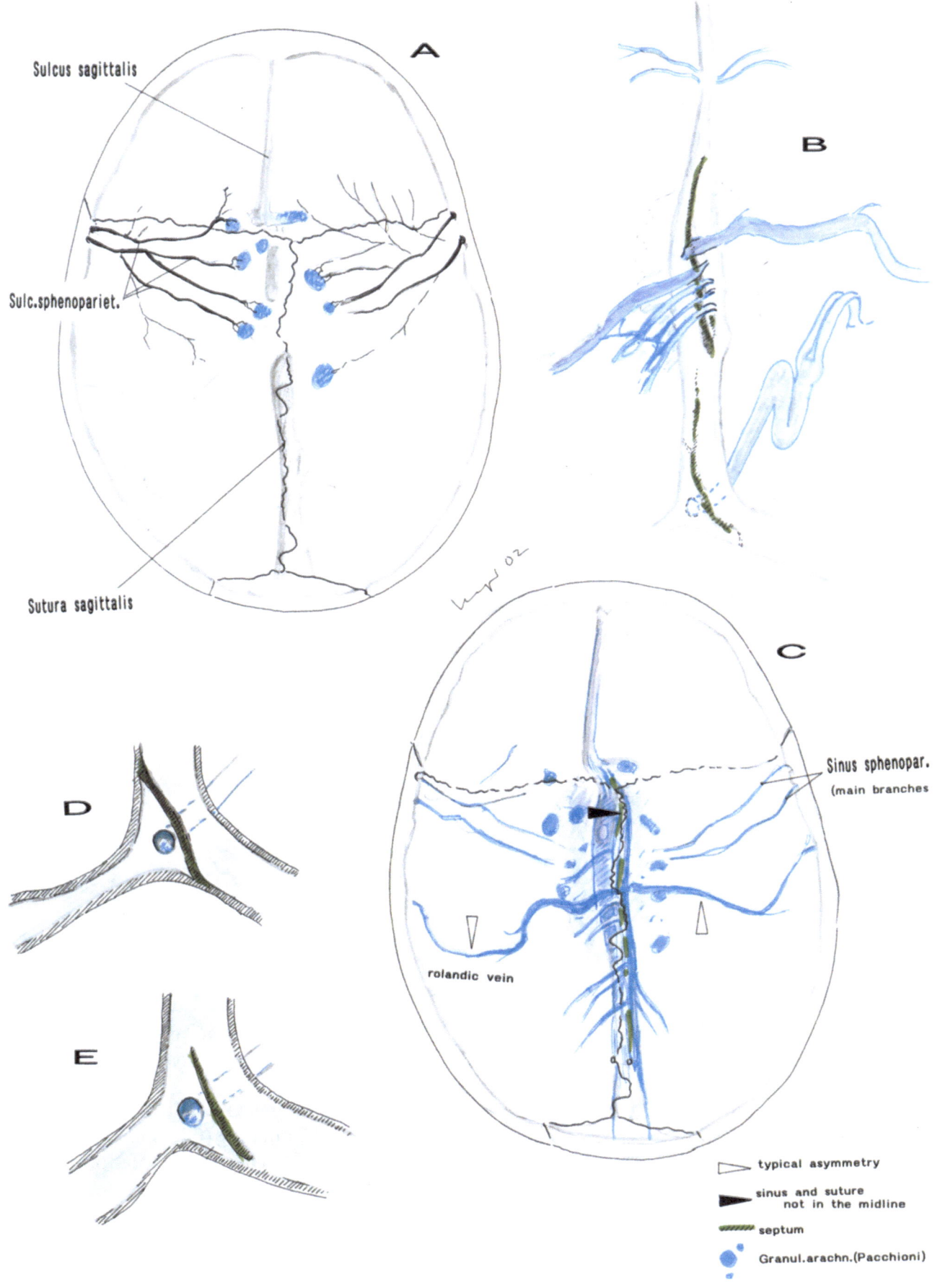

A

Sulcus sagittalis

Sulc.sphenopariet.

Sutura sagittalis

B

C

Sinus sphenopar.
(main branches

rolandic vein

D

E

▷ typical asymmetry

◀ sinus and suture
not in the midline

▨▨▨ septum

● Granul.arachn.(Pacchioni)

Fig. 108. Anatomical dissections and cast dissections, copies

A   Vv. meningeae mediae (Sinus sphenoparietalis) are typically configurated, but asymmetric

B   Suturae and Sulci are not in accordance. This finding is known for nearly 100 years

C   Typical findings: **Multiple septa of the sinus** maximal 8 cm long (O'Connel, 1934, cit. Berg-quist, 1975) in horizontal (platform-like) and in sagittal levels

Clinical aspects:
See Fig. 107

FIG. 108

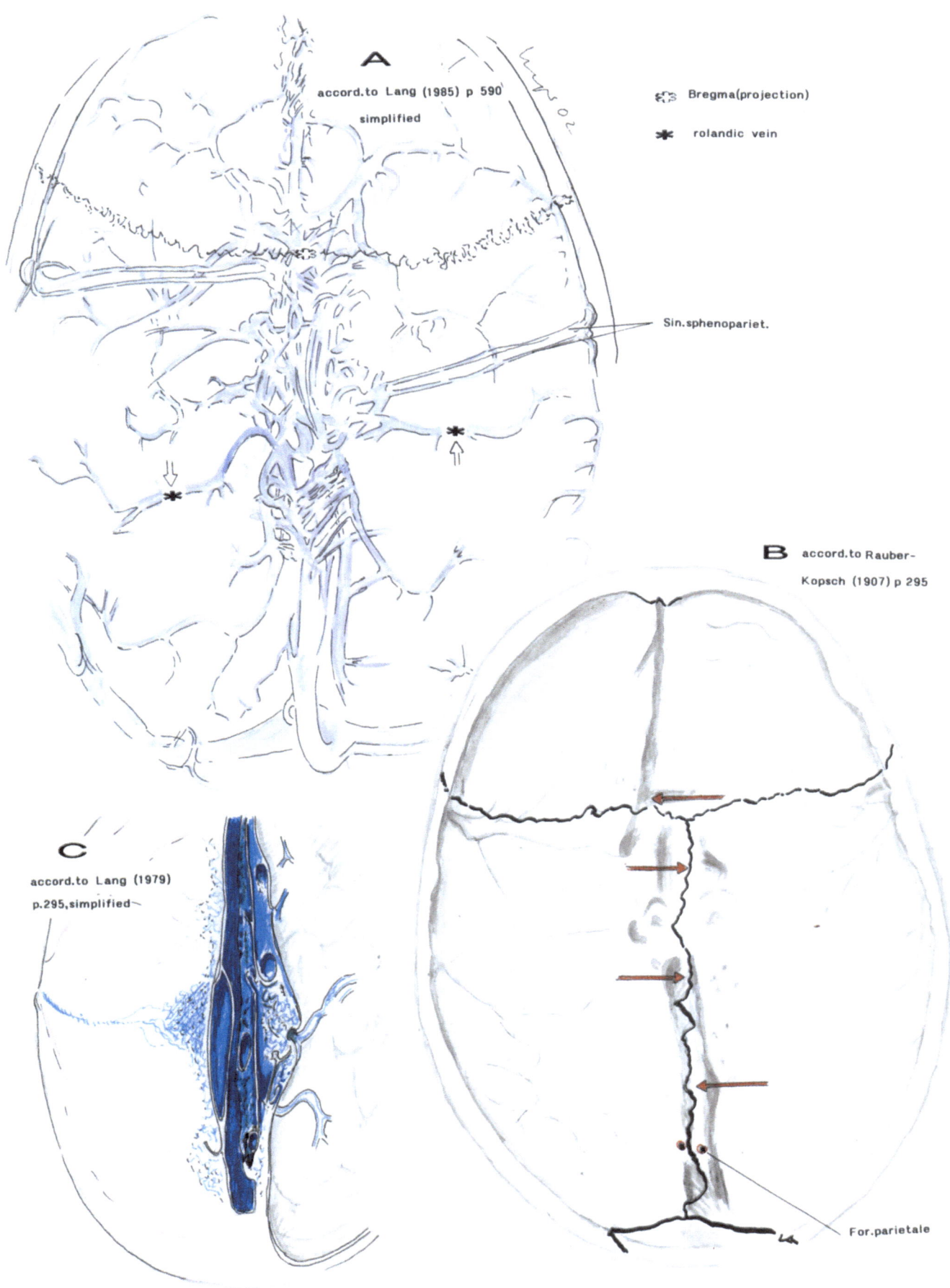

A

accord.to Lang (1985) p 590

simplified

Bregma(projection)

rolandic vein

Sin.sphenopariet.

B accord.to Rauber-

Kopsch (1907) p 295

C

accord.to Lang (1979)

p.295,simplified

For.parietale

Fig. 109.

A   Cast dissection of Stephens and Stilwell (1969) p 135, simplified

B   Sectional enlargement
- Sagittal longitudinal septum (septa)
- Bridging veins are mostly connected with a smaller part * of the sinus, which underlies the wider segment **
- Lacunae are located between dura layers, dorsal from the bridging veins. They are connected with the sinus, but not with the bridging veins ***

a to c   Common configuration of bridging veins before entering the sinus ****

c'   Stenosis of the narrow part of the sinus (common finding)

---

* Lang 1979, 1989

** Browder and Kaplan (1976)

*** Berquist (1974)

**** Crossing veins of this type were drawn by the artist of Andreas Vesalius in 1543, Jan Calcar, a pupil of Tizian. But this drawing is unclear.

FIG. 109

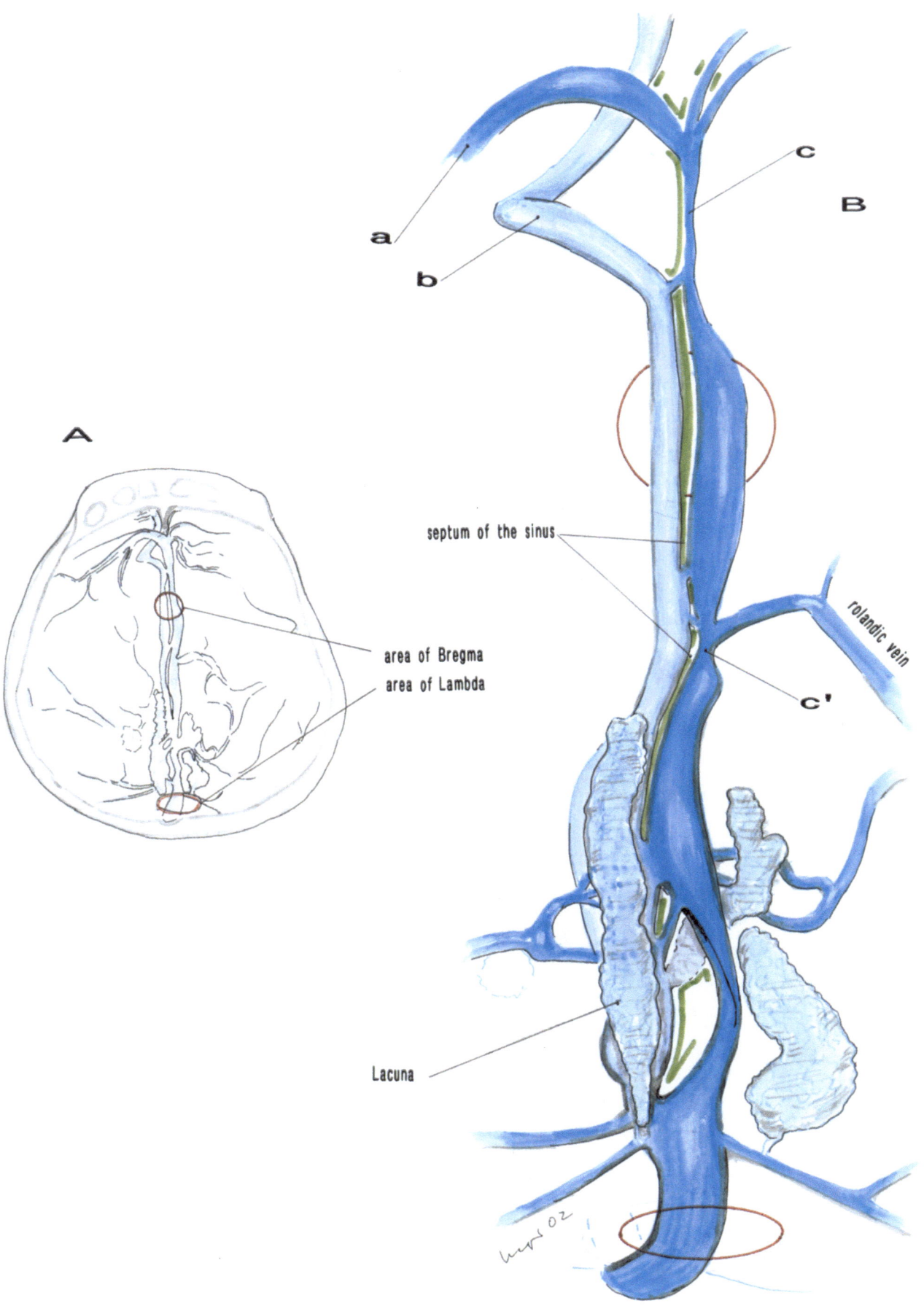

A

B

a

b

c

septum of the sinus

c'

rolandic vein

area of Bregma
area of Lambda

Lacuna

Fig. 110.

A Anatomical cast of Browder and Kaplan (1976):
  - Only few fine connections between Lacunae and the sinus
  - No connections between Lacunae and bridging veins
  - Diverticles of the sinus are rare

  Further statements of Browder and Kaplan (in agreement with Lang, 1979):
  - If the sinus is duplicated, then bridging veins enter the smaller part of the sinus

B Schematic presentation of the findings of numerous authors

Abbreviations

a bridging vein
b mushroom-like fibrous structure
c dorsal portion of the sinus
d ventral portion of the sinus
e diverticle of the sinus
f Trabeculae
g sagittal/oblique septum
h Sinus sagittalis inf., ramificated (close to the galenic area)

---

* small connection between Lacuna and the sinus

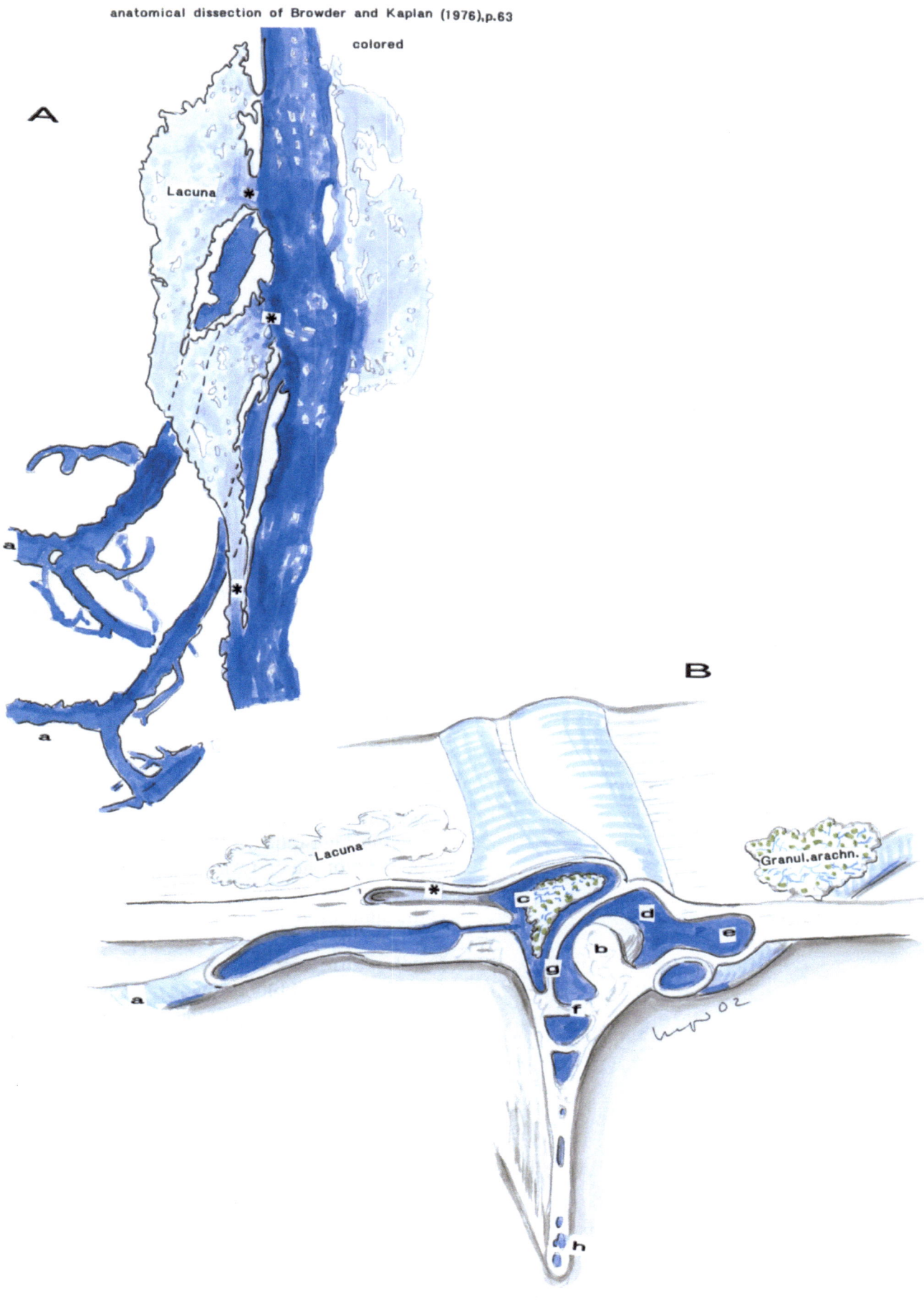

anatomical dissection of Browder and Kaplan (1976),p.63
colored

A

Lacuna

B

Lacuna

Granul. arachn.

Fig. 111. **Clinical aspects of the variability of the veins and sinuses: Ultrasonic findings during surgery**

Bloodstream of bridging veins and of the sinus

The same anatomical ocation may present a variable velocity and direction of bloodflow, dependent on

- position of the head (stenosis of a jugular vein during turning of the neck)
- position of the body, horizontal or vertical (Pacchioni's granulations filling out the lumen of the sinus or not)

Further findings:

A   Velocity of bloodflow is reduced in Lacunae
B   Irregular velocity in the same cerebral vein
C   Reduction of velocity in the larger portions of the sinus
D   Increased velocity in the smaller portions of the sinus, especially, if connected with numerous briding veins
D' Increased velocity in vertical position, if the lumen of the sinus is filled with Pacchioni's granulations
E   In horizontal position, the granulations are retracted, and the sinus lumen becomes wide

FIG. 111

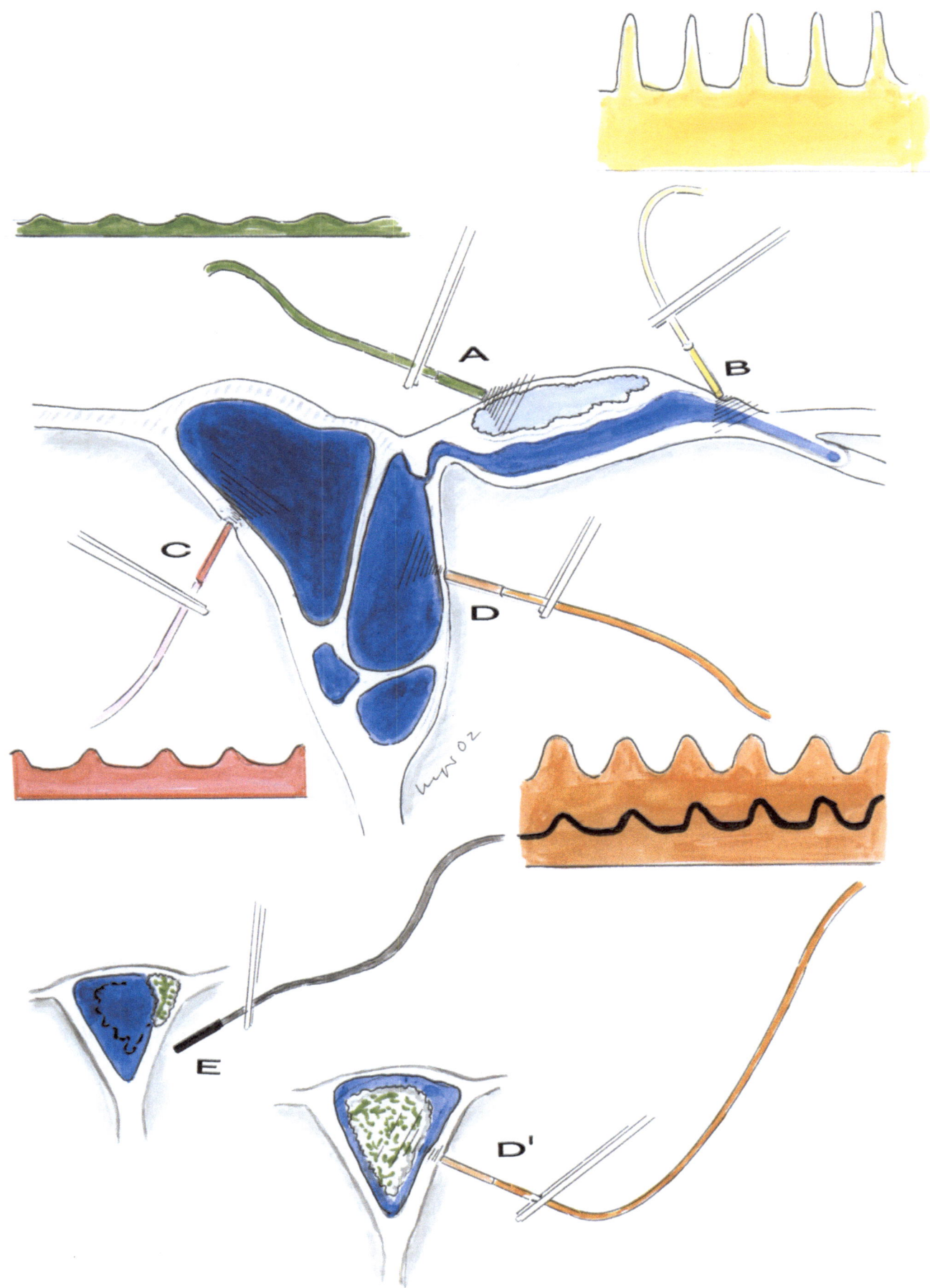

Fig. 112. These copies of an anatomical cast looks like modern computer-assissted three-dimensional presentations

A   Fine vessels accentuated and colored
B   As A, sinus and large veins highlighted and co-lored

* Widening of fine vessels artificial and/or Lacunae? Or anastomotic veins of the sinus?

Clinical aspects:
During microsurgery, the widened fine vessels, if they are located between dural sheets, may be distinguished from bridging veins before enter-ing the sinus, and from Lacunae. Bridging veins should be preserved, if possible. Lacunae may be ligated, if they obstruct the microsurgical approach

FIG. 112

* widening of the veins

A

hypothetical

outside of the skull

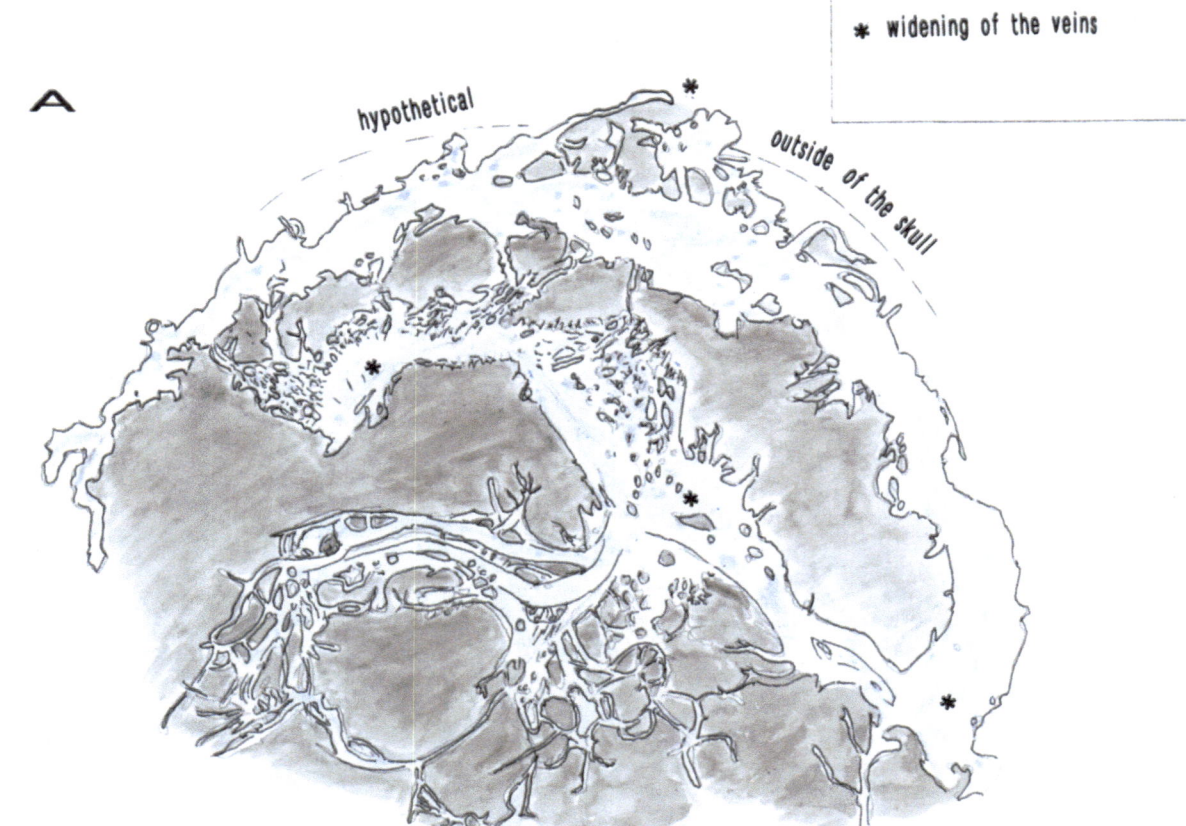

according to a cast (photograph) of Browder and Kaplan (1976) p 96

modified

B

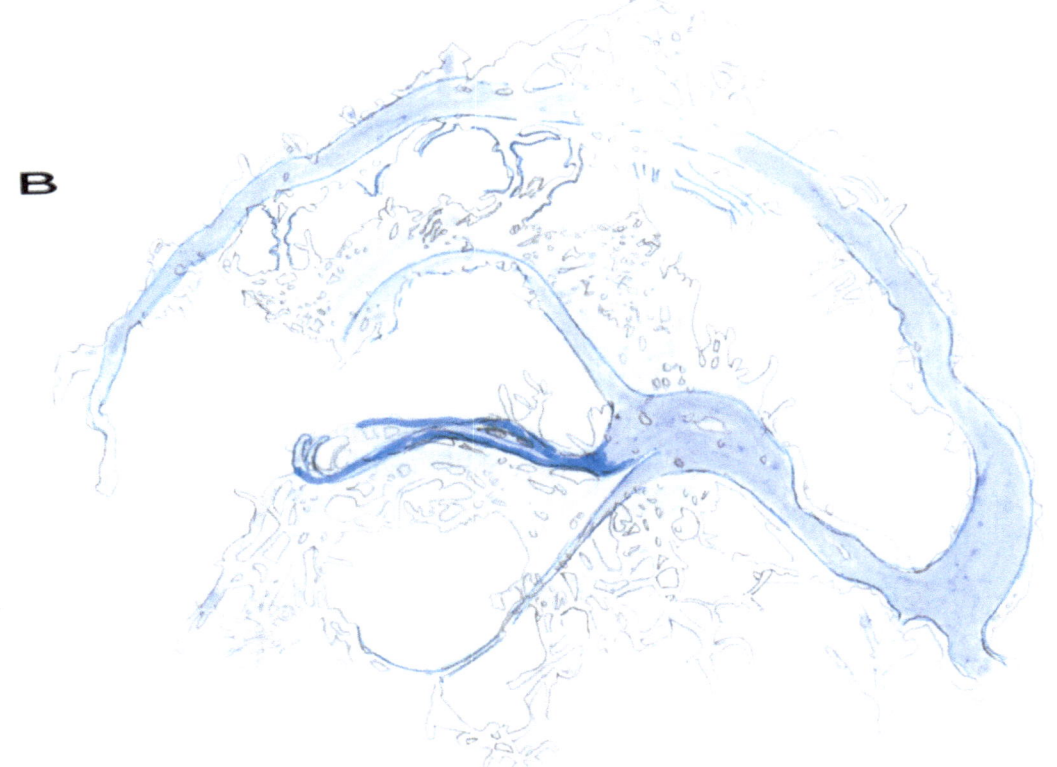

Fig. 113. **Variants of Confluens sinuum** (Torcular Herophili), and adjacent veins, common variants

A Septum divides Sinus sagittalis sup. and Confluens sinuum
B Left Sinus transversus duplicated
C Circular variant of Confluens sinuum

Clinical aspects:
See Fig. 112

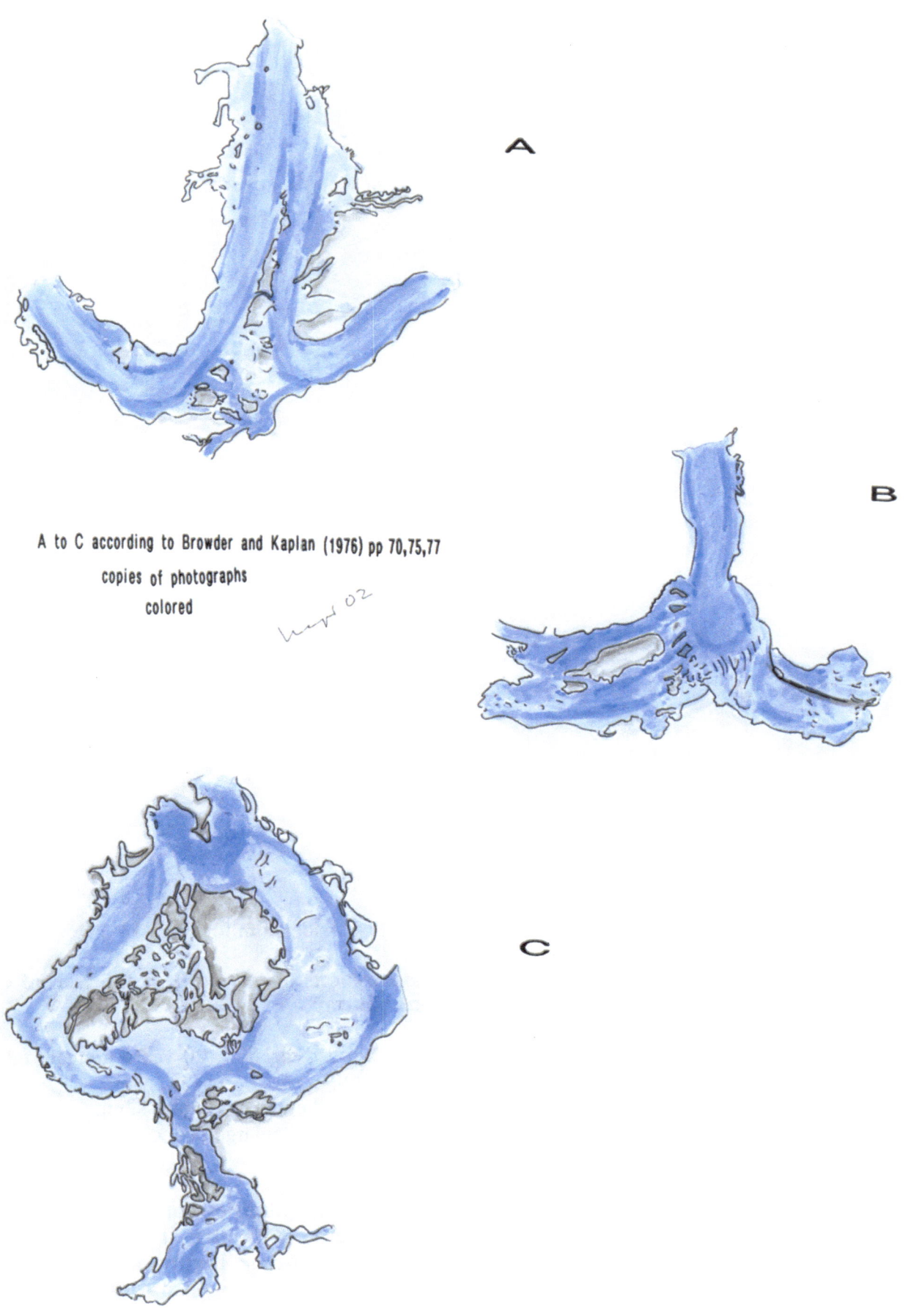

A

B

C

A to C according to Browder and Kaplan (1976) pp 70,75,77
copies of photographs
colored

Fig. 114. Addendum for Fig. 113

Impressions of the skull by the wall of Confluens sinuum

A   Skull dissection (20 years, female, tuberculosis)

B   Hypothetical Confluens sinuum (blue colored)

C   Sectional enlargement of A

Clinical aspects:
Normally, trepanations at the area of Confluens sinuum are not problematic. But Meningiomas, AVM's and other highly-vascularized lesions may mask the shape of Confluens sinuum

FIG. 114

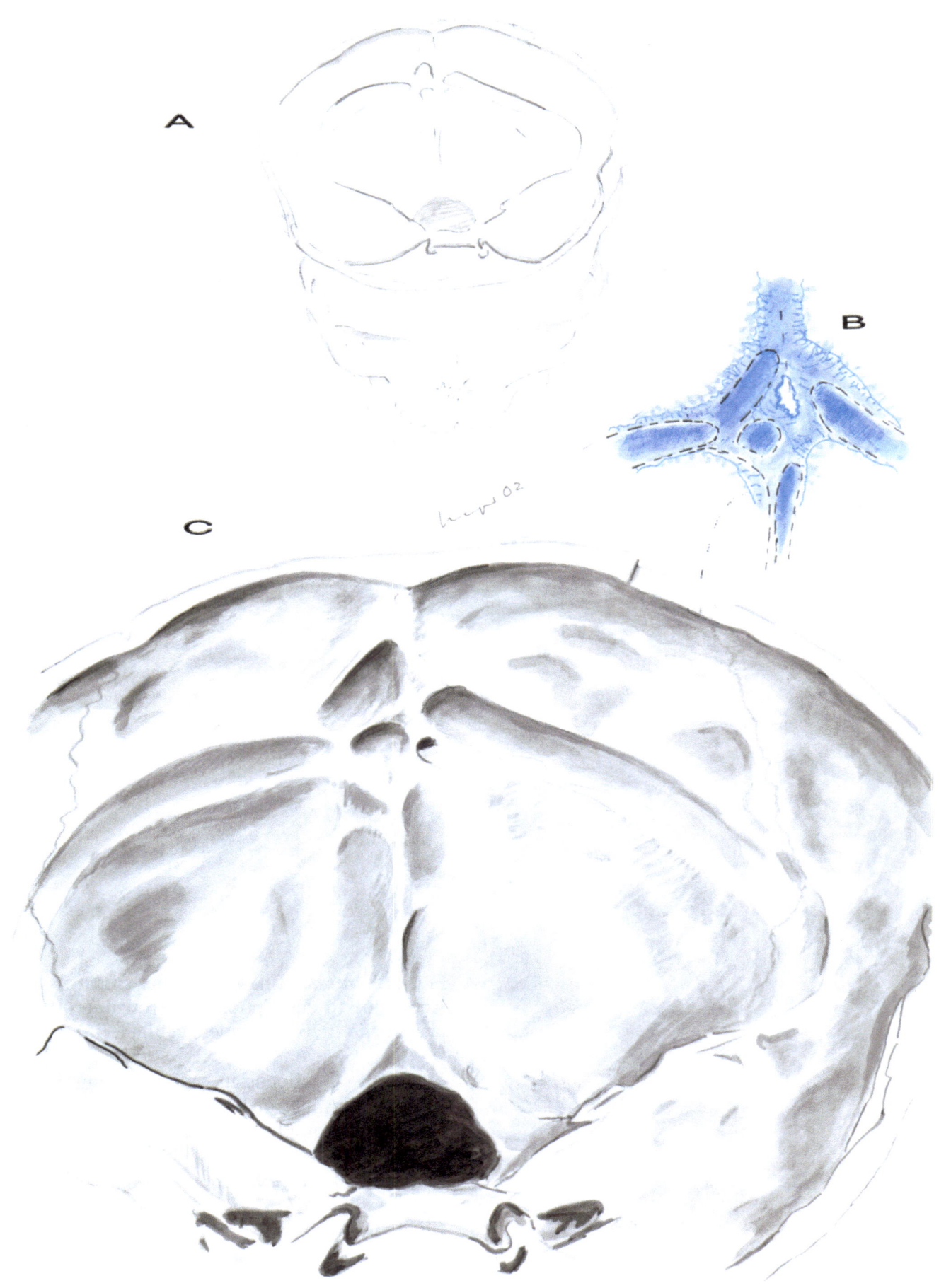

Fig. 115. **Variants of Confluens sinuum**
According to descriptions of Browning (1953), Kaplan et al. (1976), Browders et al. (1975)

A   Confluens and adjacent veins, simplified
B   Adhesion builds an incomplete septum
C   Complete septum and anastomotic channel (arrow)
D to **F**   Further variants

Clinical aspects:
See Fig. 116

FIG. 115

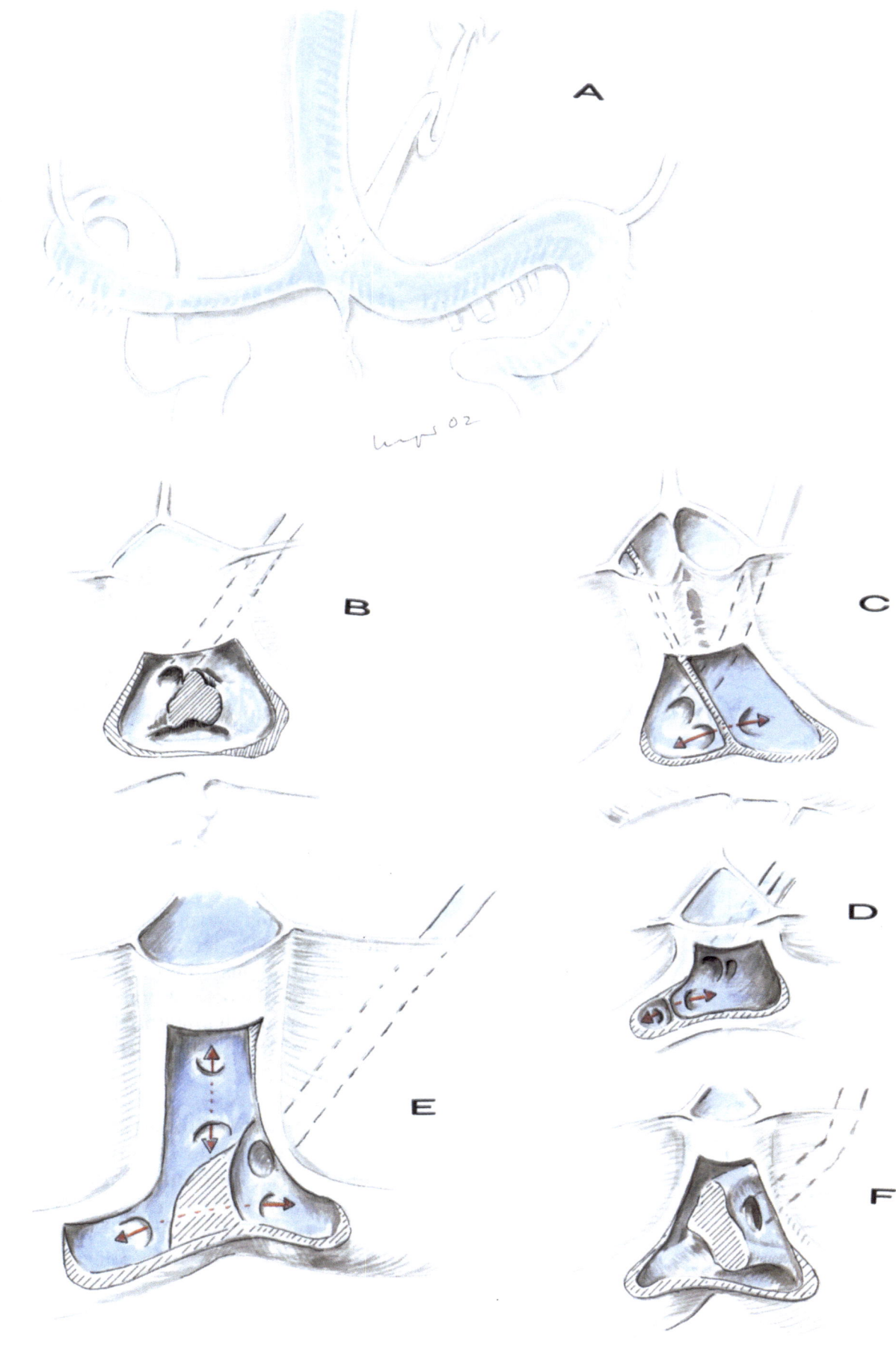

Fig. 116. Continuation of Fig. 115

G No septum
H Symmetrical Confluens (rare) with adhesion
I As E Fig. 115, without channels
J Sinus rectus duplicated
K Complete separation of the two cerebral systems of veins
L Both systems are draining in the right Sinus transversus. The left Sinus transversus is feeded by the numerous veins of the cranial base, especially from the petrous bone, V. Labbé and from superior cerebellar/tentorial veins

Clinical aspects:
Most important are K and L, which are presentable using neuroradiological methods. Sinus rectus is mostly connected with the left side of Confluens. After its ligation, the bloodflow in the inner cerebral veins will be diverted into numerous tentorial veins. As the bloodstream of tentorial veins is very slow, there is danger of thrombosis. Ligations of Sinus rectus should be avoided, if possible.

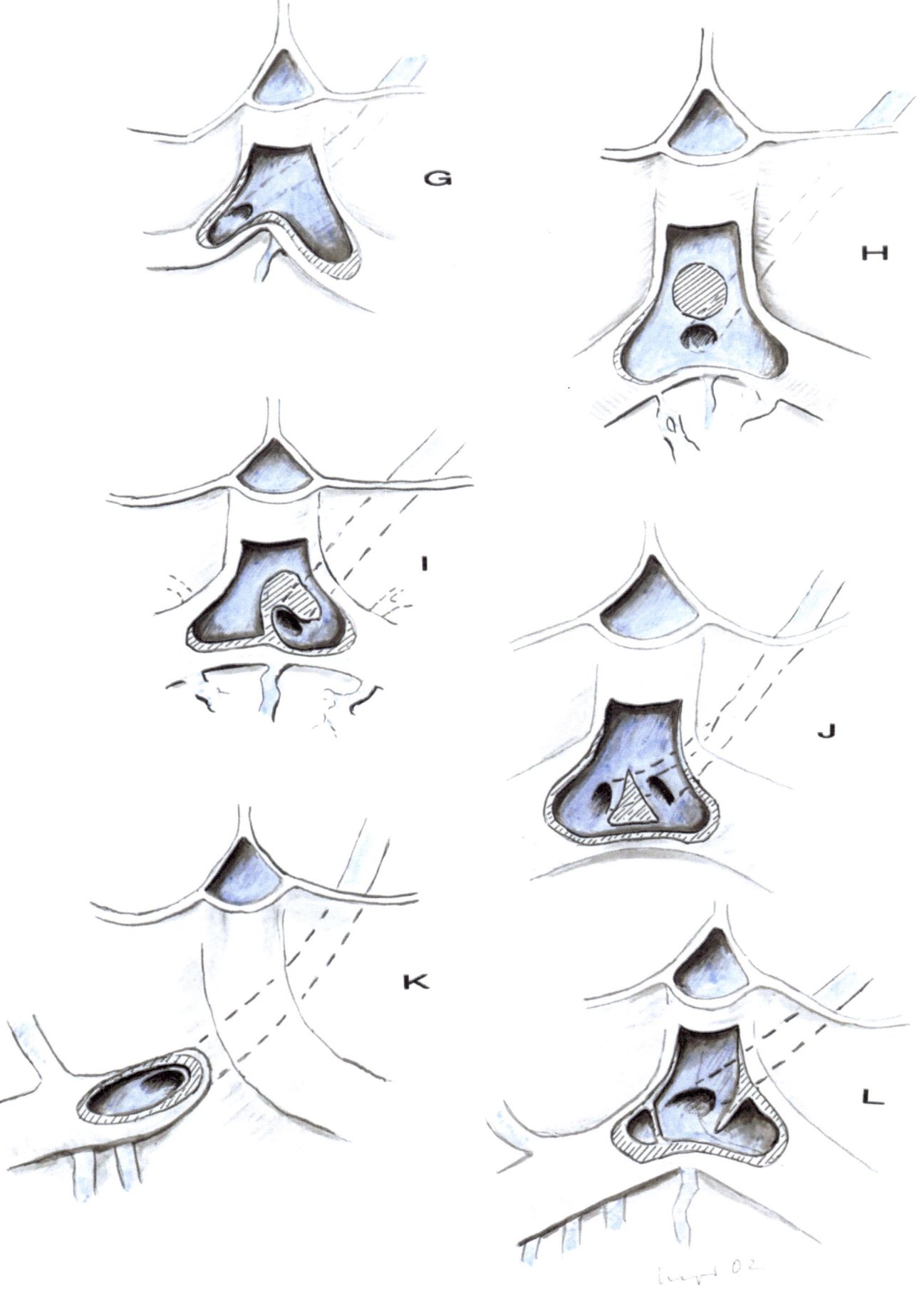

Fig. 117. **Performance of a cast of the sinus(es) without deformation of the sinus wall,** using a low-pressure injection of formaldehyd*

---

* Danninger, thesis Freiburg/Br. (1979)

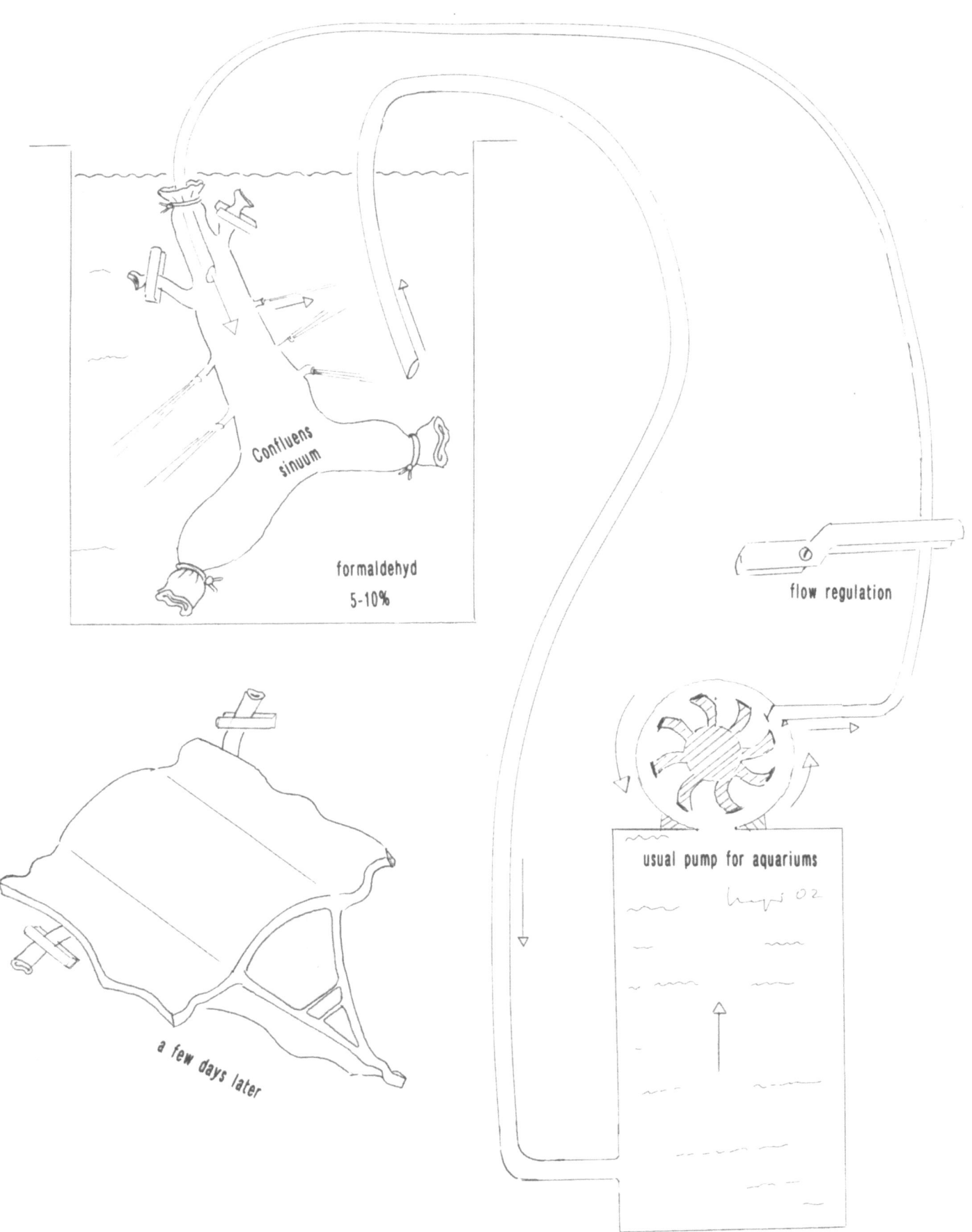

Confluens sinuum

formaldehyd
5-10%

a few days later

flow regulation

usual pump for aquariums

Fig. 118. **Bridging veins of Sinus sagittalis inf., 2 common variants** (Stephens and Stilwell (1969), p 145 and 149, Mc.Cord et al. (1972)

In both types the sinus begins in the frontal segment of Falx. A bundle of bridging veins, originating from Corpus callosum (or one singular bridging vein) are present at the anterior segment of the sinus. Further bridging veins in the occipital portion of the Sinus were not present.

Clinical aspects:
These bridging veins are draining not only Corpus callosum, but also it drains the cingulum Allocortical areas are endangered during midline-approaches – caution –

FIG. 118

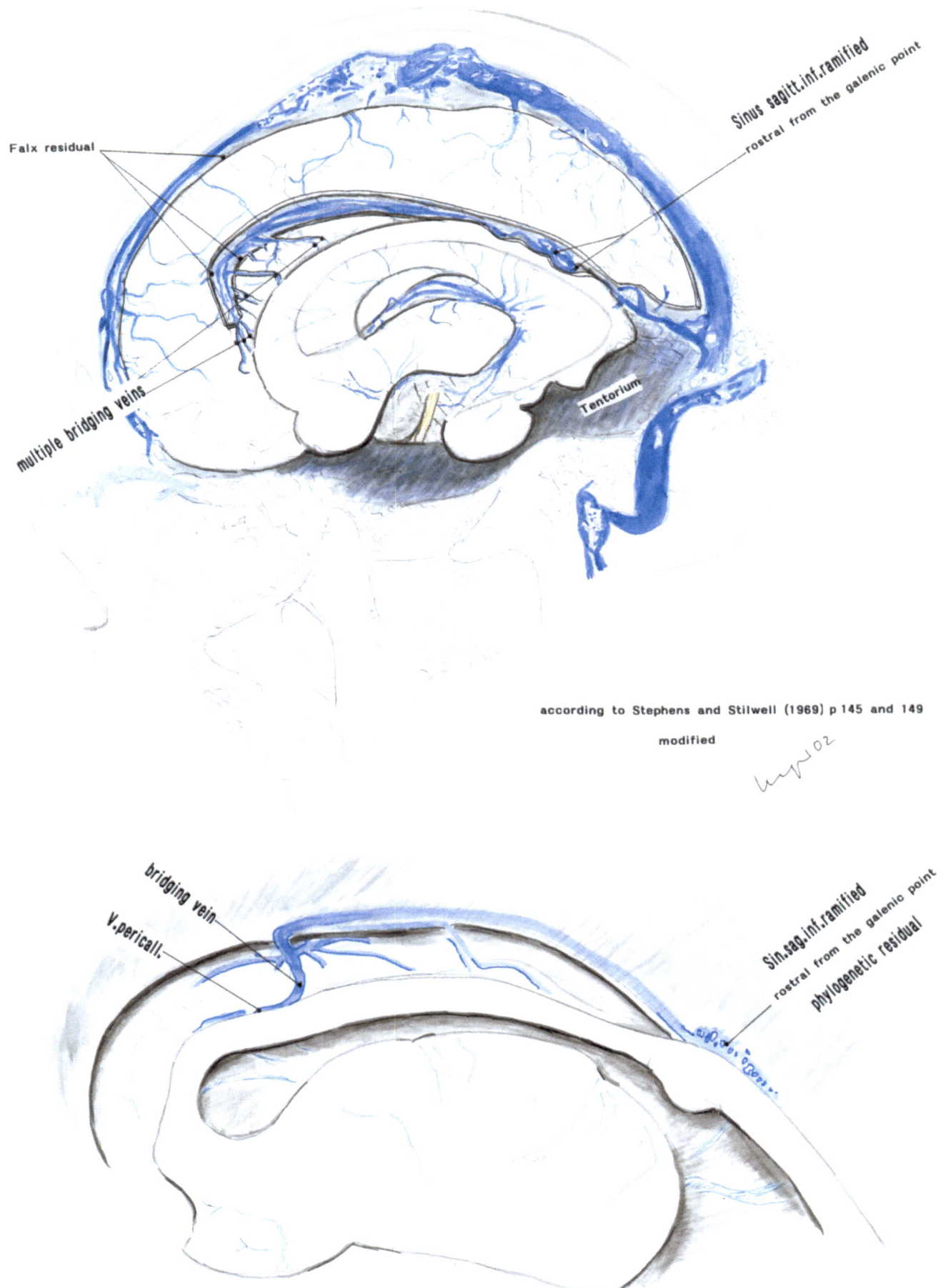

Falx residual

Sinus sagitt.inf.ramified
rostral from the galenic point

multiple bridging veins

Tentorium

according to Stephens and Stilwell (1969) p 145 and 149

modified

bridging vein

V.pericall.

Sin.sag.inf.ramified
rostral from the galenic point
phylogenetic residual

Fig. 119. **Sinus rectus, variants**

A   Normal finding
Note the ramifications of Sinus rectus and Sinus sagittalis inf.
B   Sinus rectus absent (rare)
Drainage of the inner cerebral veins into the inferior and superior sagittal sinus (normal pattern in mammals)

Clinical aspects:
If Sinus rectus is absent, then resection of Falx may be dangerous

FIG. 119

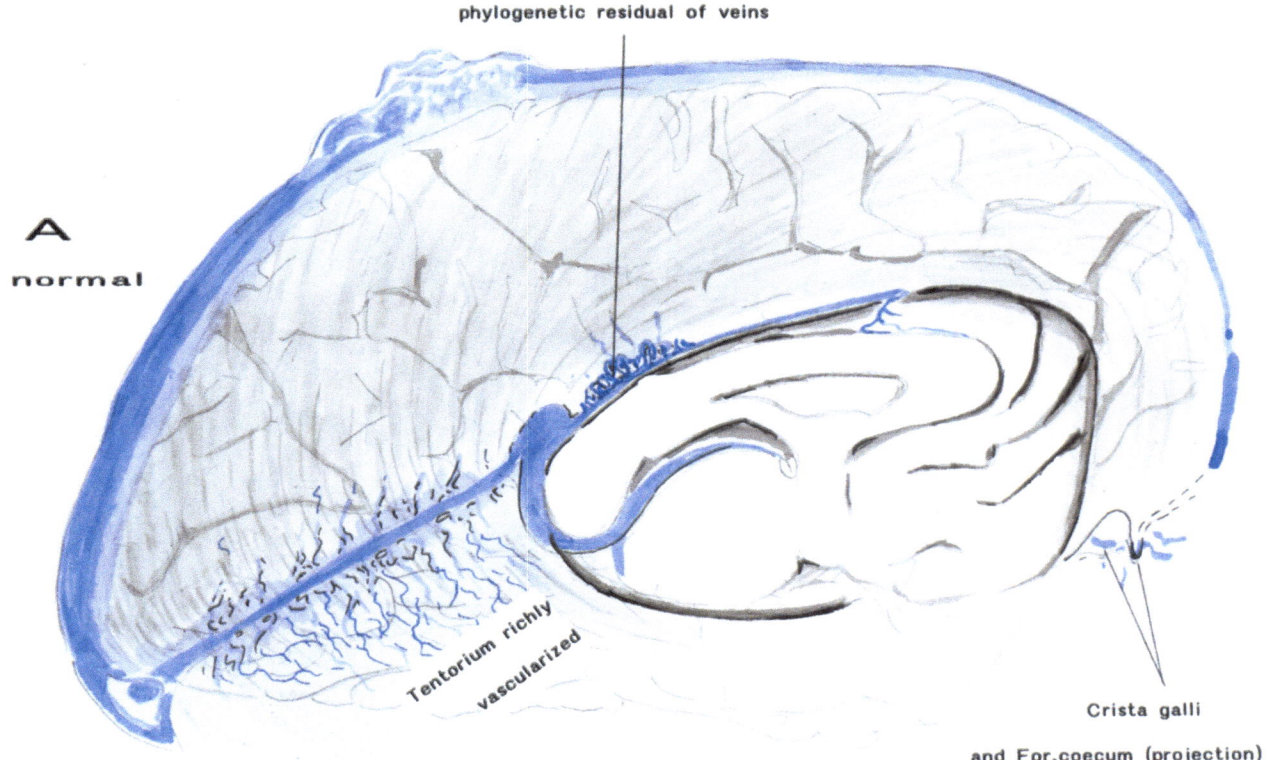

phylogenetic residual of veins

A
normal

Tentorium richly vascularized

Crista galli

and For.coecum (projection)

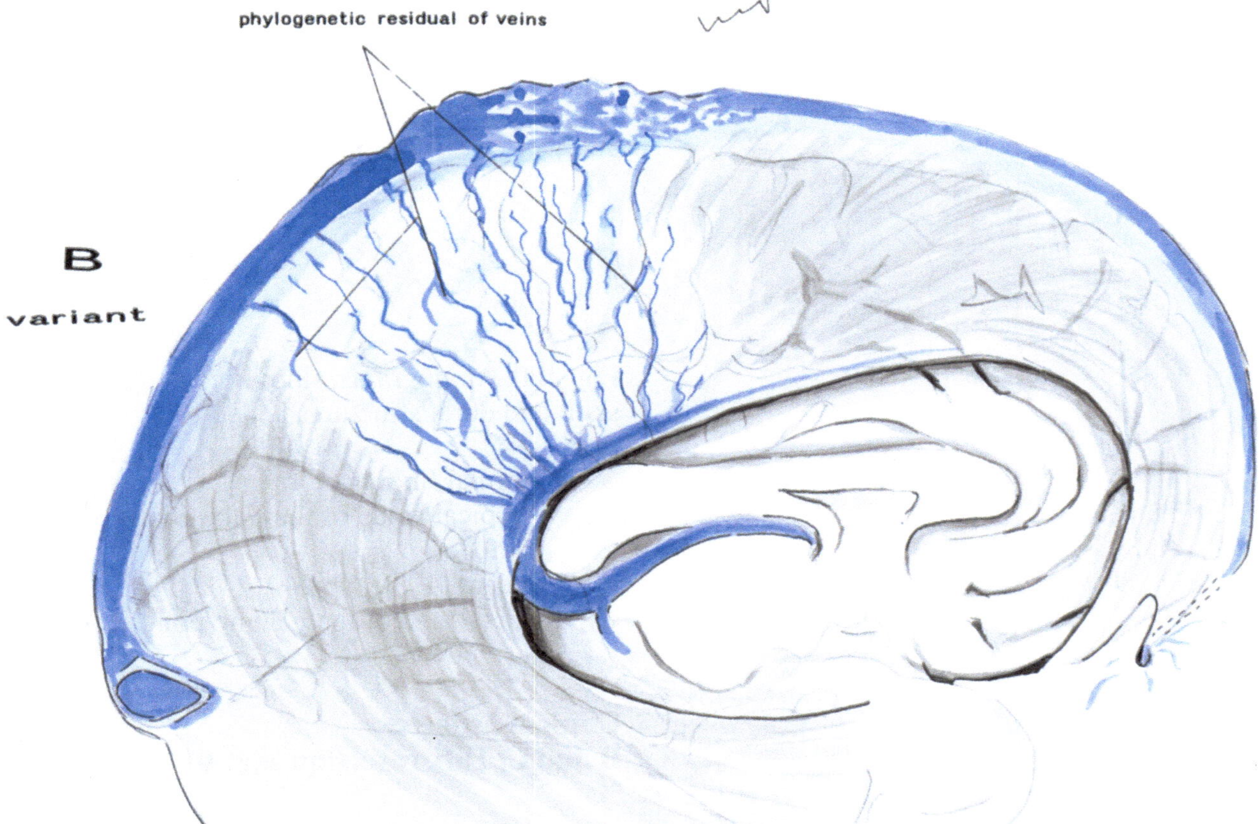

phylogenetic residual of veins

B
variant

Fig. 120. Angiograms, simplified and colored skull and dural structures added

A Normal finding
B Atypical sinus of the Falx. Sinus rectus absent (rare)
Seeger (1980) pp 240, 241*
C Dorsal size of the galenic point, common variant

Clinical aspects:
B and C: Falx resection may be dangerous

---

* according to P. Danninger, thesis, Freiburg (1979)

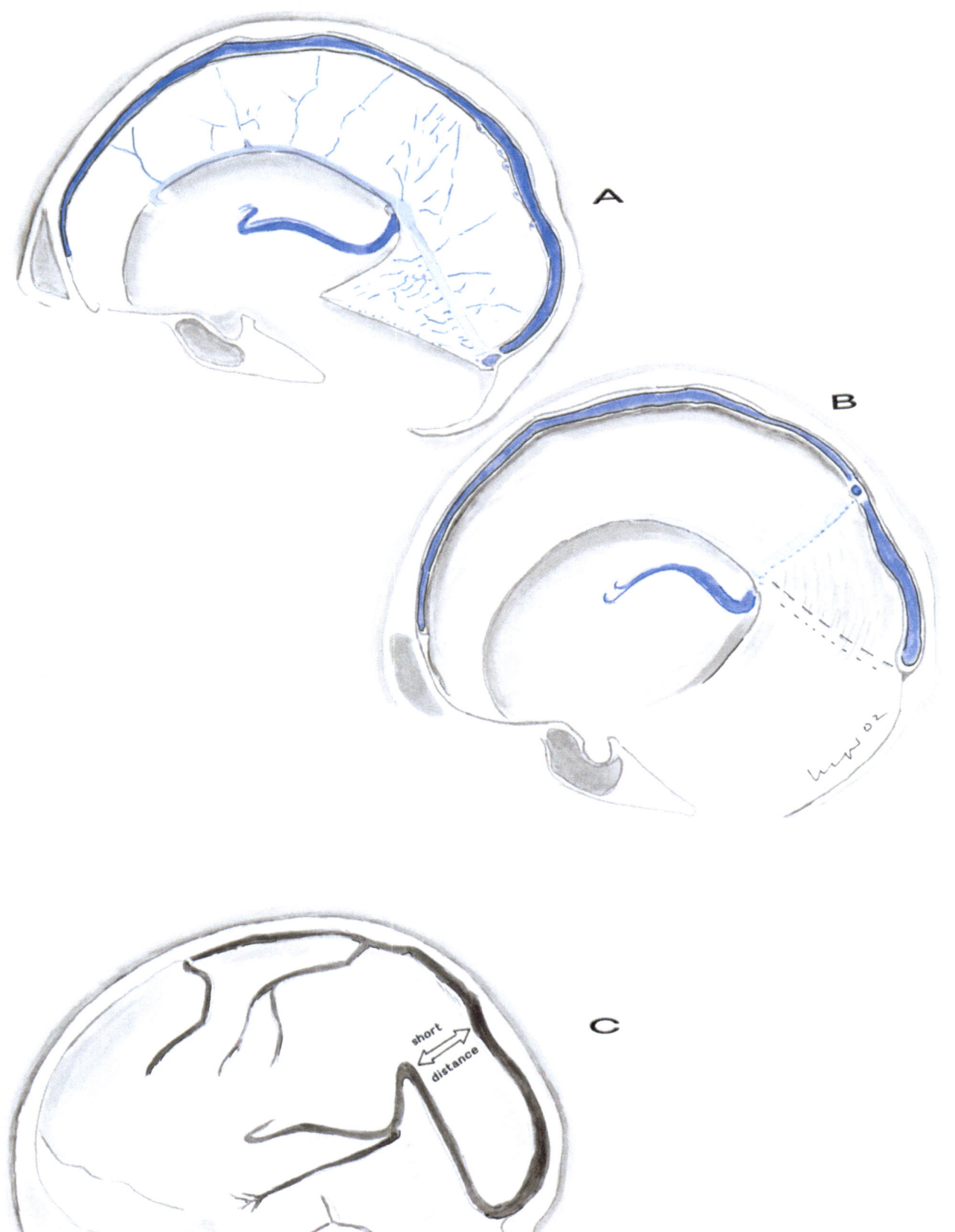

A

B

C

short
distance

Fig. 121. **Further clinical aspects** at usual findings

a　　　**Danger for the frontal segment of bridging veins, which are draining Cingulum**

b and *c*　　**The high venous vascularization has to be considered at resection of Falx or Tentorium**

FIG. 121

microsurgical approach

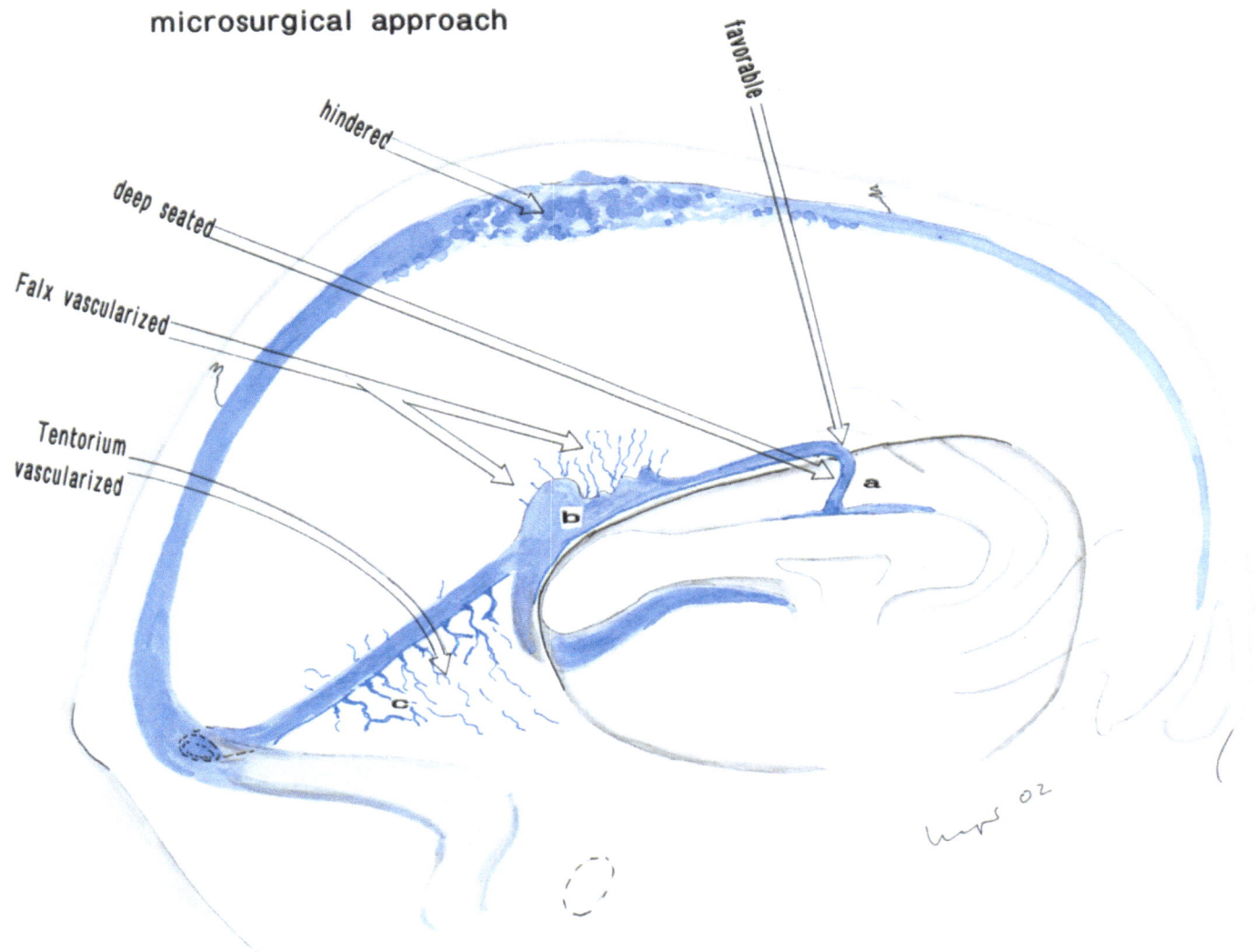

Fig. 122. **Sinus rectus, septa**

Its numerous variants are illustrated by multiple septa (approximately 5 to 15 mm long, sagittal and platform-like), adhesions and variants of its connection with Confluens sinuum

A	Drawing according to photographs of cadaver dissections and x-ray-findings of Bergquist (1975)
B	Sectional enlargement from Fig. 112
C	Platform-like septum, sketch
D	As C, multiple
E and **F**	Duplications of Sinus rectus and its connections with Confluens sinuum
B to **F**	According to Bergquist (1975), and Browder and Kaplan (1976)

Clinical aspects as Figs. 115 and 116

FIG. 122

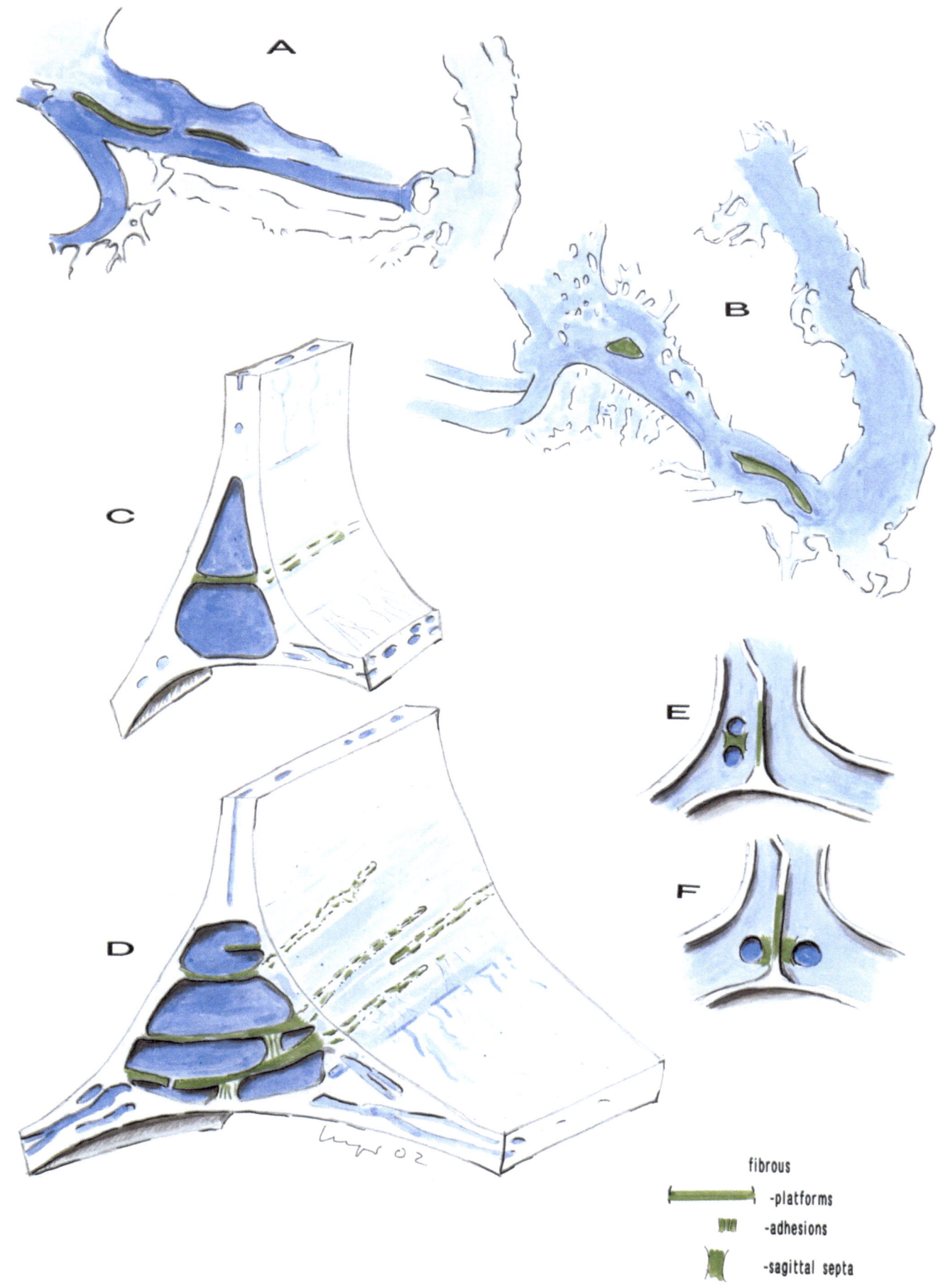

fibrous
-platforms
-adhesions
-sagittal septa

Fig. 123. **Veins of the lateral ventricle** (except Cornu inf.)
Common findings, schematic presentation

A   Dorsal and basal veins
B   Dorsal veins
C   As A, relationship with basal ganglia; a lateral ventricle, b Nucleus caudatus, c Thalamus, d lateral atrial veins. Medial atrial veins omitted

Clinical aspects:
See Fig. 126

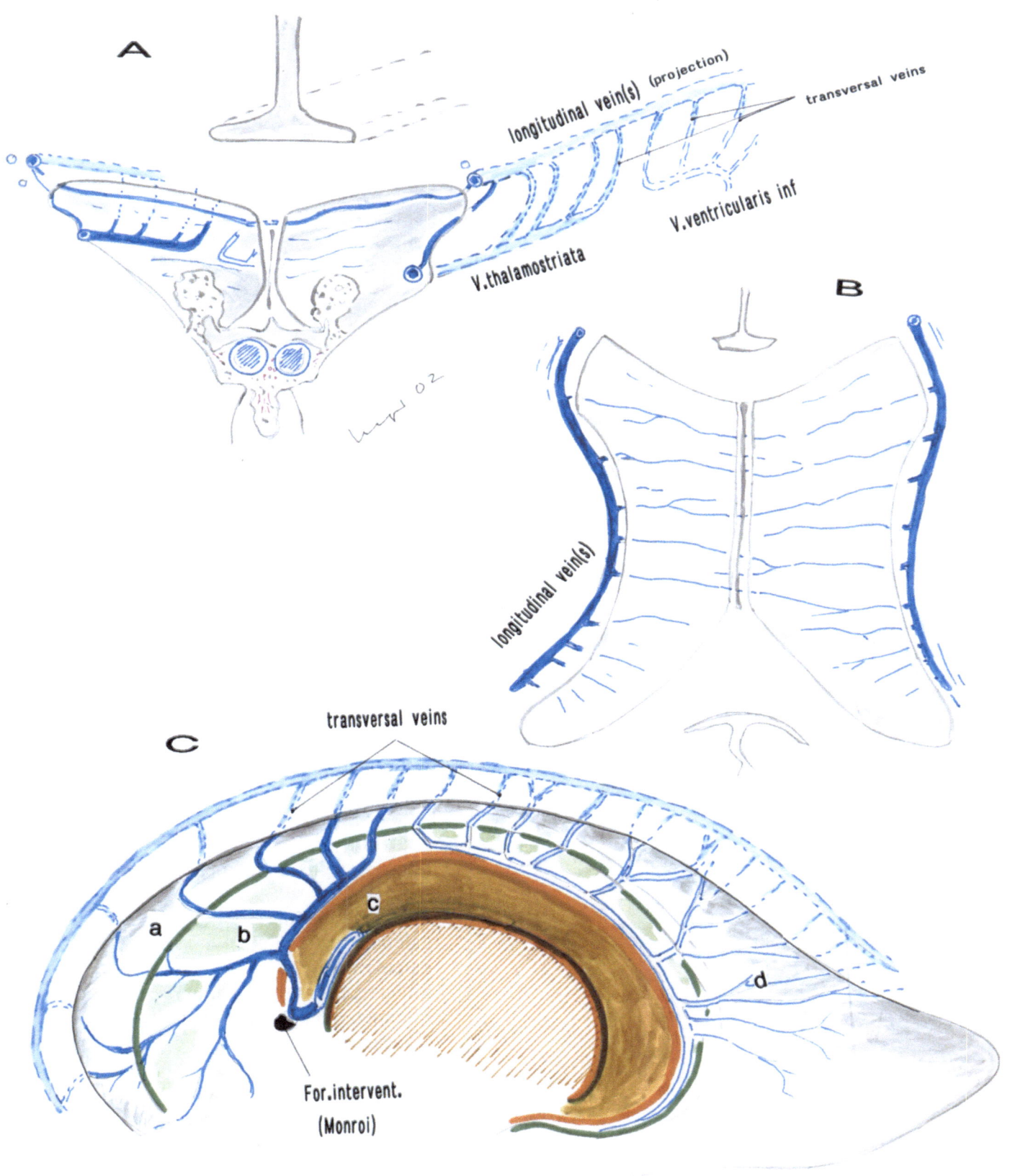

A

longitudinal vein(s) (projection)

transversal veins

V.ventricularis inf

V.thalamostriata

B

longitudinal vein(s)

transversal veins

C

transversal veins

a

b

c

d

For.intervent.
(Monroi)

Fig. 124.

A   V.thalamostriata and V. ventricularis inf. are not connected
- usual finding -
(connection of these veins is a common variant*)

B   **Usual finding**

Clinical aspects
See Fig. 126

Abbreviations

a   bridging veins are mapping the begin of Sinus sagitt. inf.
b   Sin. sagittalis inf.
c   galenic area
d   galenic vein
e   Sinus rectus (straight sinus)
f   Foramen interventriculare (Monroi)
g   V. cerebri int.
h   V. basalis (Rosenthal)
i   longitudinal vein (projection)
j   as i
k   Rami transversi
l   Rami transversi
m   medial atrial veins
n   lateral atrial veins

---

* According to the thesis of G. Hölzel (1977), Freiburg

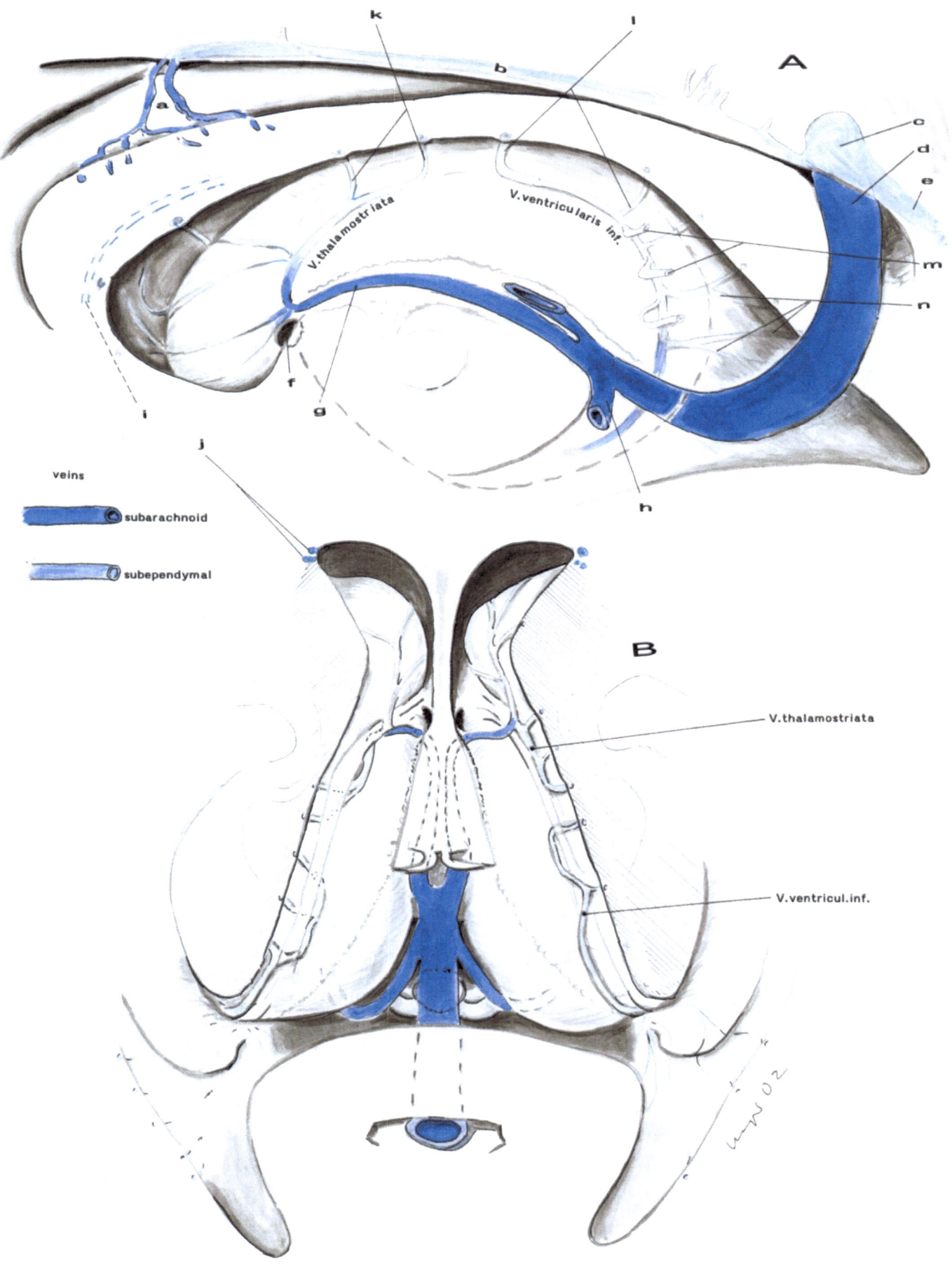

veins

subarachnoid

subependymal

A

V.thalamostriata

V.ventricularis inf.

B

V.thalamostriata

V.ventricul.inf.

Fig. 125. **Further variants**

A   Hyperplasia of V. cerebri int. (proximal, left-sided) V. ventr. lat. directa: common variants
B   Hypoplasia of V. cerebri int. (proximal, left-sided), combined with V. ventriculi lat. directa
C   V. thalamostriata (Angulus venosus) distant from Foramen interventriculare (Monroi)

Clinical aspects
Variants should be considered during microsurgical dorsal approaches to the 3rd ventricle.

Abbreviations
a   V. septi pellucidi
b   Vv. nuclei caudati
c   V. thalamostriata
d   Foramen interventriculare (Monroi)
e   V. cerebri int. (hypoplastic anterior from j)
f   V. ventriculi lat. directa
g   plexus vein
h   dorsal thalamic vein(s)
i   V. basalis (Rosenthal)
j   atypical connection of r and k
      favorable for microsurgical approaches to Cornu inf.
k   both Vv. basales (Rosenthal)
l   galenic vein
m   V. mesencephalica lat.
n   V. mesencephalica dorsalis
o   Connection of m and n
p   V. petrosa sup.
q   Rr. transversi
r   V. ventricularis inf.

---

* According to the publications of Salamon and Huang, Wolf,Huang and Wolf (see Seeger, 1984)

FIG. 125

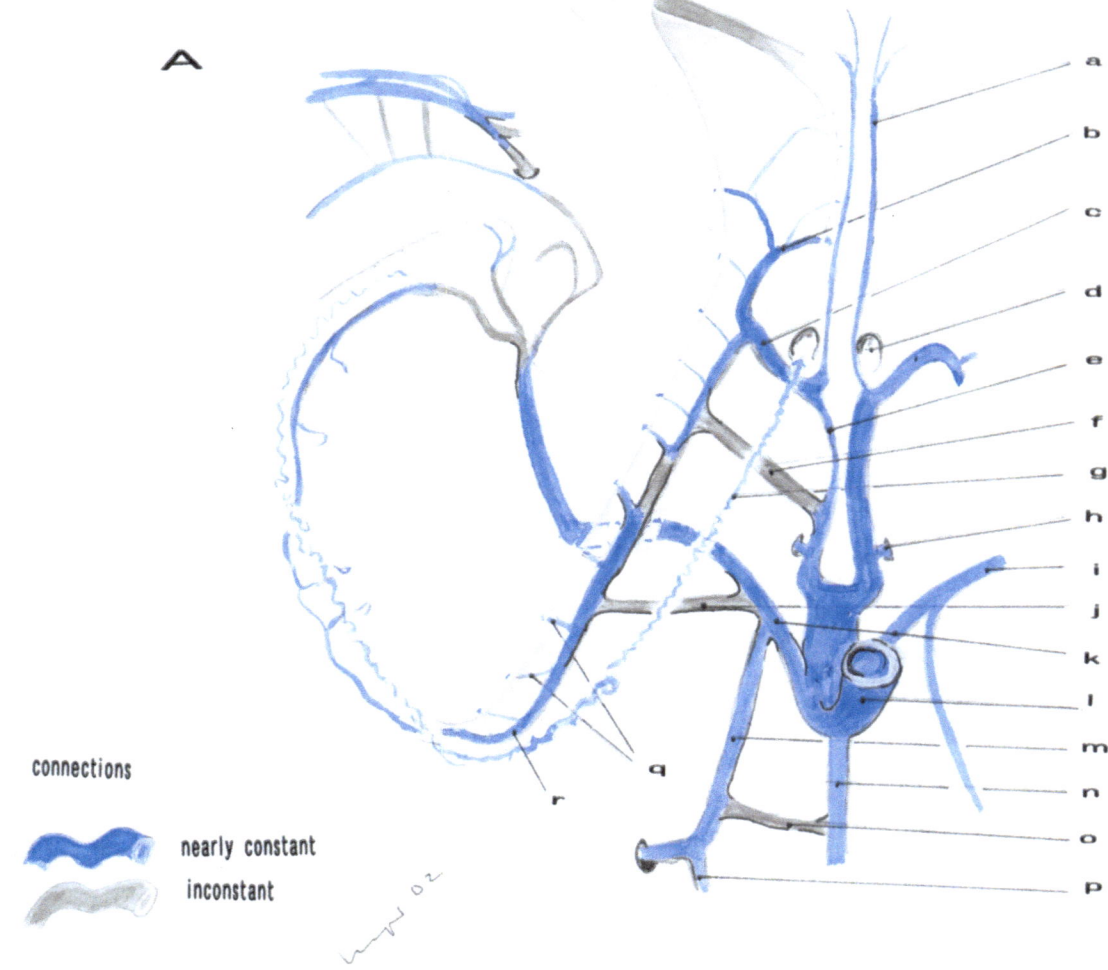

A

a
b
c
d
e
f
g
h
i
j
k
l
m
n
o
p

q

r

connections

nearly constant

inconstant

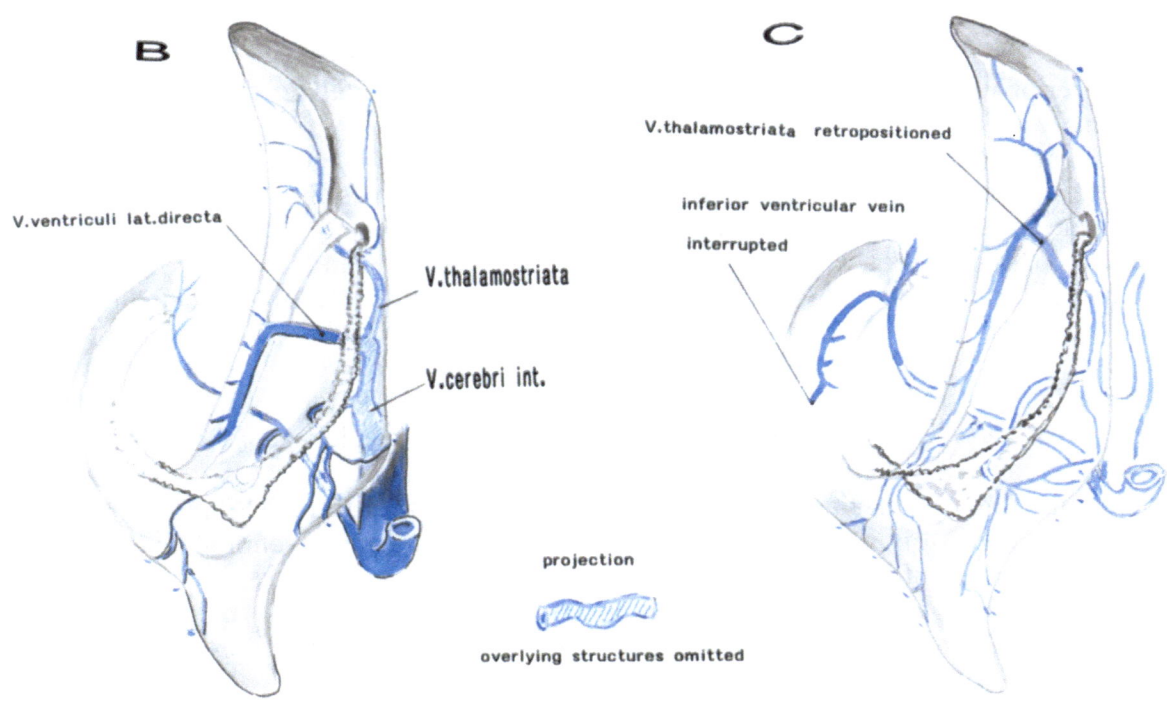

B

V.ventriculi lat.directa

V.thalamostriata

V.cerebri int.

C

V.thalamostriata retropositioned

inferior ventricular vein

interrupted

projection

overlying structures omitted

Fig. 126. Complete substitution of the anterior segment of V. cerebri int.

by V. ventriculi lat. directa*

Clinical aspects:

Usually, the posterior wall of the foramen of Monro contains Angulus venosus (transition of V. thalamostriata and V. cerebri int.). Sometimes the rostral (anterior) segment of V. cerebri int. is hypoplastic, the posterior segment is thickened by a variant, which connects V. thalamostriata and/or V. ventricularis inf., far distant from the foramen of Monro, with V. cerebri int.:

– V. ventriculi lateralis directa –

Here microneurosurgical **approaches to the third ventricle are favorable**. The posterior wall of the foramen of Monro can be splitted and widened in a posterior direction without damaging the inner cerebral veins. V. septi pellucidi is not complicating the approach. It may be shifted into a lateral or medial direction, but it should not be interrupted, because it is a draining vein of Columna fornicis and Septum pellucidum. The surgical widening of the foramen may

be hindered, if the anterior segment of V. cerebri int. is not completely aplastic, as seen in Fig. 125

Abbreviations

a	V. thalamostriata
b	R. transversus
c	**V. septi pellucidi (can be pushed aside by microsurgery)**
d	one of numerous septal veins, running to Fornix and Plexus chor.
e	contralateral V. cerebri int.
f	galenic vein
g	V. basalis (Rosenthal), small variant
h	Plexus chor.
i	dorsal choroid branches** dividing close to Habenula (Plets, 1969)
j	Caput nuclei caudati
k	Thalamus
l	Columna fornicis
m	Foramen intervenriculare (Monroi)

---

* for MRT-findings see Seeger (2000)

** i1 to i4: dentrifications according to Stephens and Stilwell (1969), pp 65, 66, 97, 115 to 119

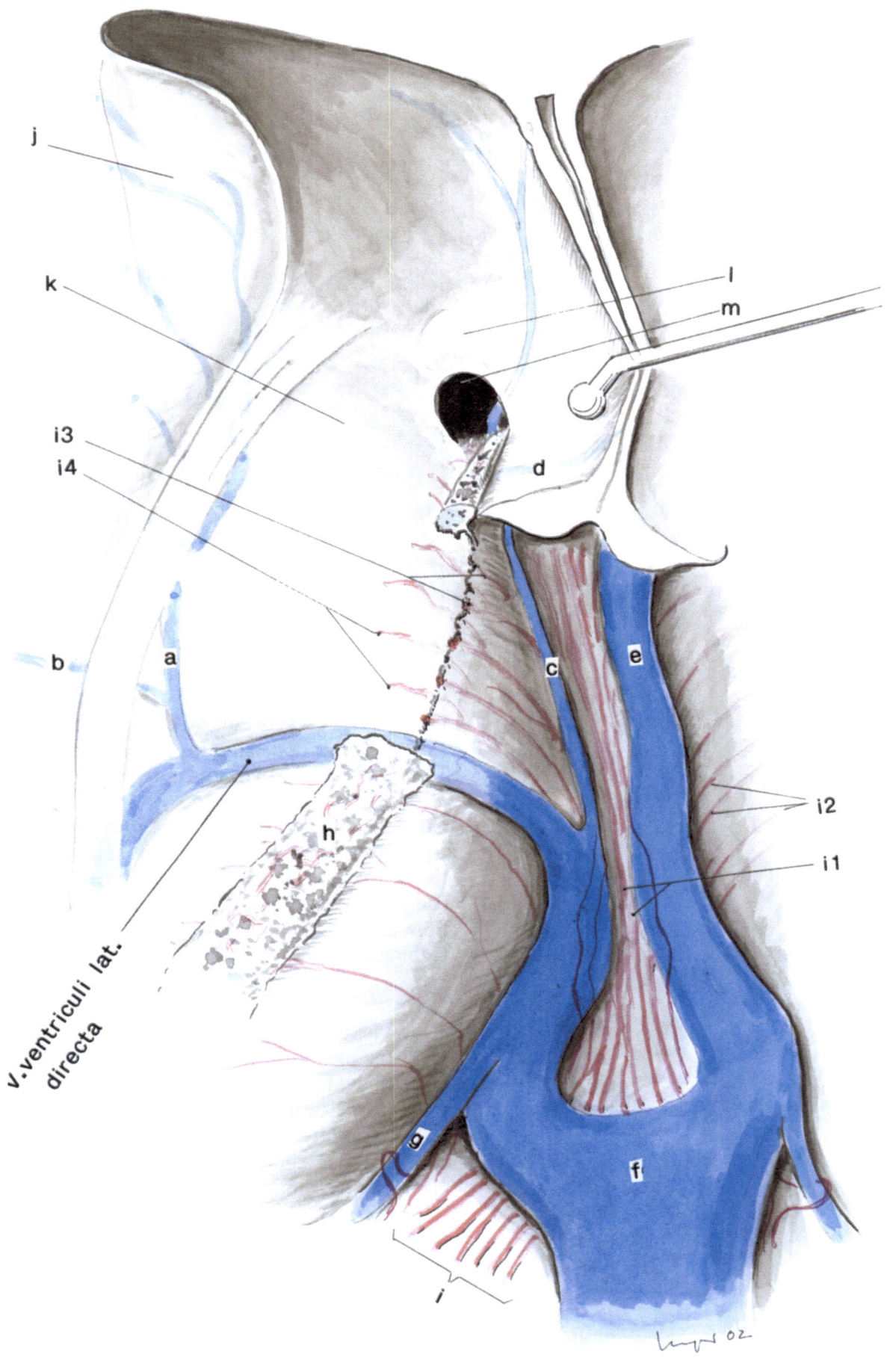

j

k

i3

i4

b   a

v.ventriculi lat.
directa

h

l

m

d

c   e

i2

i1

g   f

i

Fig. 127. **Angulus venosus and a wide medial septal vein are enclosed by the posterior wall of Foramen interventriculare Monroi, covered by a thin layer of Ependyma, which also envelopes Thalamus and Fornix as a common membrane***

Clinical aspects:
Ependyma must be stepwise transected, until the veins are visible. Now a further splitting of the posterior wall of the foramen has to be avoided: Danger for brain damage by venous infarctions.

Abbreviations

a　Plexus of the third ventricle not connected with the plexus of the lateral ventricle

b　dilated **medial septal vein**

c　**anterior end of the plexus of the lateral ventricle**

d　inner cerebral vein

e　plexus omitted for demonstration of Taenia chorioidea

f　Columna fornicis

g　V. thalamostriata

---

* based on a microsurgical finding (Seeger, 1988, p 263)

FIG. 127

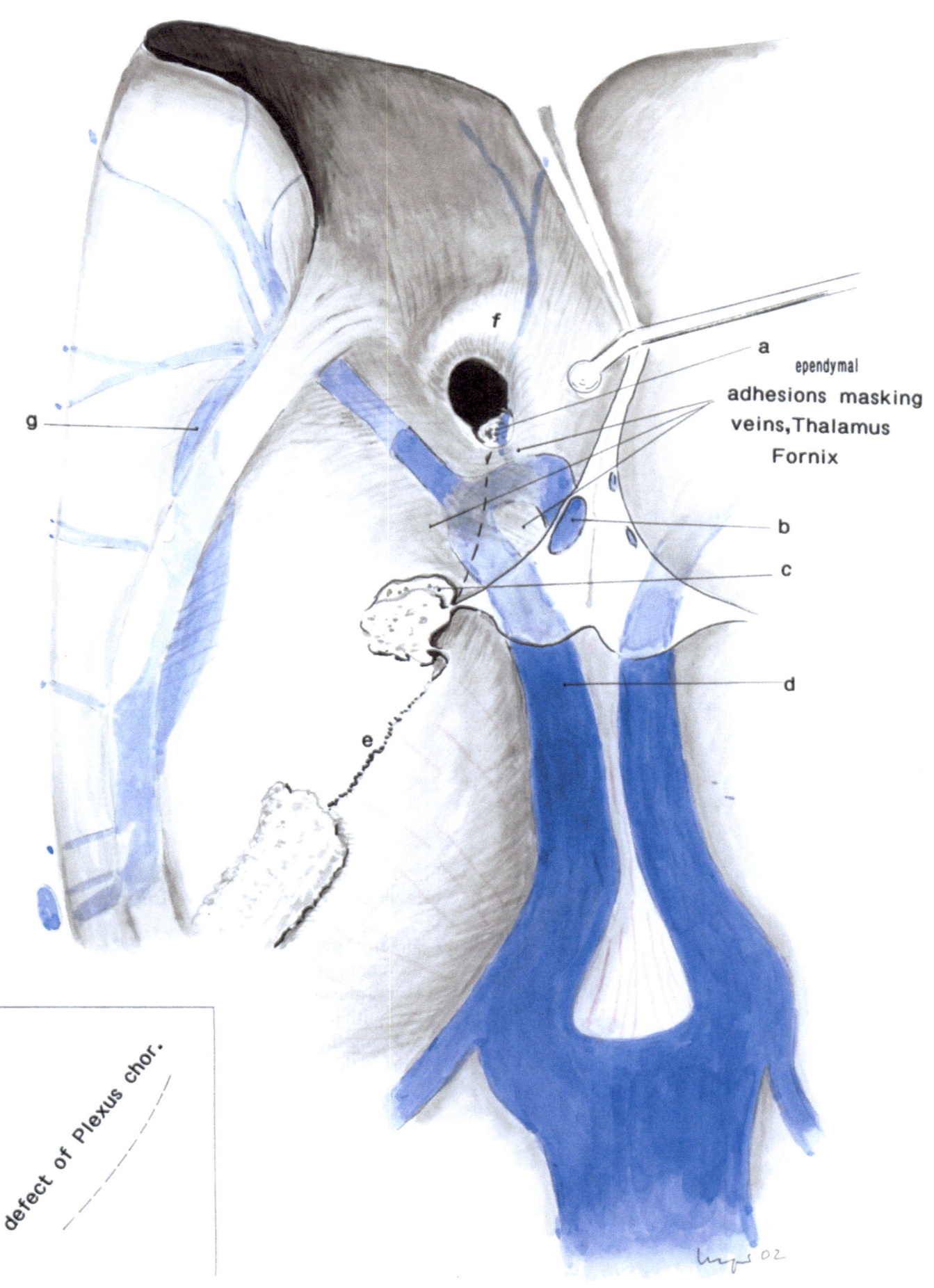

f

a

ependymal
adhesions masking
veins, Thalamus
Fornix

g

b

c

d

e

defect of Plexus chor.

Fig. 128. **Further variants which can complicate microsurgery, common findings**

- Heterotopy of plexus segments, combined with atypical epi- and subependymal thalamic veins
- Hyperplasia of Taenia chorioidea, mostly combined with hyperplasia of Velum interpositum. In this drawing the velum is omitted for a better presentation of the veins and of Thalamus
- V. ventricularis inf. is atypically connected with the inner cerebral vein

Clinical aspects
- **Microsurgical loosening of the plexus from Thalamus is difficult**
- **Thalamic veins should be preserved, as well as V. ventricularis inf.**

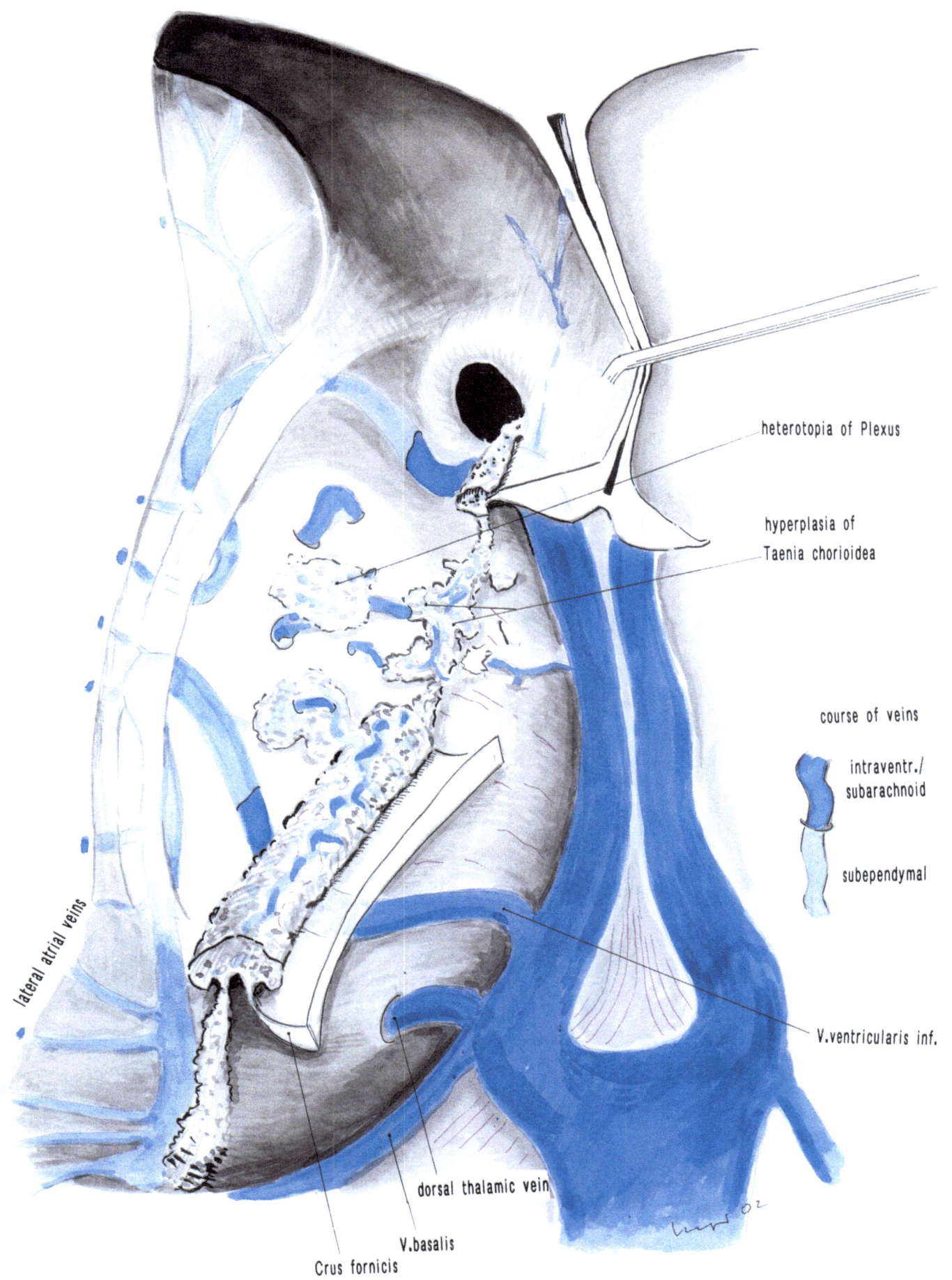

lateral atrial veins

heterotopia of Plexus

hyperplasia of
Taenia chorioidea

course of veins

intraventr./
subarachnoid

subependymal

V.ventricularis inf.

dorsal thalamic vein

V.basalis

Crus fornicis

Fig. 129. Atypical drainage of the inner cerebral veins* (rare variant)

- V. magna Galeni and inner cerebral veins are not recognizable
- Combination of Sinus rectus with a venous network of the Falx (see Fig. 119)
- Main drainage of the inner cerebral veins by using a dilated **V. cerebellaris praecentralis**

Clinical aspects:
This is a variant, no pathological finding (no venous angioma)

---

* Angiogram of the Department of Neurology Ulm (Prof. Kornhuber) in 1984

FIG. 129

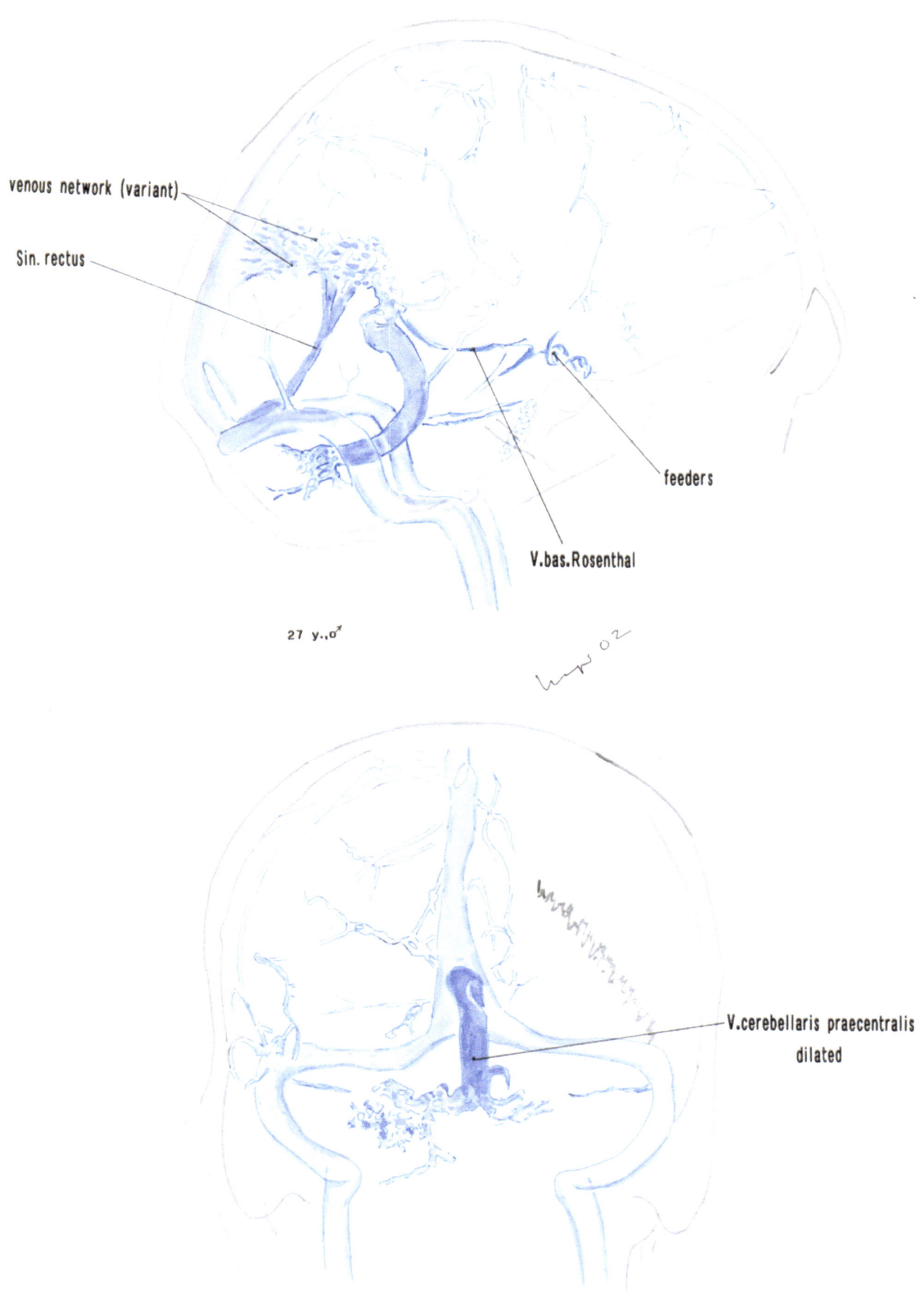

venous network (variant)

Sin. rectus

feeders

V.bas.Rosenthal

27 y.,♂

V.cerebellaris praecentralis
dilated

Fig. 130. **Cavum Vergae** (Verga 1851, according to Lang, 1981)

A　　　MRT, normal finding, simplified, for comparison with B
B　　　Cavum Vergae
C and **D**　Indian ink copies. Sketch for topogram added. View from an anterior direction

Abbreviations
a　　Cavum Vergae
b　　Plexus chor.
c　　layer af Septum pelluc.
d　　Adhaesio interthalamica

For clinical aspects see Figs. 132 and 133

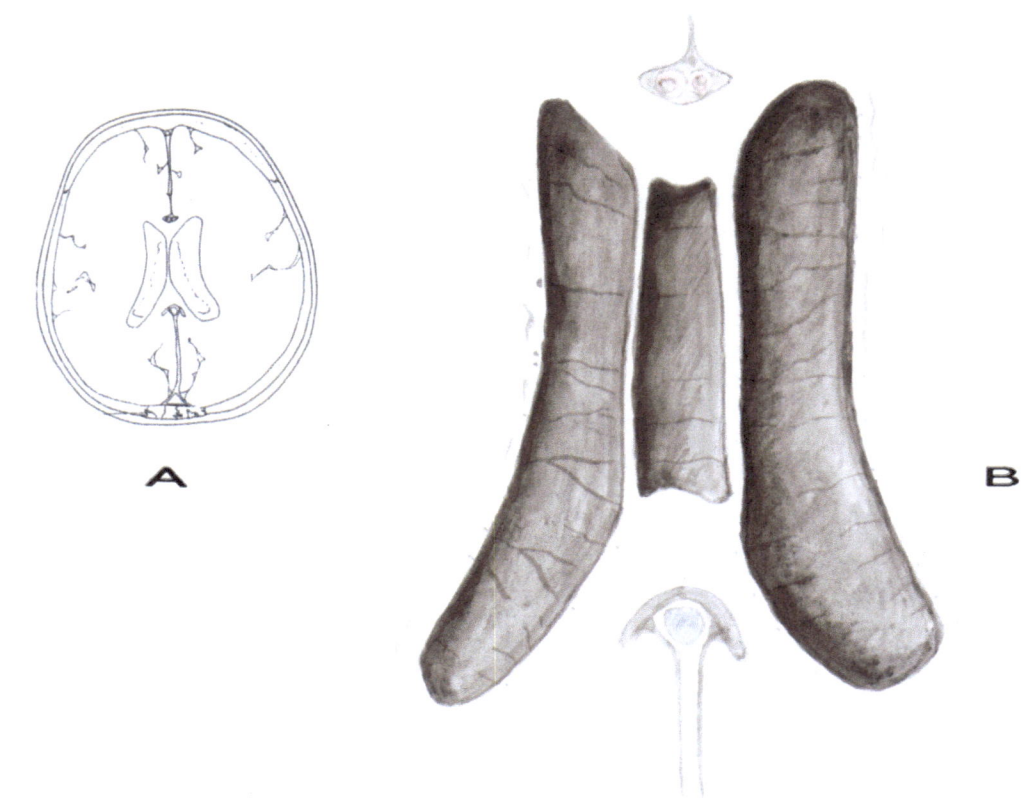

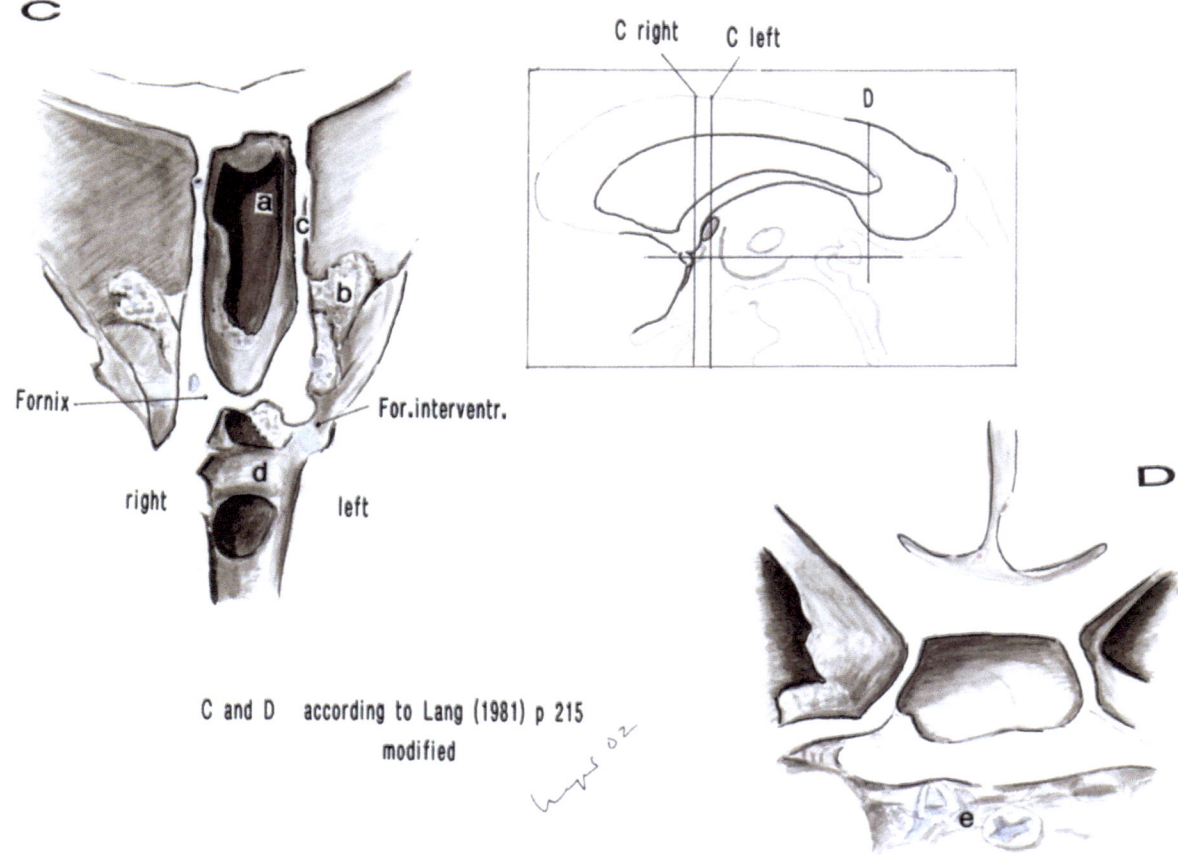

A

B

C

C right   C left

D

Fornix

For.interventr.

right   left

C and D   according to Lang (1981) p 215
modified

D

Fig. 131. Continuation of Fig. 130

Ontogenetic aspects

A   Early stage of development: Separation of Commissura fornicis from Corpus callosum (light arrow)

B   Later on: Commissura fornicis is shifted into a basal direction. Fenetrations of the wall of the Cavum with connection of Cavum Vergae with the lateral ventricle (mostly leftsided, Lang, 1981), and Fissura transversa (black arrows). These are usual connections of Cavum Vergae.

FIG. 131

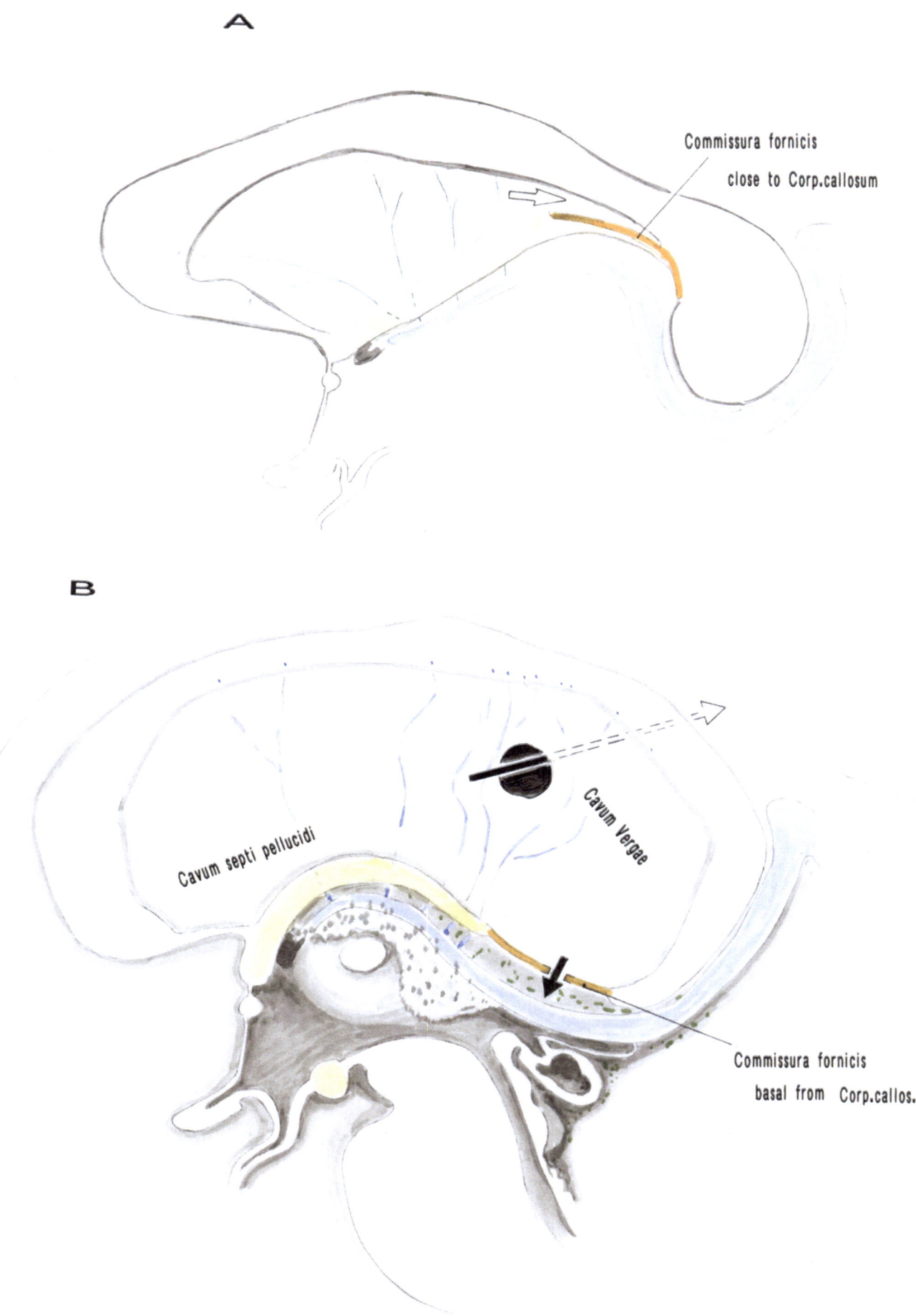

A

Commissura fornicis
close to Corp.callosum

B

Cavum septi pellucidi

Cavum Vergae

Commissura fornicis
basal from Corp.callos.

Fig. 132. **Illustration of clinical aspects**

A   Normal findings: **Long microsurgical route**
B   Cavum Vergae: **Short microsurgical route**

Yellow colored: Allocortical (=limbic) structures
Arrows (red colored): Microsurgical routes

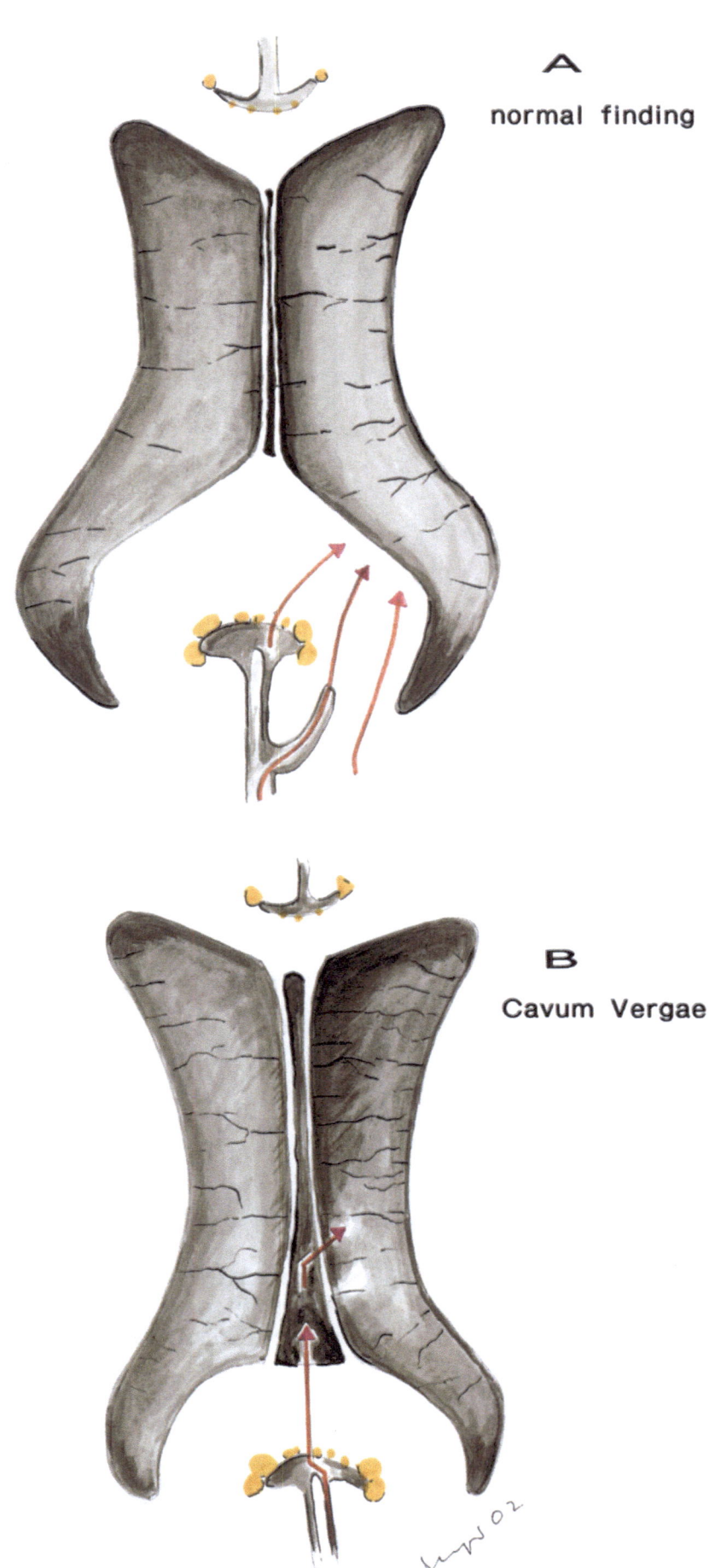

A
normal finding

B
Cavum Vergae

Fig. 133. Continuation of Fig. 132

Interfornical approach* transpassing Cavum Vergae is more favorable than transpassing the narrow Cavum septi pellucidi

Yellow colored: Allorocortical (=limbic) Striae
Arrow (red colored): Microsurgical route

---

* Glioma II of Plexus chor. of the third ventricle, female, 41 y., according to an MRT, 3/12 2002; microsurgical extirpation by Priv. Doz. Dr. Vera van Velthoven, Department of Neurosurgery Freiburg/Br., chairman Prof. Dr. J. Zentner

FIG. 133

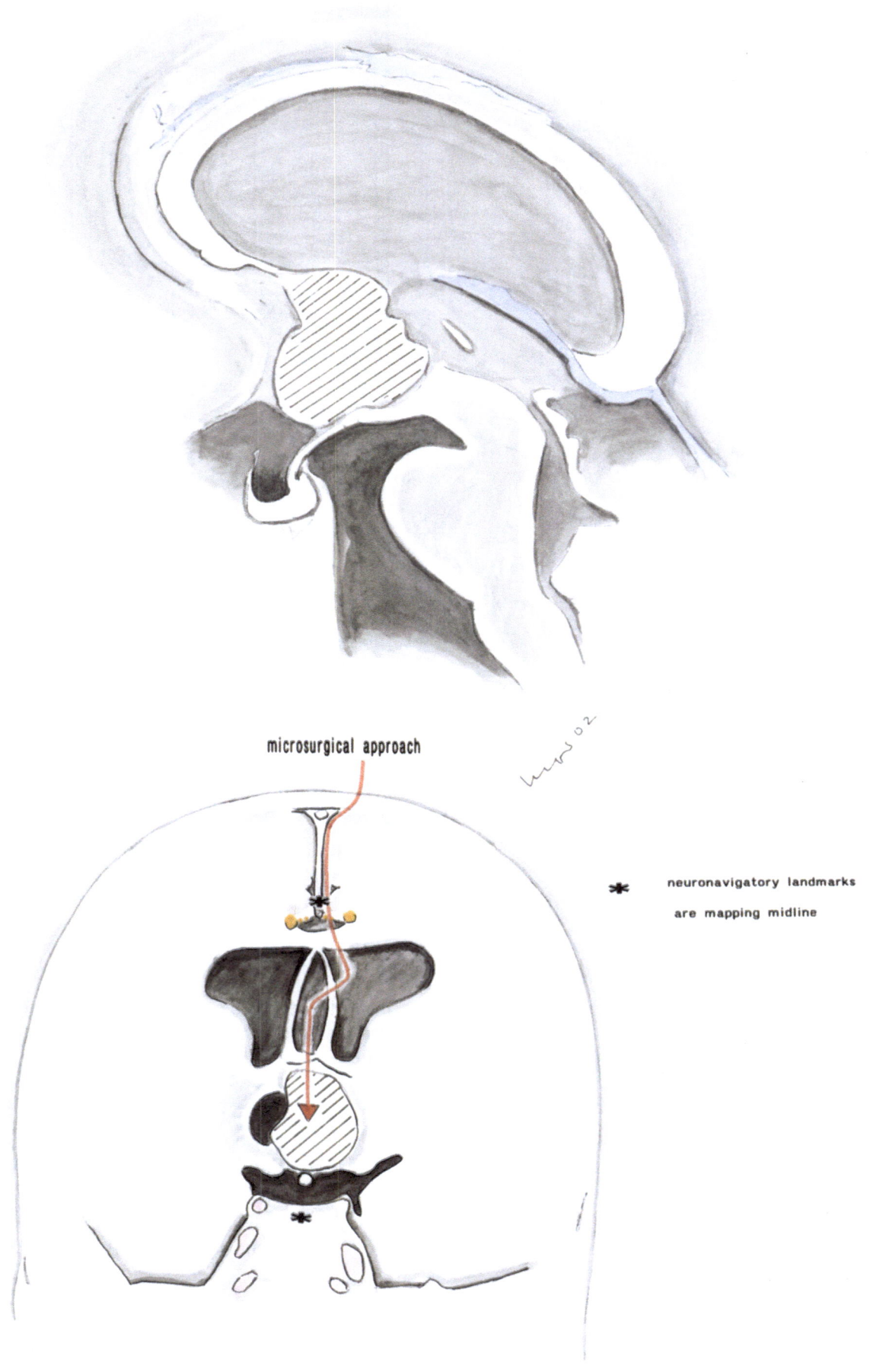

microsurgical approach

✱ neuronavigatory landmarks
are mapping midline

Figs. 134 to 140. **Configuration of the 3rd ventricle and adjacent structures** according to MRT-findings

Fig. 134

A Small and long configurated 3rd ventricle, Crus fornicis basally located, Adhaesio interthalamica absent (common variant)
B Usual findings (for comparison with A)

Clinical aspects:
A: Favorable for microsurgical interfornical approaches

FIG. 134

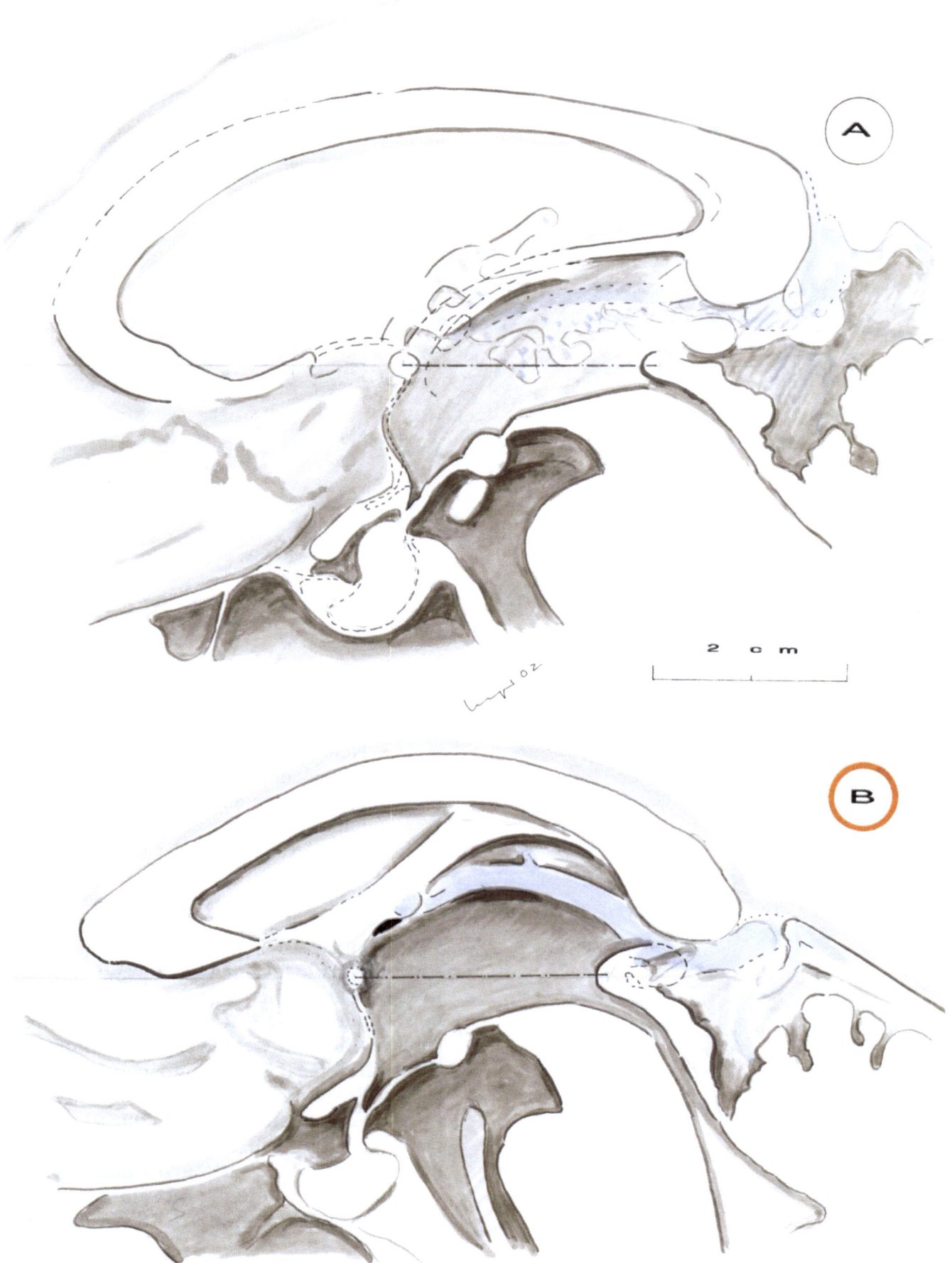

2 cm

Fig. 135

C • Senile atrophic brain (male, 65 years). Com-
missura ant. distant from Rostrum, Cisterna
tecti wide
• Note thickened variant of Rostrum, size and
atrophy of pituitary gland
D • Commissura ant. close to Rostrum and Co-
lumna fornicis
• Well developed medial septal veins
• Cisterna tecti narrow
• Note thickened variant of Rostrum

Clinical aspects:
Anterior Callosotomy (e.g. in epilepsy surgery)
may be dangerous. Typically, Callosotomy will
be stopped at the transitional area between the
thick calibrated Genu corporis callosi and the
thinwalled Rostrum. C and D present a thick-
walled Rostrum. If the callosotomy is continued
in a too far posterior direction, as long as the
rostrum fibers are thickwalled, then Fornices
and Commissura ant. are endangered. This
complication would be prevented by using a
neuronavigatory landmark (Seeger and Zentner,
2002)

FIG. 135

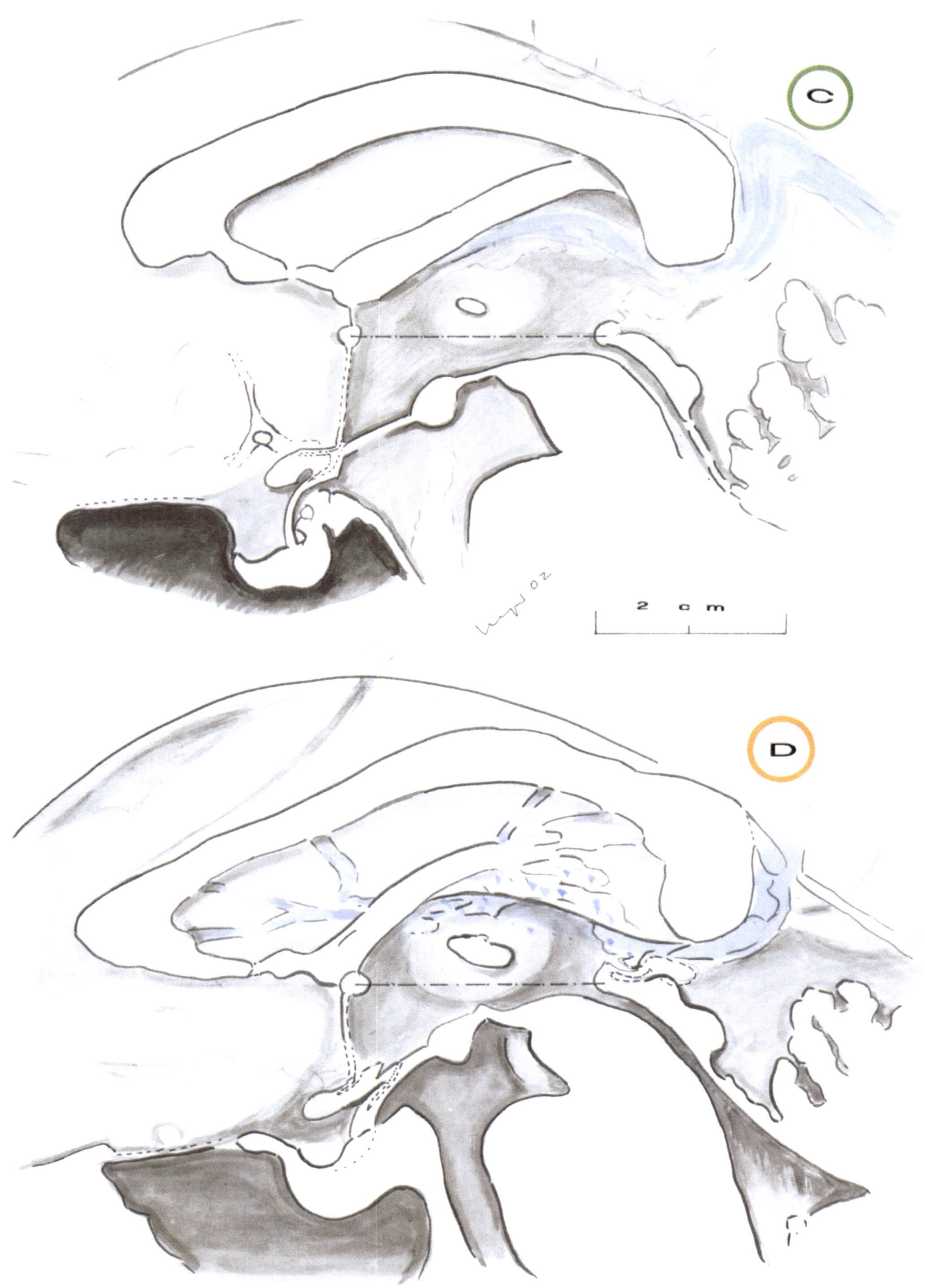

2 c m

C

D

Fig. 136.

E   Hyperplasia of Plexus chor. of the third ventricle (common finding)
Note variant of Genu corporis callosi

Colored shapes of the third ventricle and adjacent structures of A to E for comparison

Clinical aspects:
Ad E: If Plexus chorioideus is adherent to the wall of the 3rd ventricle, then microsurgery will be problematic.
Note: Adhesio interthalamica absent (common variant).
It cannot be used for landmark

Ad colored projections A to E: High-grade variability of the distance measurements of Splenium corporis callosi and Cisterna tecti. This should be considered during Yasargil's infratentorial-supracerebellar microsurgical approach to the 3rd ventricle and adjacent structures

FIG. 136

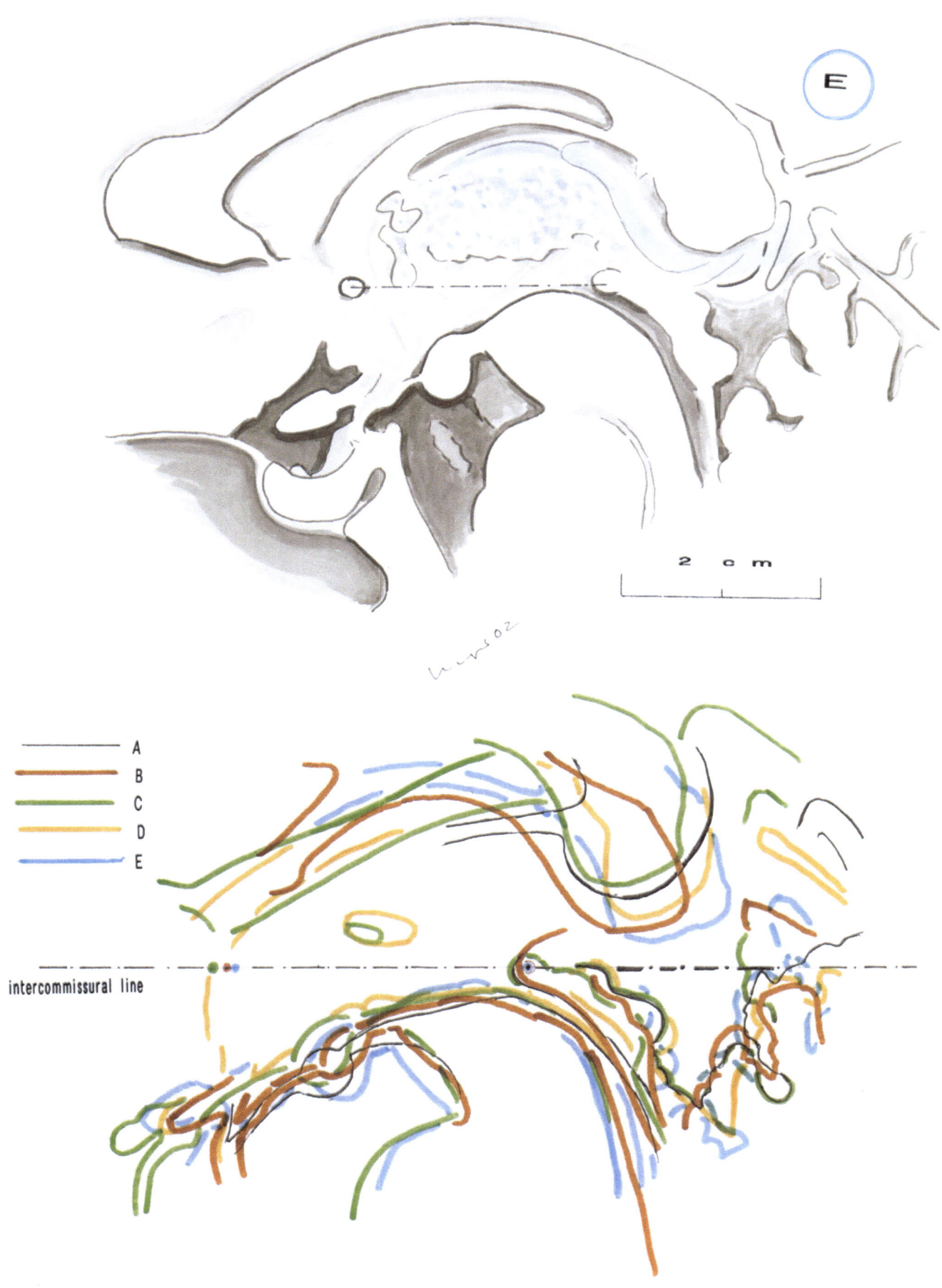

E

2 cm

A
B
C
D
E

intercommissural line

Fig. 137. Addendum for Fig. 134

a   Wide distance between Splenium and Tectum
     In sagittal slices of MRT's definition of the ga-
     lenic point may be inaccurate
b   Small distance between Splenium and Tectum

Clinical aspects:
If the distance is small, then arachnoid layers
must be resected more extensively than in nor-
mal conditions in microneurosurgery.

distance measurement

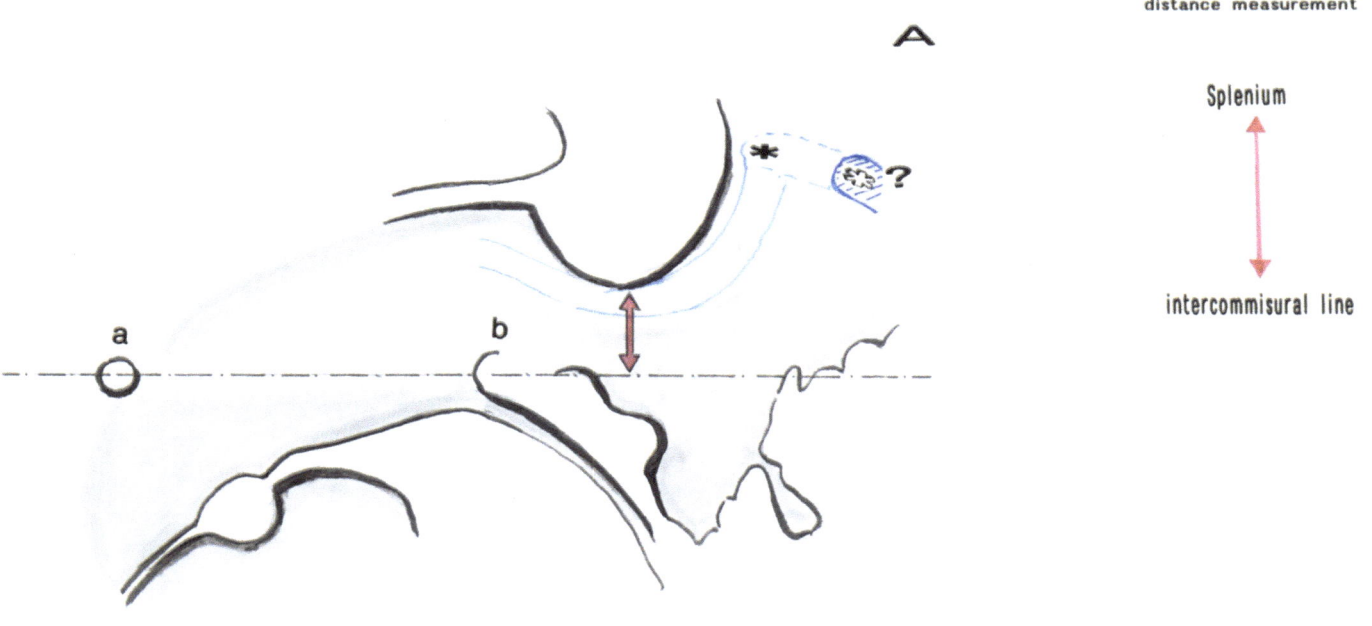

A

Splenium

intercommisural line

a   b

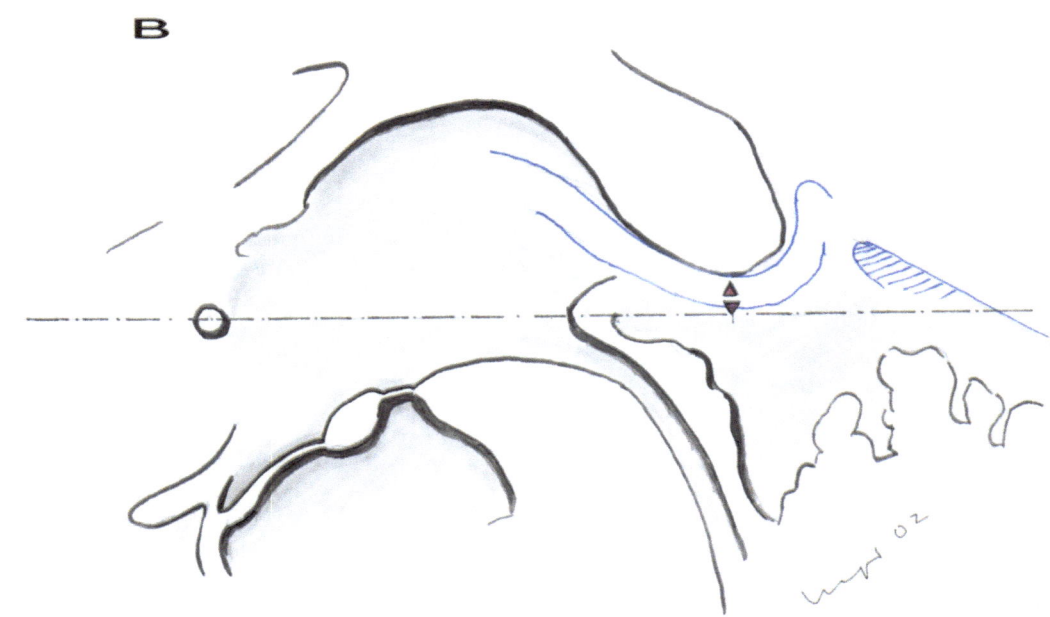

B

Fig. 138. Addendum for Fig. 135 E and 136 E

FIG. 138

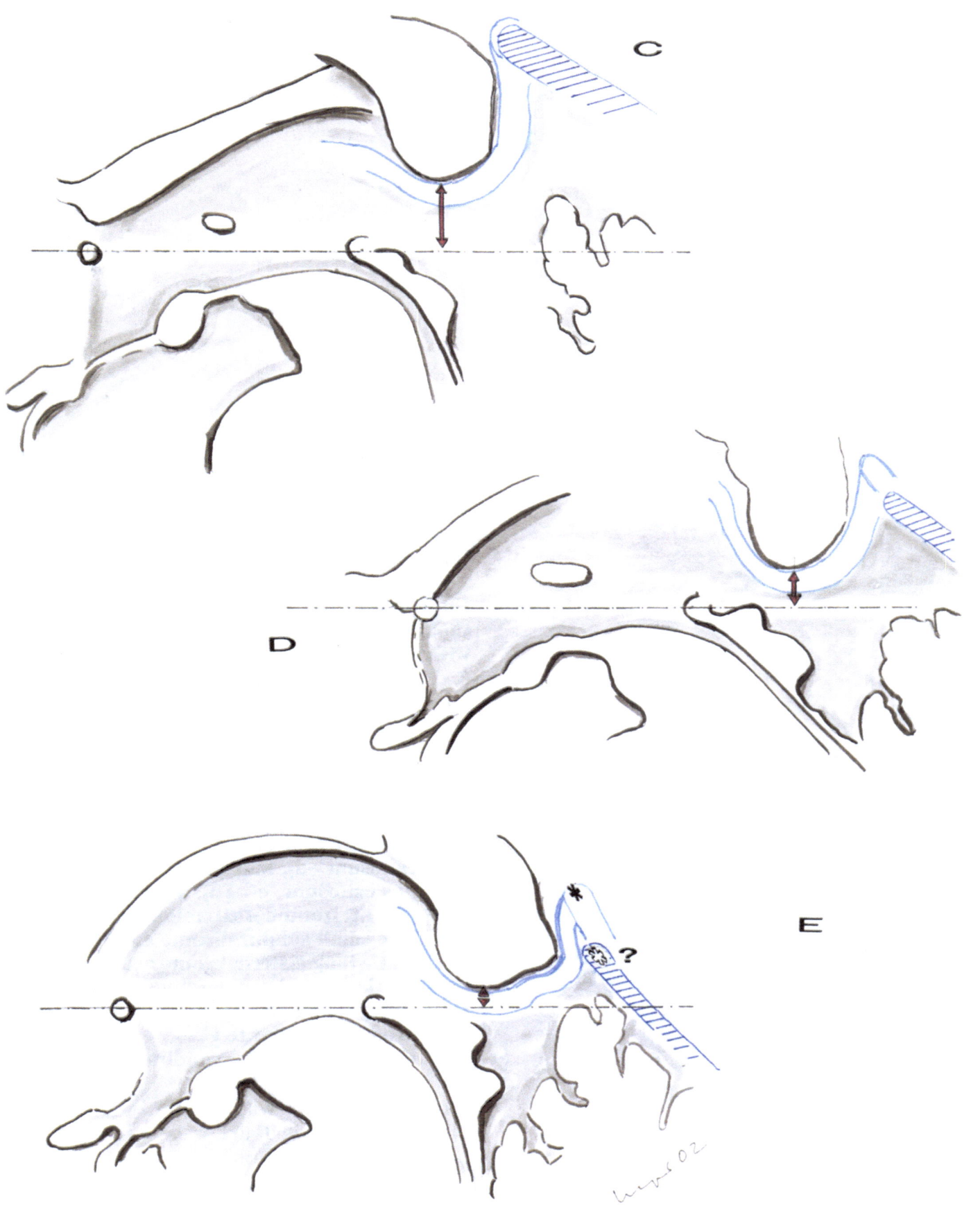

Fig. 139. Addendum for Fig. 134 to 136

Variants of the **configuration of Corpus callosum** and its relationship with Tectum, defined by the inter-commissural line

Clinical aspects:
Comparison of intraoperative ultrasonic scans with preoperative neuronavigatory landmarks

A **Favorable for frontobasal interhemispheric microsurgical approaches to Lamina terminalis and to third ventricle. Favorable for infratentorial supracerebellar approaches to third ventricle (A: Usual topographical conditions)**

B **Similar to A**

C **Note the wide distance of Falx and Corpus callosum –d-, and the thickening of Corpus callosum –e-. This should be considered at frontodorsal midline approaches The small subdural route and the problematic long cisternal route are unfavorable for the frontobasal midline approach**

D **Unfavorable for frontobasal midline approaches similar to C**

E **Favorable for dorsal midline approaches**

a to e for comparison with a' to e'
f for comparison with f' to f''''
Blue colored: Relationships of Corpus callosum, Commissura ant., and Lamina terminalis
Red colored: Configuration of Splenium-Crus-fornicis-complex
Yellow colored: Inferior margin of Falx

FIG. 139

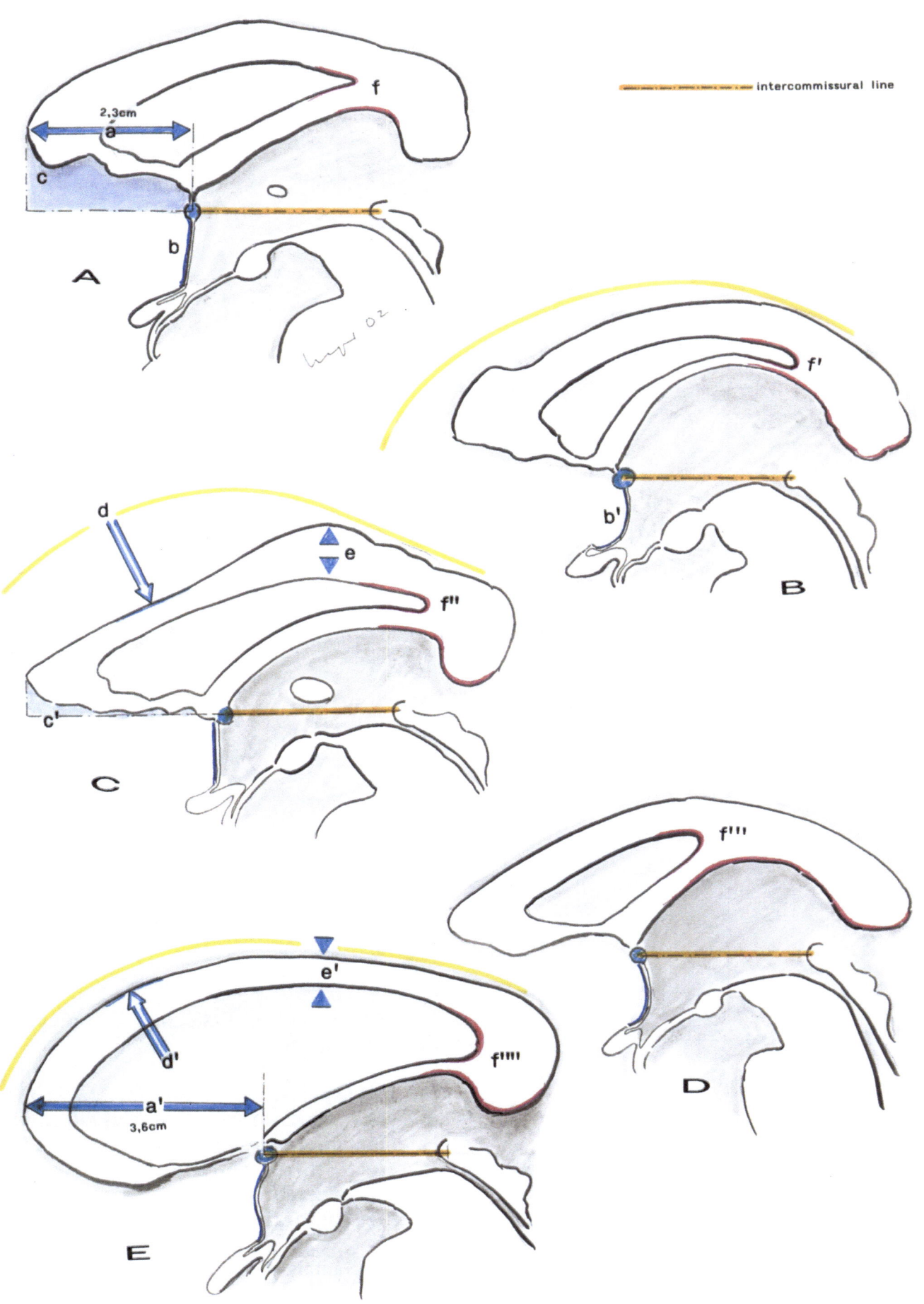

intercommissural line

Fig. 140. **Variants of Cisterna tecti and surrounding CSF-spaces**

Clinical aspects:
A   **Favorable for infratentorial supracerebellar microsurgical approach to Cisterna tecti, third ventricle, and Fissura transversa cerebri**
B   **Less favorable**

- Intercommissural line

FIG. 140

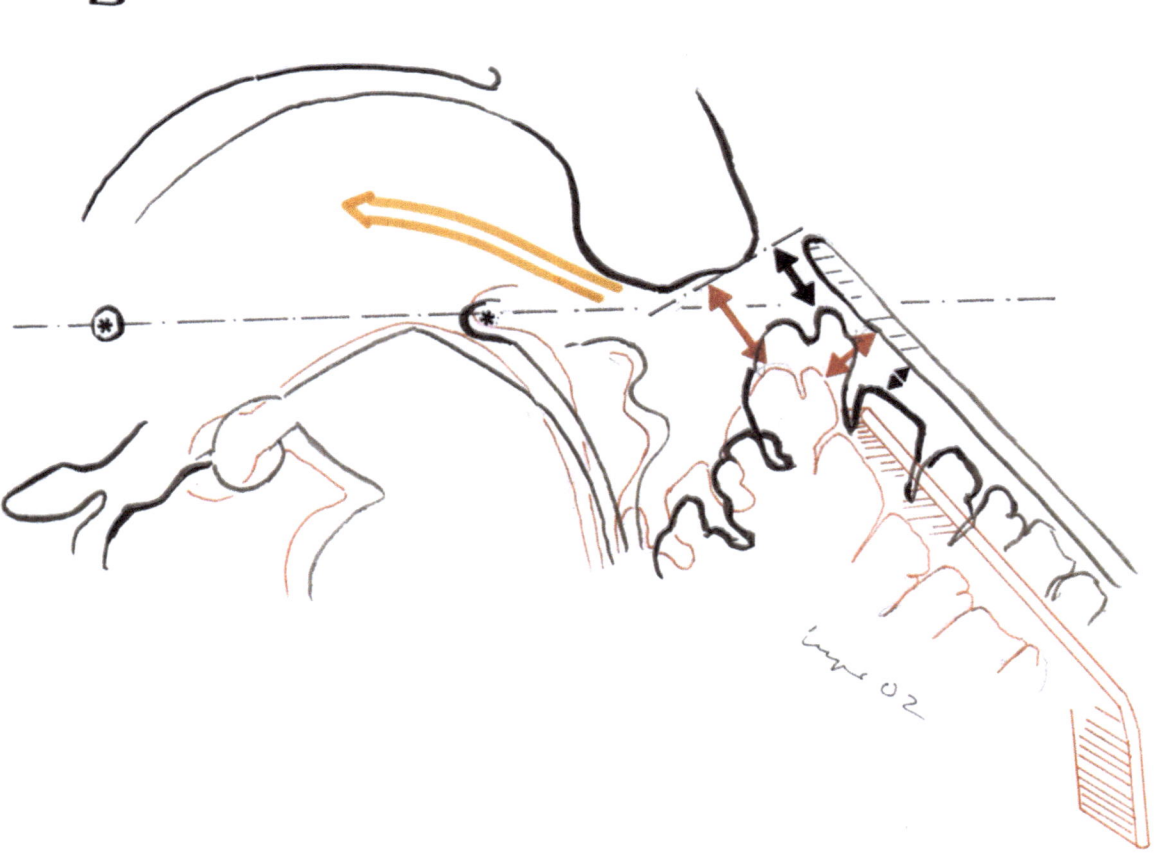

A

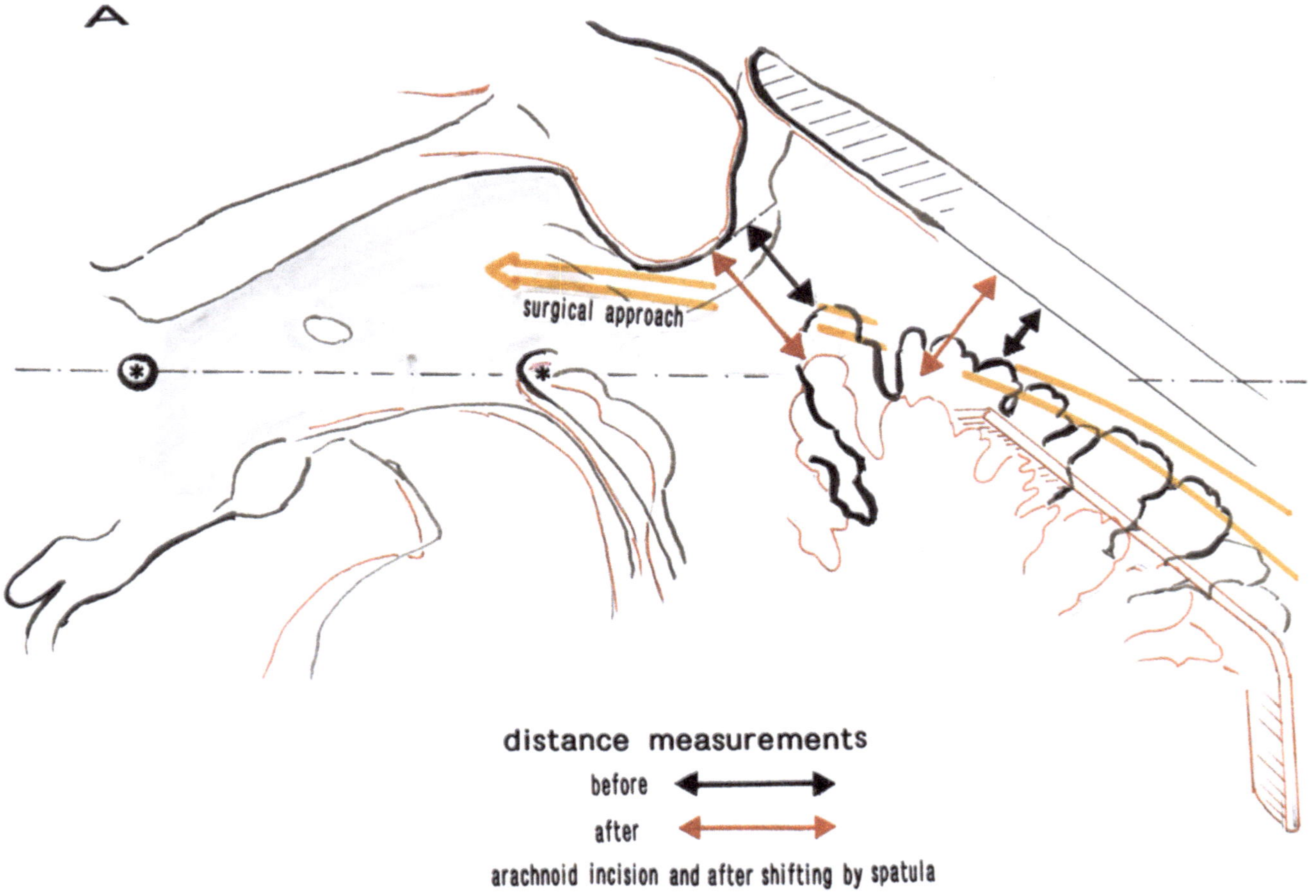

surgical approach

distance measurements
before
after
arachnoid incision and after shifting by spatula

B

Fig. 141. **Variant of the wall of the third ventricle**
which may be confused with Adhesio interthalamica

Clinical aspects:
Preoperative MRT should be considered

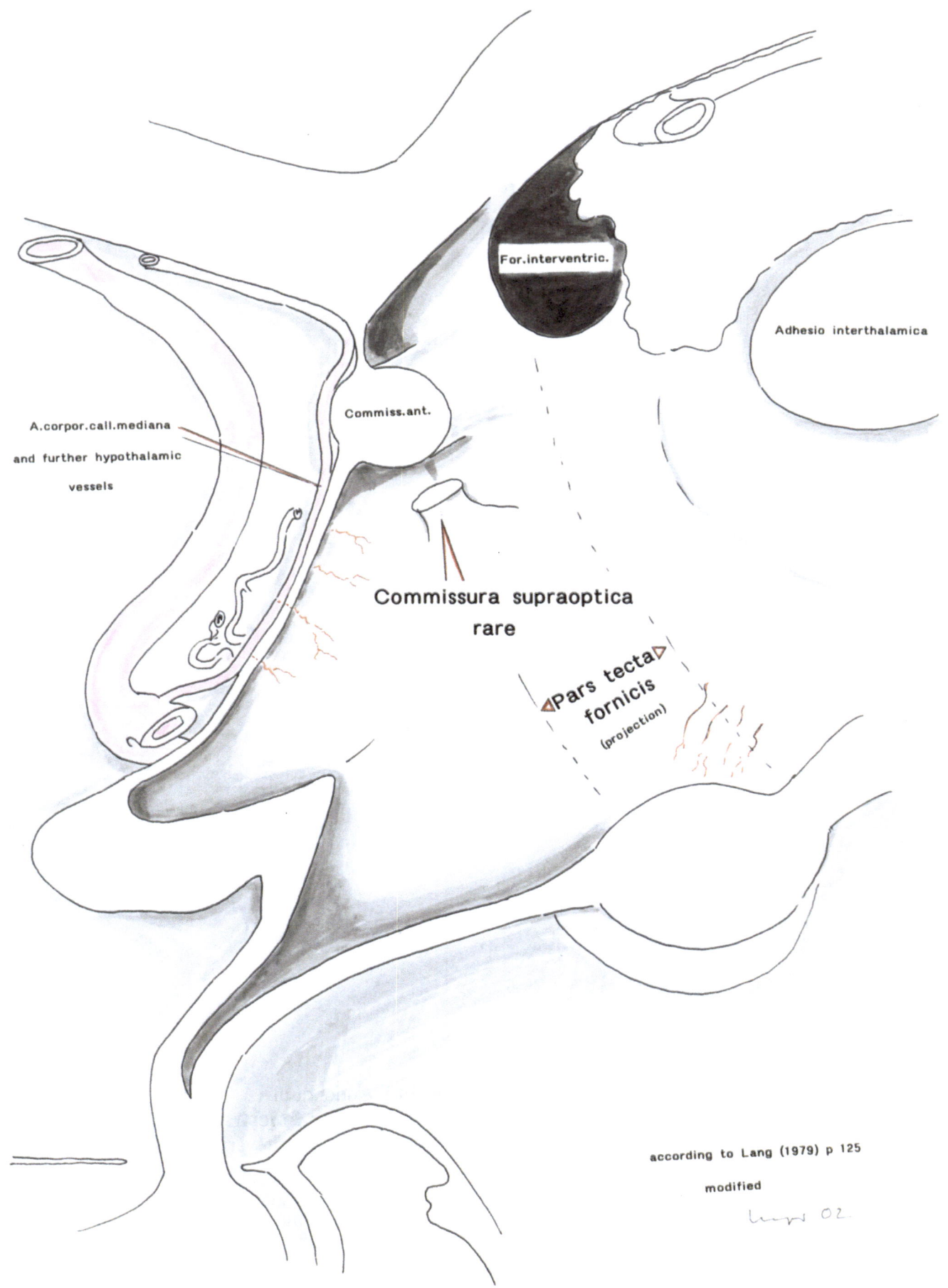

For.interventric.

Adhesio interthalamica

Commiss.ant.

A.corpor.call.mediana
and further hypothalamic
vessels

Commissura supraoptica
rare

▷Pars tecta▷
fornicis
(projection)

according to Lang (1979) p 125

modified

Fig. 142. Addendum
Transventricular **fenestration of the floor of the third ventricle** and variants of the bifurcation of A. basilaris

A   Usual finding: Surgery can be complicated by adhesions
B   More favorable for surgery

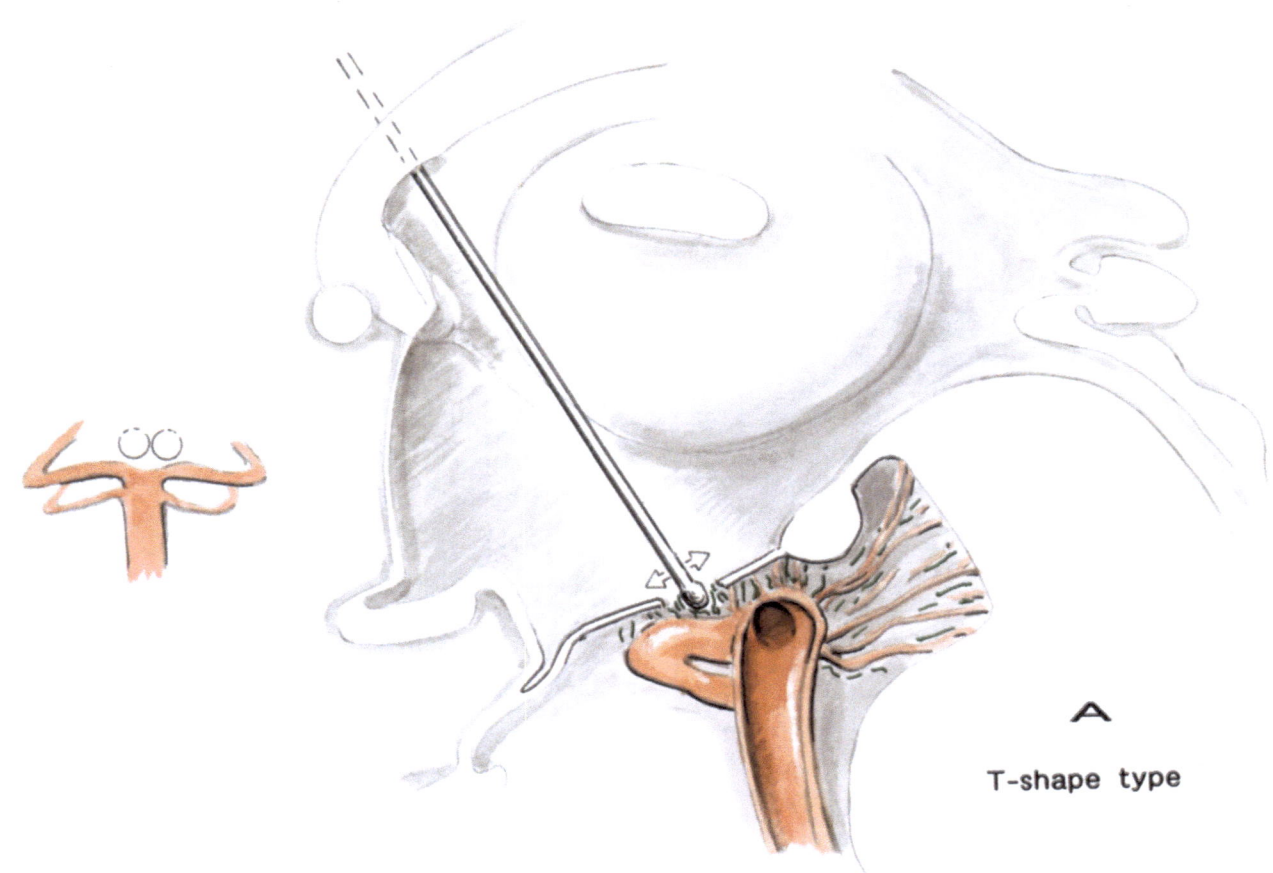

A

T-shape type

Ventriculo–Cisternostomy

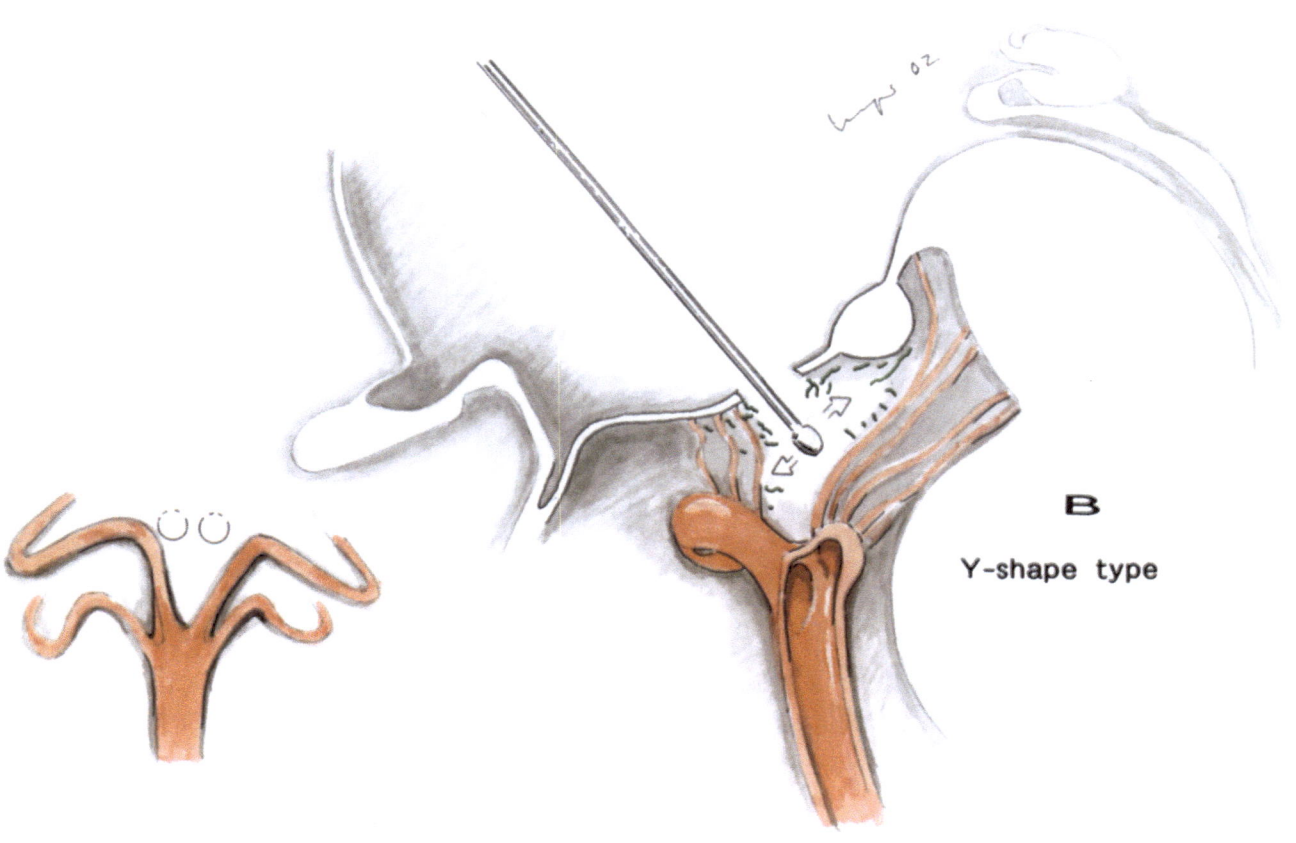

B

Y-shape type

Chapter 5
Variants of the infratentorial areas
(Figs. 143 to 177)

Figs. 143–145.

Suboccipital microsurgical approaches

Schematic presentation

---

* Bony landmarks

FIG. 143

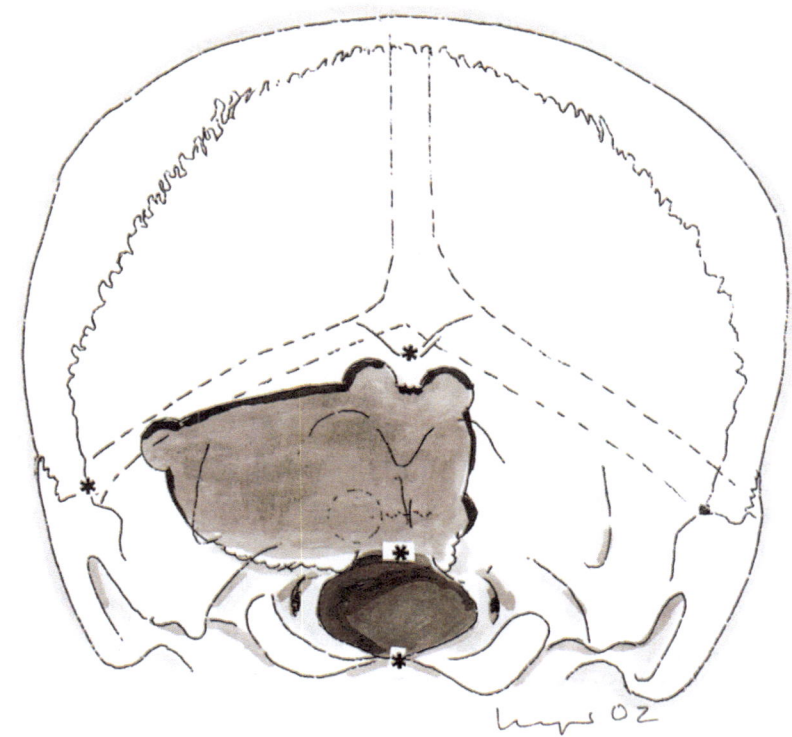

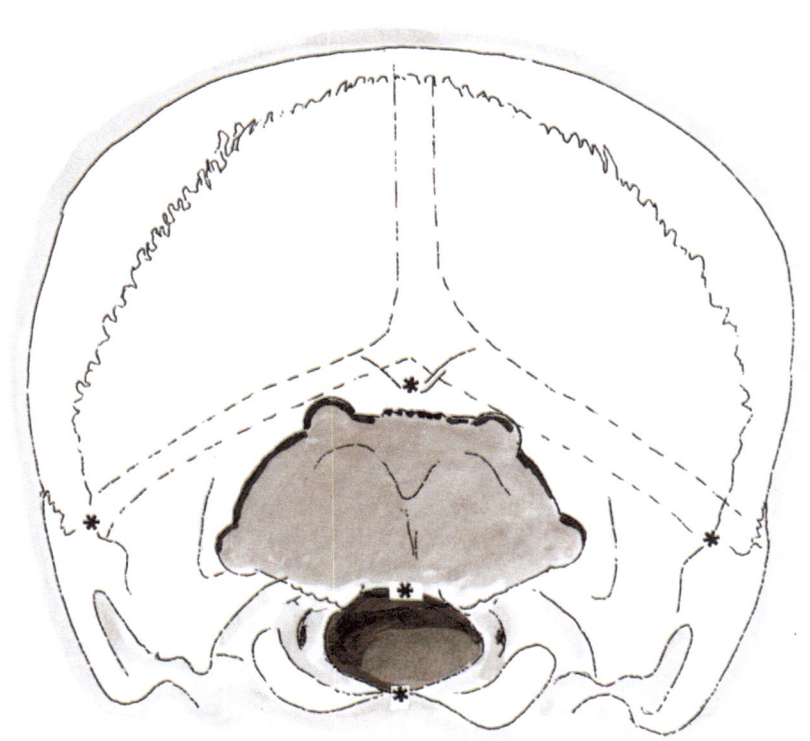

FIG. 144

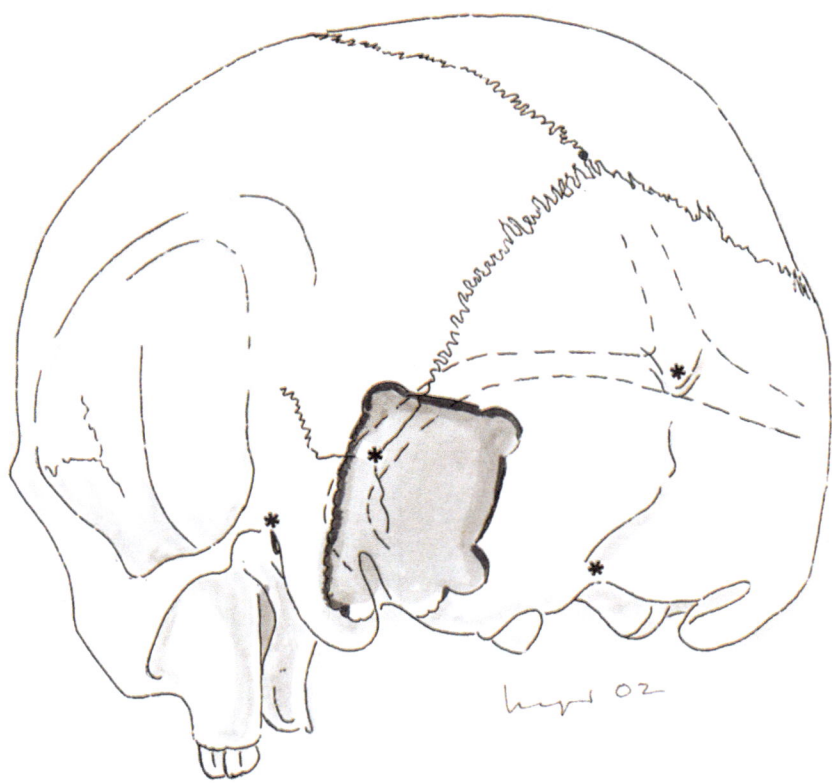

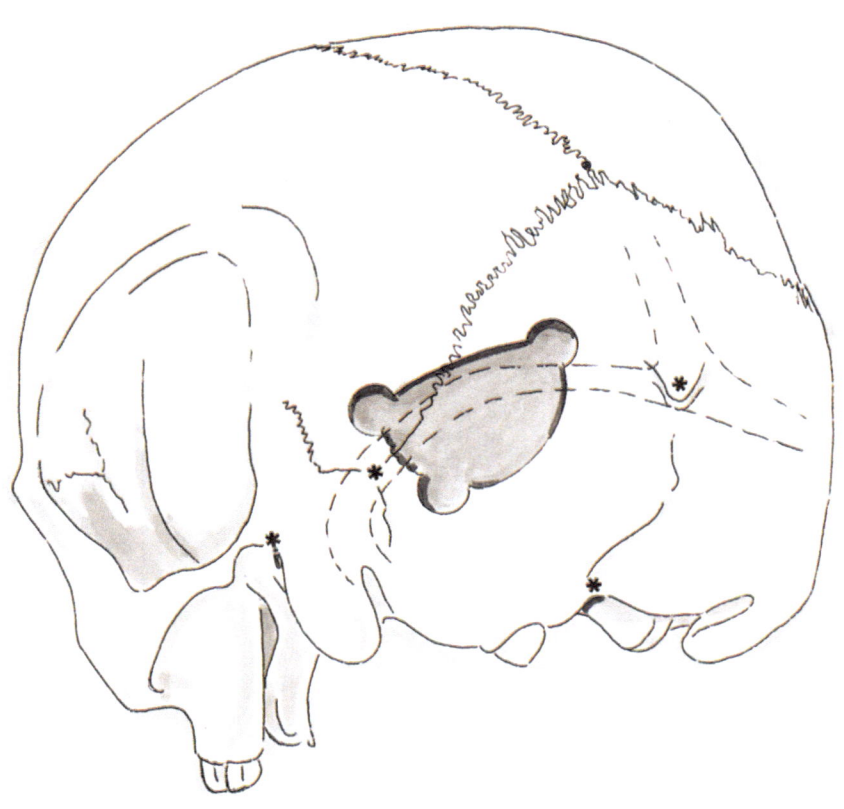

FIG. 145

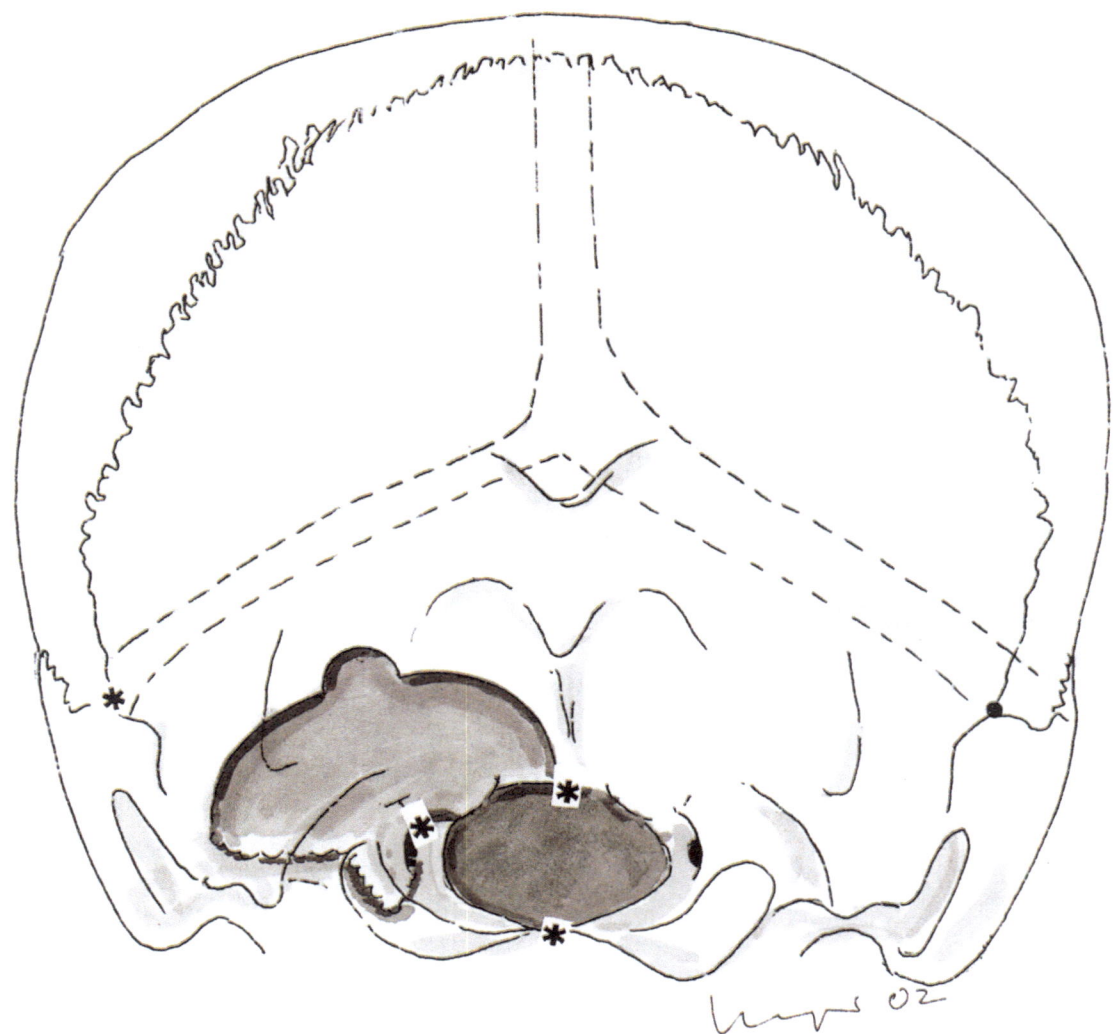

Fig. 146. **Variants of Atlas**

A   Defect of Arcus atlantis post.
B   Defect of Arcus atlantis post. and ant.
A' and **B'** Microsurgical approaches to the areas, in
          which the defects are located. Schematic
          presentation
C   Sulcus arteriae vertebralis transformed to
          a foramen by an congenital bony bridge
          bilaterally

Clinical aspects:

Ad A:
**Midline incision of the muscles of the neck can
endanger dura and medulla**

Ad B:
**Definition of the midline during Microsurgery
may be confused. Neuronavigatory definition of
Clivus** (Basion) **can be recommended**

Ad C:
**Bony protection of A. vertebralis is defect
Note: Preoperative CTs (and MRTs) should be
considered**

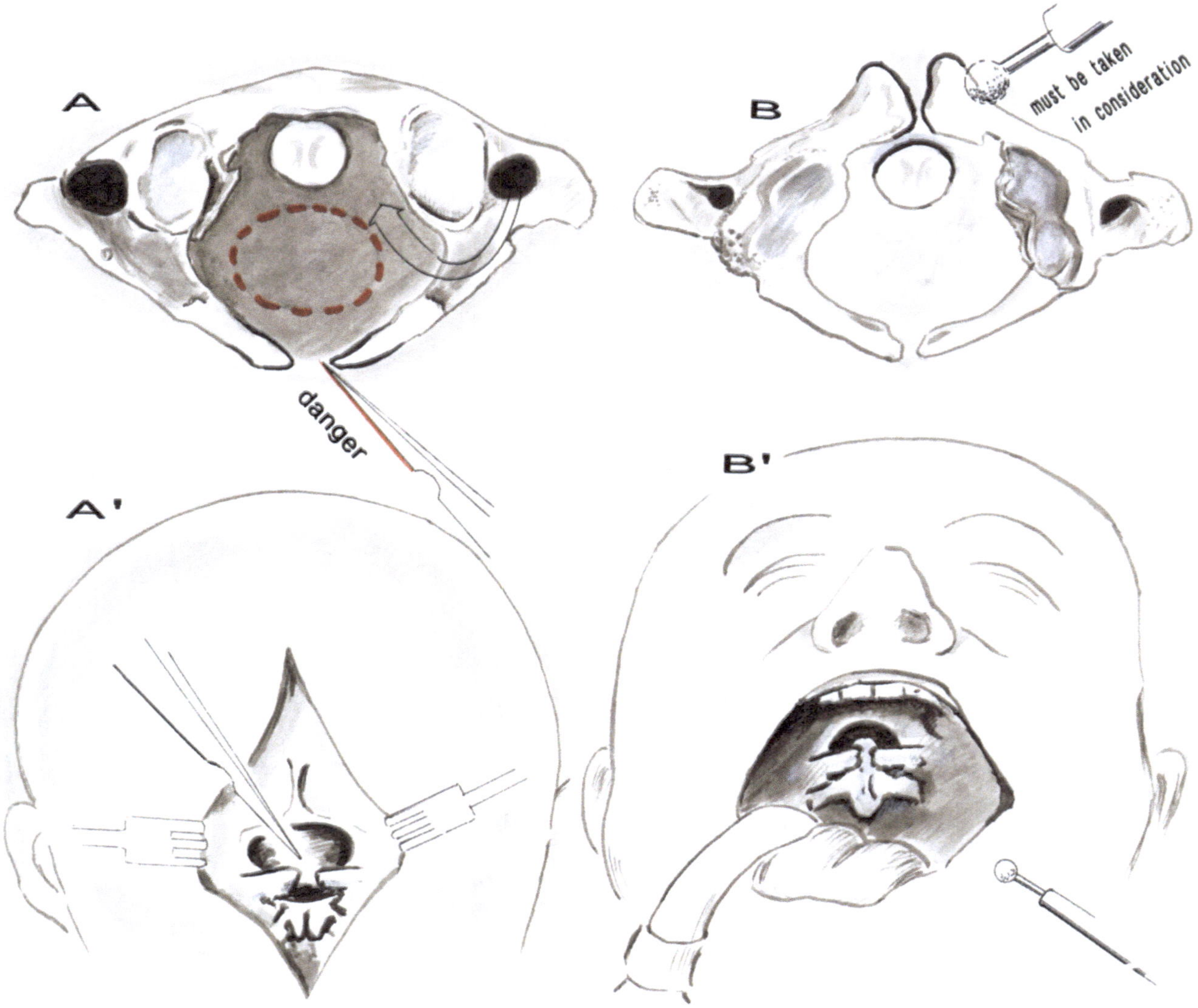

A,B,and C according to Lang (1979)p 318,319,modif.

Fig. 147. **Sinuses of Fossa cranii post., variants,** according to Lang (1979) p 234 ff

Clinical aspects:
- **Danger for unexpected bleeding during surgery**
- **Danger for unexpected increase of venous pressure**
- **Sometimes x-ray-findings are helpful for the diagnosis by looking for bony sulci, before the typical angiographic and MRT-findings are analysed**

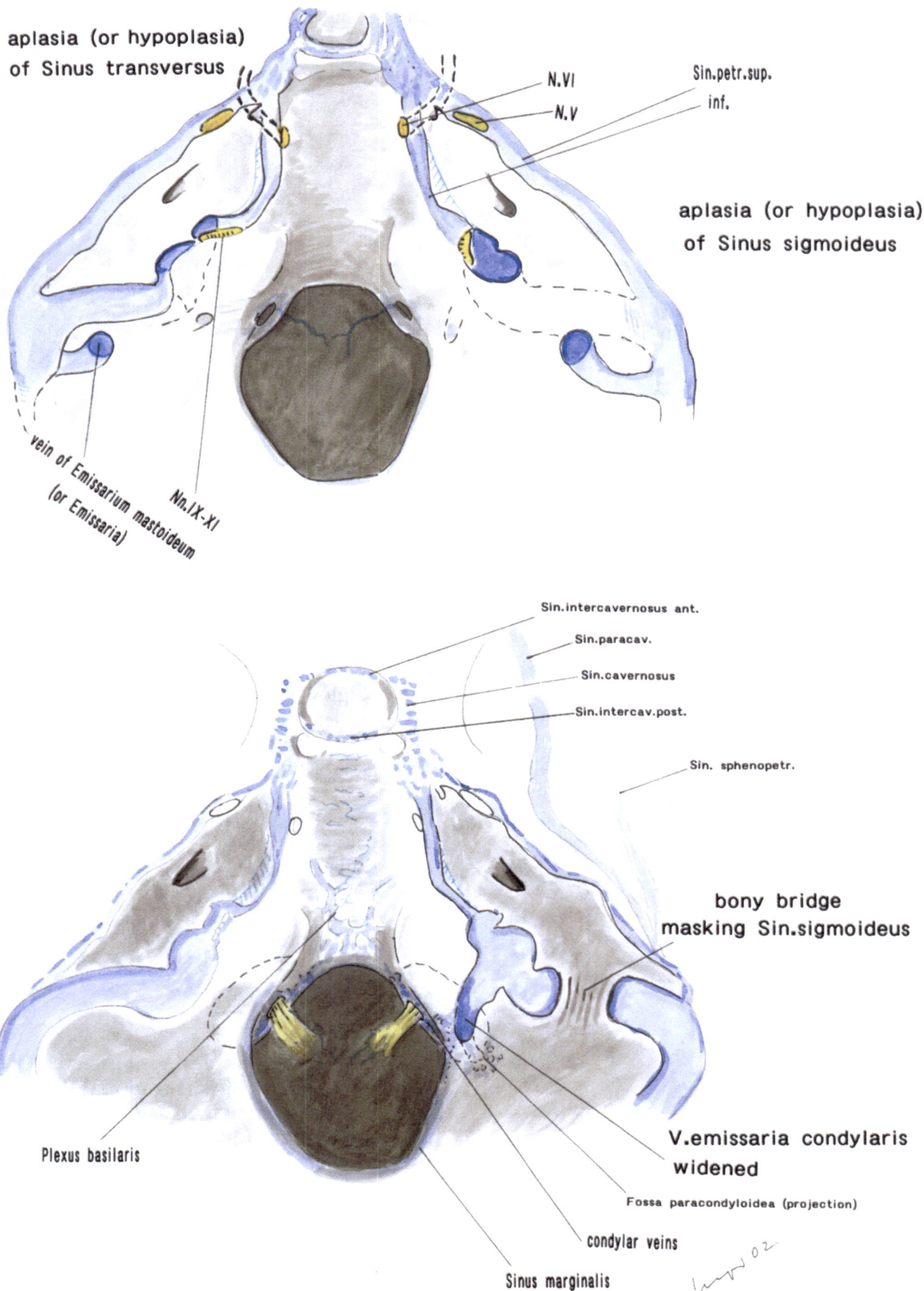

aplasia (or hypoplasia) of Sinus transversus

N.VI

N.V

Sin.petr.sup.

inf.

aplasia (or hypoplasia) of Sinus sigmoideus

vein of Emissarium mastoideum (or Emissaria)

Nn.IX-XI

Sin.intercavernosus ant.

Sin.paracav.

Sin.cavernosus

Sin.intercav.post.

Sin. sphenopetr.

bony bridge masking Sin.sigmoideus

V.emissaria condylaris widened

Fossa paracondyloidea (projection)

Plexus basilaris

condylar veins

Sinus marginalis

Fig. 148. **Variants of bony channels of cranial nerves**

A   N. glossopharyngeus and its bony channel are located far distant from Nn.X and XI and its channel (= For. jugulare)
B   As A. Dura and sinuses drawn in

Abbreviations
a   Nn. X–XI
b   Bulbus sup.
c   Canalis condylaris (project.)
d   Clivus
e   Apex petrosi
f   Sulcus petrosus sup
g   Porus acust. int.

Clinical aspects:
Normally, the group of cranial nerves IX to XI is located on an interforaminal line, which connects Porus acusticus int. (posterior margin), For. jugulare (anterior segment) and Canalis n. hypoglossi (anterior margin). The unexpected location of a separate Can. glossopharyngeus puts the nerve at risk during microsurgery. Similar problems might occur with a duplication of Canalis n. XII.

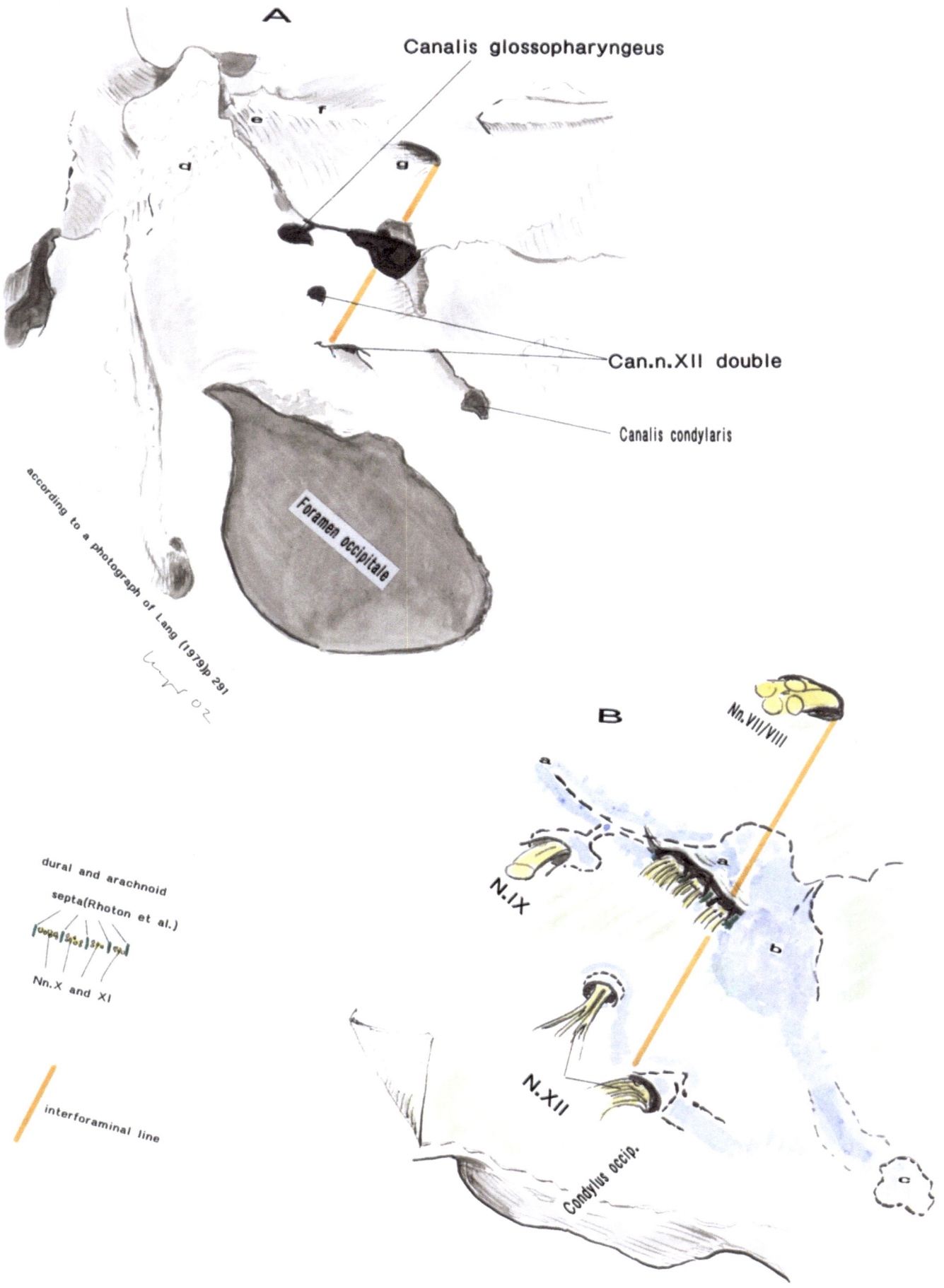

A

Canalis glossopharyngeus

Can.n.XII double

Canalis condylaris

Foramen occipitale

according to a photograph of Lang (1979)p 291

B

Nn.VII/VIII

N.IX

N.XII

Condylus occip.

dural and arachnoid

septa(Rhoton et al.)

Nn.X and XI

interforaminal line

Fig. 149. Continuation of Fig. 148

A   Usual findings (f, 48 y.)
       1 to 6 and 1' to 6': Corresponding points
B   Multiseptal variants of For. jugulare (m, 65y)
       For further variants see Lang J (1985) pp 134 ff
C   Combination of different variants (anonymous)
       • No Sulcus sigmoideus on the right side
       • Atypical and wide bony sulcus is an indication for a wide
       Sinus occipitalis which is a substitute for the aplastic Sinus sigmoideus
       • asymmetric Foramen occipitale
D   Normal and duplicated For. jugulare (nerval and venous segments are always separated, except Sin. petrosus inf.)

For clinical aspects see Figs. 147 and 148

FIG. 149

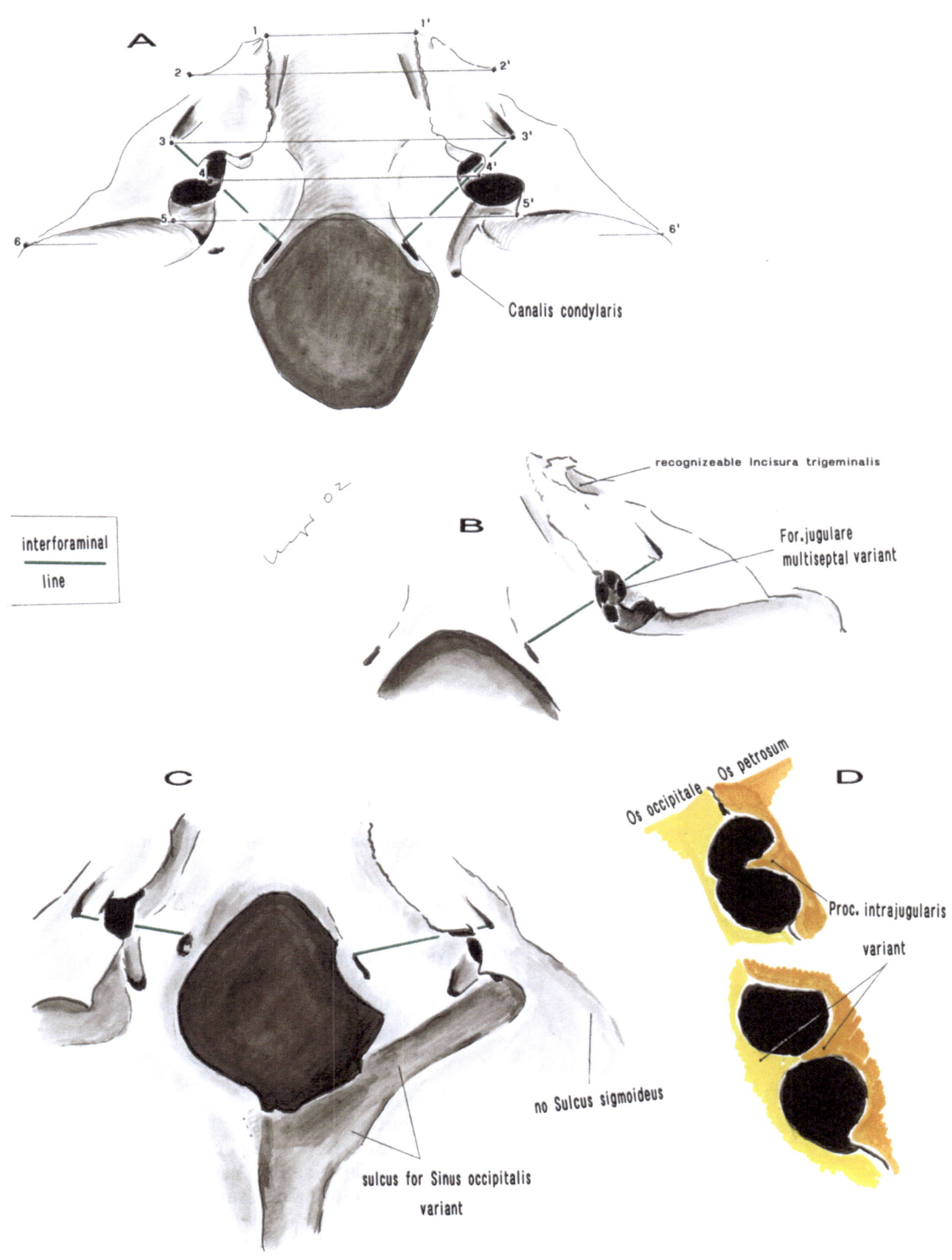

A

Canalis condylaris

interforaminal
line

B

recognizeable Incisura trigeminalis

For. jugulare
multiseptal variant

C

D

Os occipitale   Os petrosum

Proc. intrajugularis

variant

no Sulcus sigmoideus

sulcus for Sinus occipitalis
variant

Fig. 150. Continuation of Fig. 149

Double variant of N. hypoglossus
Cisternal and intraosseous segment

A   Anatomical dissection
B   As A, simplified

Yellow colored: Cranial nerves
Green colored: Dural and arachnoid layers
Grey colored: Bony structures (petrous and occipital bone)
Red colored: A. vertebralis

Clinical aspects:
Variant of N.XII may be confused with other cranial nerves during operation.
Danger for bleeding of the numerous dural or osseous veins, which surround N.XII and the accessory N.XII

FIG. 150

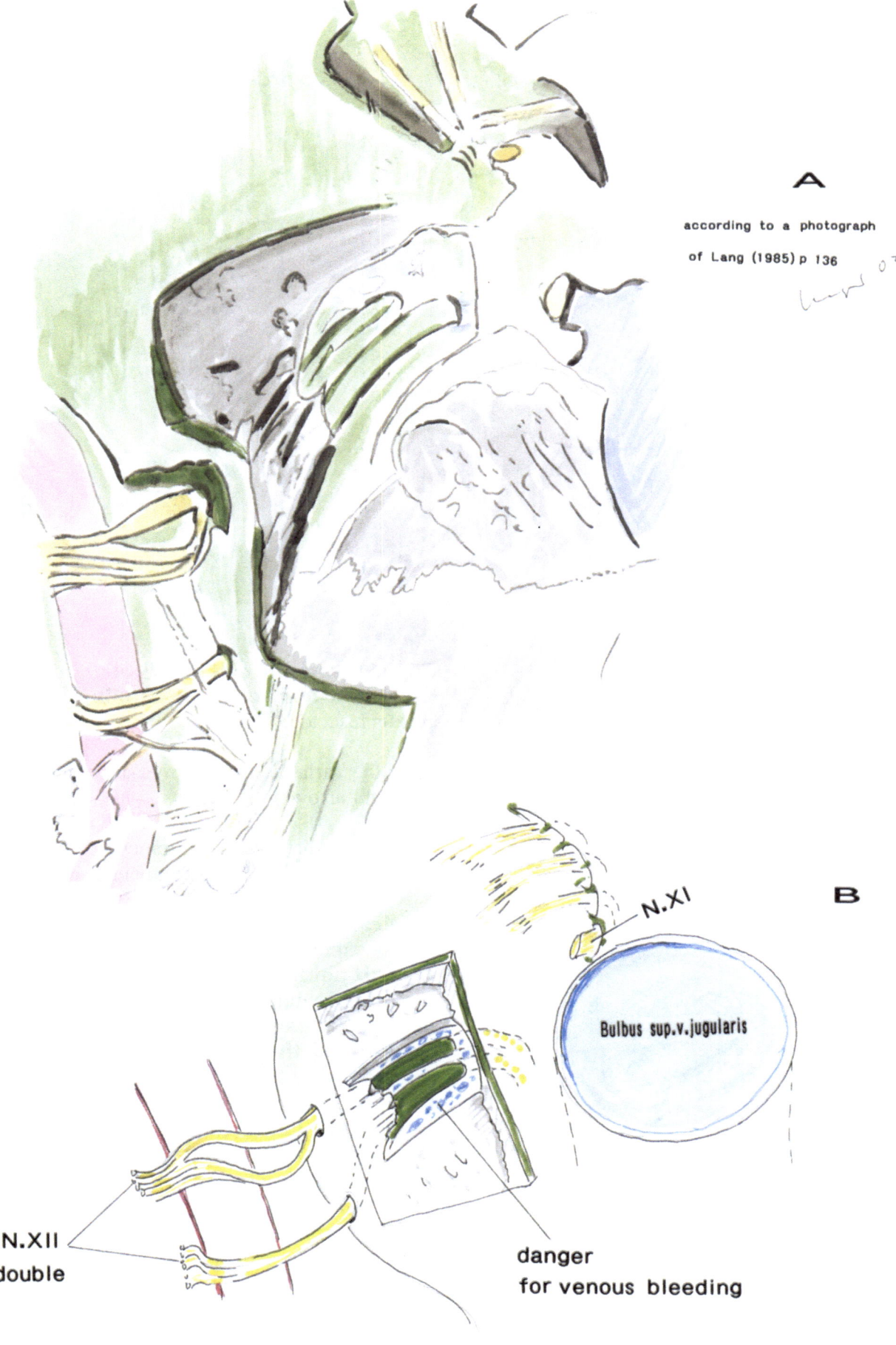

A

according to a photograph

of Lang (1985) p 136

B

N.XI

Bulbus sup.v.jugularis

N.XII
double

danger
for venous bleeding

Fig. 151. **Variable exit points of cranial nerves at the brainstem, common findings.** Bochdalek's plexus resected.

Note the variable distance measurements of N.VI to XI (light arrows)

A   Short distance measurements
B   Long distance measurements

Clinical aspects:
These variants of the distance measurements of the exit points may cause confusion of identification especially of N.VII and VI (in microsurgery of Schwannoma of N.VIII), and of identification of the N.VI-VII-VIII-IX-complex (in microsurgery for petroclival meningiomas and others)
Anatomical landmarks (if there does not exist an excavation and dilation of the brainstem by a space-occupying lesion):
- **Bochdalek's plexus is located superior from N.IX**
- **Fossa paraolivaris (origin of N. VII)**
- **Floor of Recessus lateralis (relief of N.VIII)**

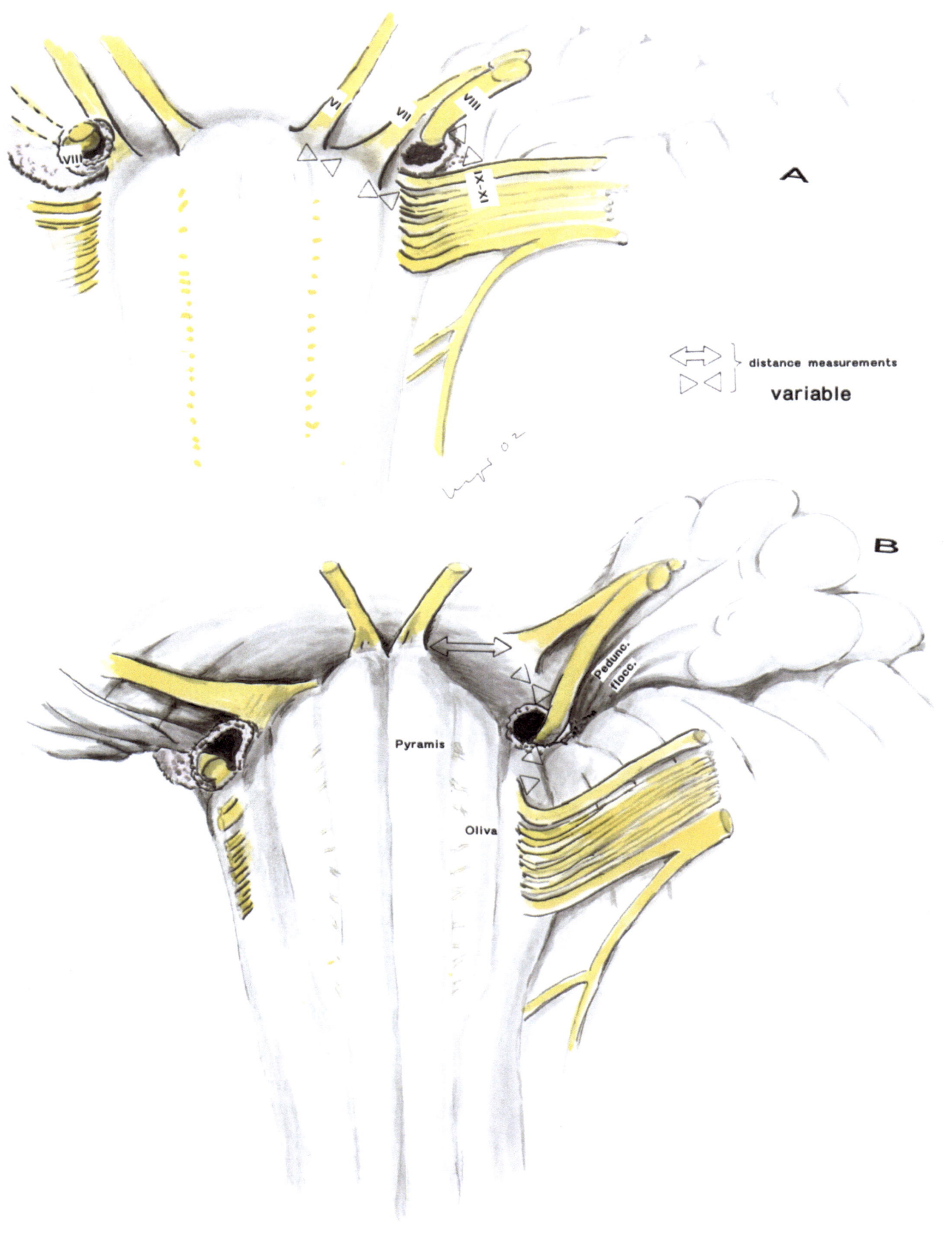

Fig. 152. **A. vertebralis and the proximal segment of PICA**

A   PICA follows A. vertebralis, right-side
Hypoplastic A. vertebralis close to the verte-bro-basilar junction, right-side

B   Inferior caudal position of the junction (common findings: Position at the pontomedullary level)

C   Aplasia of A. vertebralis close to the junction, left-side. A. basilaris follows A. vertebralis

D   Elongation of the arteries, common finding in elderly individuals (degenerative vessels; these are no congenital variants)

Clinical aspects:
Combination of variants and degenerative elongations with aneurysms must be considered in microsurgery and interventional neuroradiological methods

Further well-known aspects: Janetta's compression syndromes of cranial nerves and basal arteries.

FIG. 152

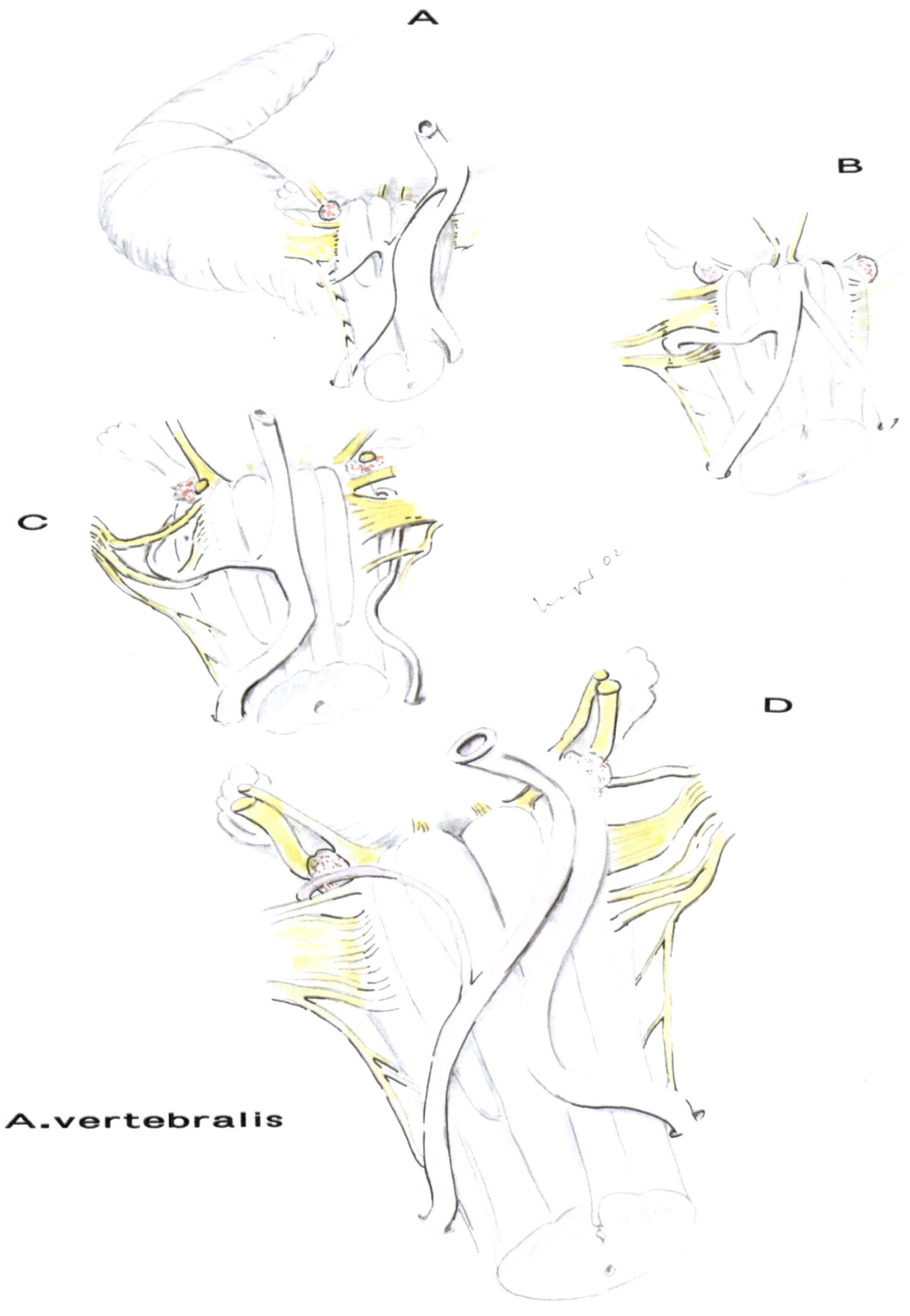

A

B

C

D

A.vertebralis

Fig. 153. **Further variants of basal arteries and caudal cranial nerves**

A Fenestration of the left A. vertebralis, containing some fibers of N. hypoglossus
The contralateral origin of PICA is a rare variant
B N. hypoglossus, common variants
B' Crossing of N. abducens and PICA, common variants

Clinical aspects:
These variants must be taken into consideration during microsurgery for petroclival and craniospinal meningiomas. Craniospinal meningiomas are located in a tumor-dilated subdural space. A. vertebralis and N. hypoglossus are penetrating the tumor, not other vessels or cranial nerves

Abbreviations
fl Flocculus
pl Bochdalek's plexus (caudal position)
ts Tonsilla cerebelli
* Nn.IX to XI (shifted in a cranial direction, according to a photograph of Lang)

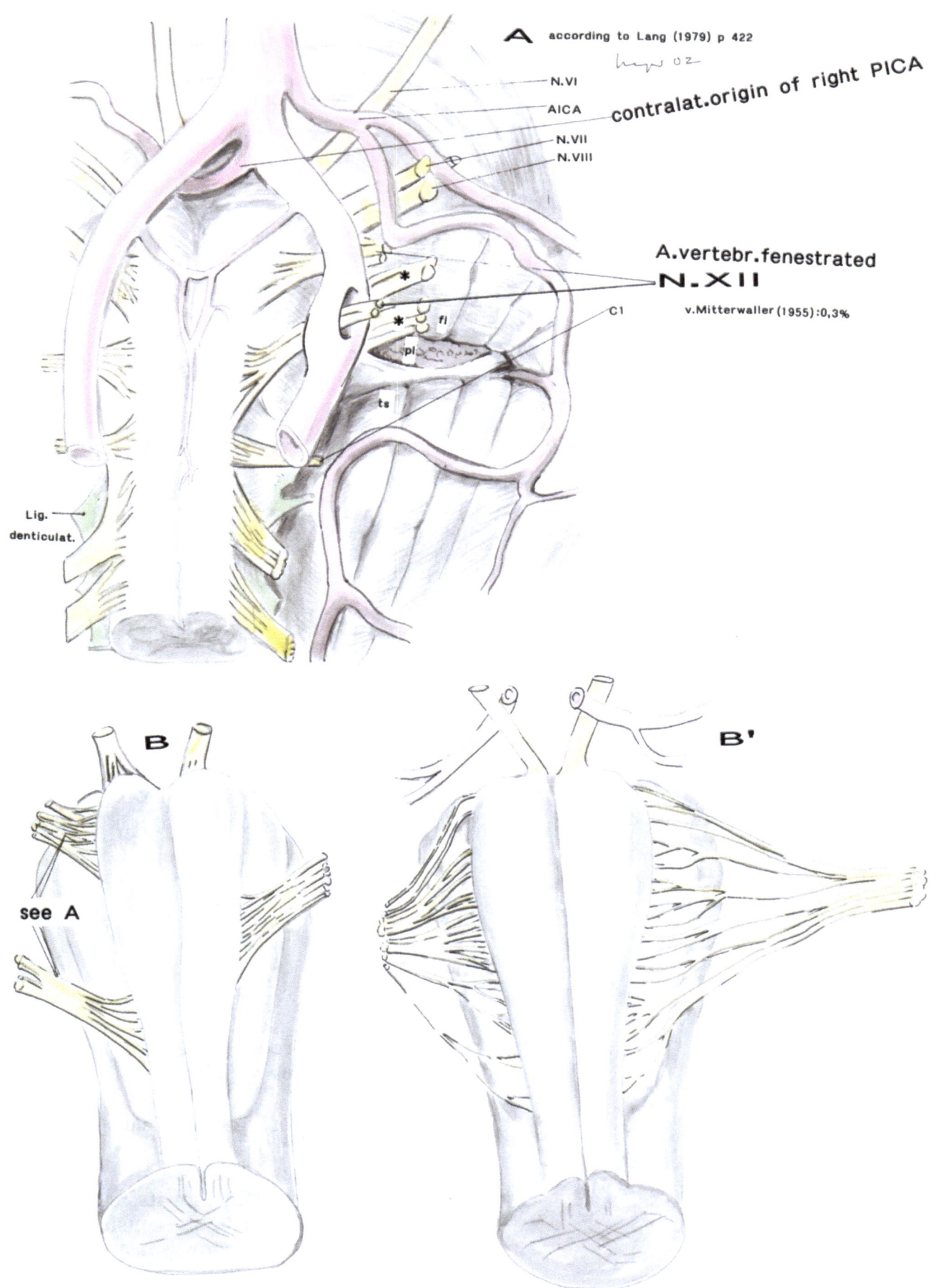

A according to Lang (1979) p 422

N.VI

AICA

contralat.origin of right PICA

N.VII
N.VIII

A.vertebr.fenestrated
N.XII

v.Mitterwaller (1955):0,3%

C1

*

*

fi

pl

ts

Lig.
denticulat.

B

B'

see A

Fig. 154. **PICA, relationship with cranial nerves IX to XI, common variants**

Note the position of Lig. denticulatum (green colored)

Clinical aspects:
Relationship of PICA and its aneurysms with cranial nerves must be considered during microsurgery. Danger for aspiration and pneumonia with lower cranial nerve deficits.

FIG. 154

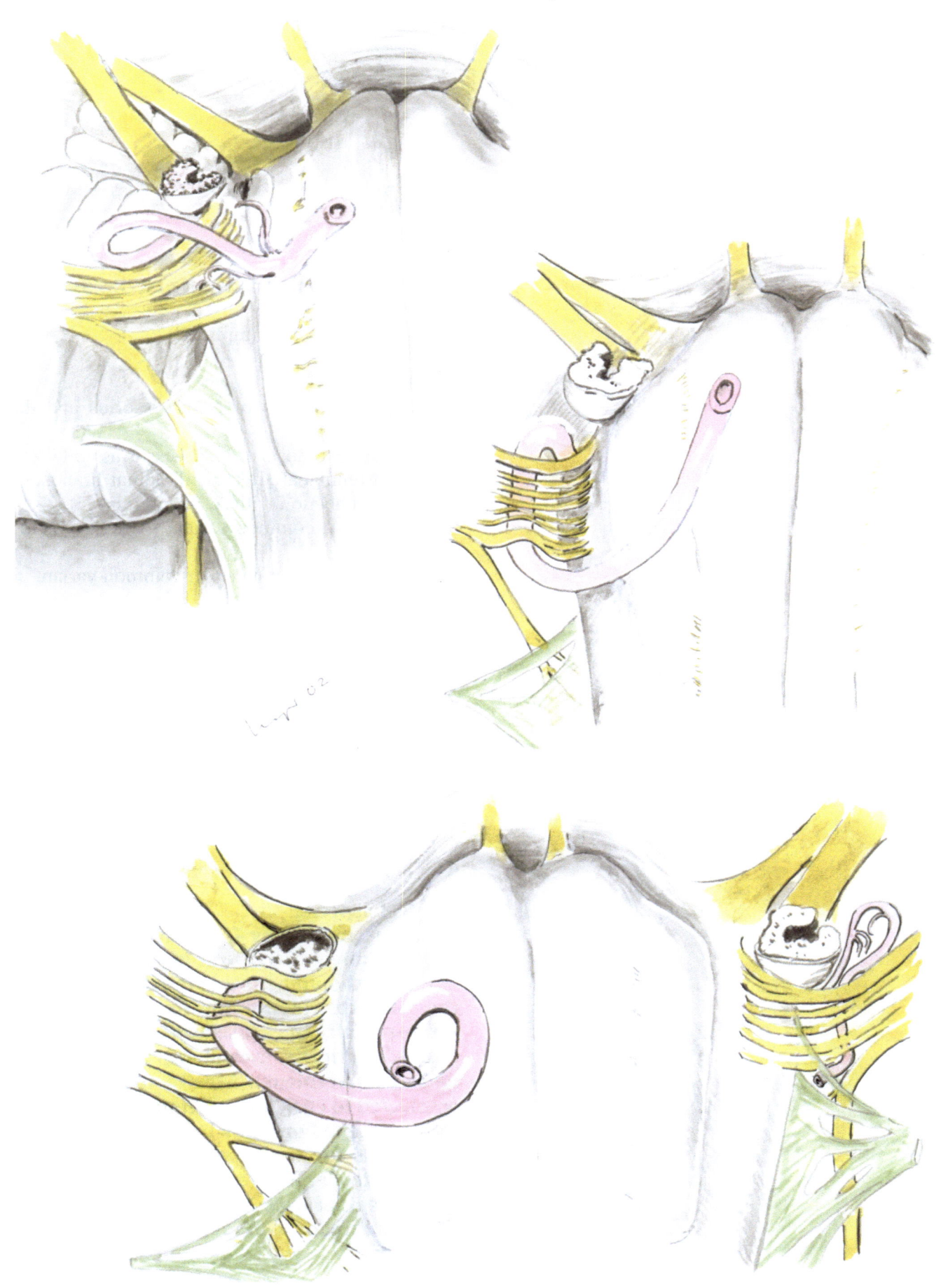

Fig. 155. **Further variants of arteries and caudal cranial nerves**
(illustration of Lang, simplified)

Variants

a   Anastomosis of N. accessorius and dorsal root of C1
b   Anastomosis of N. accessorius and C2
c   Meningeal artery with spinal branches and connection with A. vertebralis
d   PICA, common variant (dotted: typical finding)
e   Ganglion spinale C1, common variant, right side
         (left: C1 exclusively motoric nerve, normal finding)
f   see d
g, h, i see c

Further abbreviations:
j   A. vertebralis
k   Craniospinal articulation
l   Processus articularis sup. of Atlas
m   Sulcus a. vertebralis of Atlas
n   Ligamentum denticulatum

Clinical aspects:
Even the numerous fine anastomotic vessels, which connect extradural arteries and the spinal cord, should be preserved during surgery of the vertebral column and its adjacent structures, especially the muscles of the neck (especially A. cervicalis ascendens, which accompagnies N. phrenicus). Mostly, these vessels are connected to each other. Eliminations of the arteries close to Foramina intervertebralia may be dangerous for the blood-supply of the spinal cord. Extradural arteries may often accompagny the spinal roots. Sometimes they penetrate the dura directly without relationship to the spinal roots.

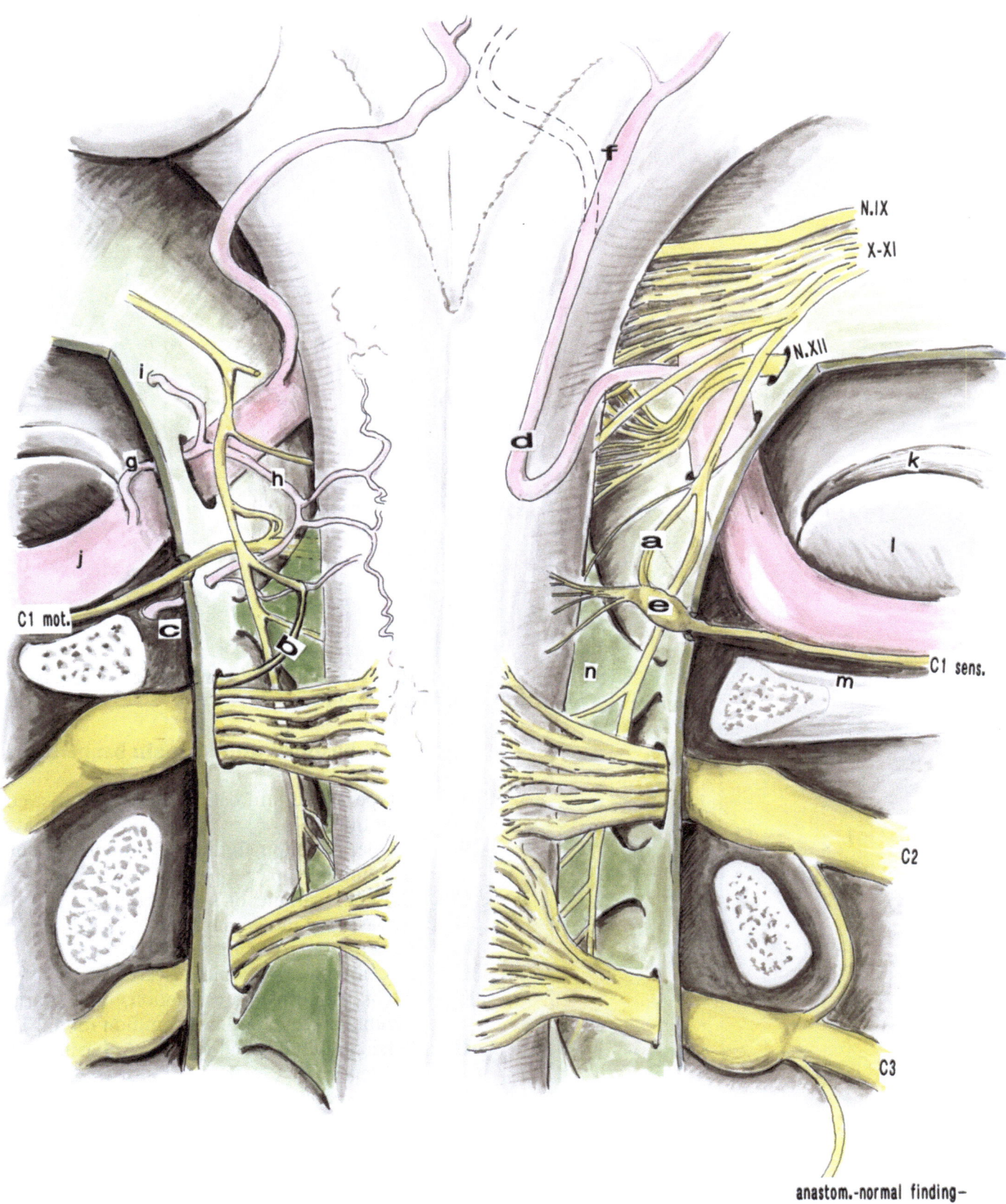

N.IX
X-XI
N.XII
f
d
k
l
a
e
C1 sens.
m
n
i
g
h
j
C1 mot.
c
b
C2
C3

anastom.-normal finding-

according to Lang (1979) p 399

modified

Figs. 156–166. **Further variants of the vertebro-basilar system**

Fig. 156. **A. basilaris and its main branches**

A   Y-type of the bifurcation of A. basilaris, common finding

B   T-type, common finding. Common variant of A. cerebelli sup.

C   Common variants of A. cerebelli sup.

D   A. basilaris, rare variant

E   A. vertebralis, common variant

F   A. cerebelli sup., common variant

G   AICA, multiple presentation, rare variants (double presentation is a common finding)

H   Persisting embryonal type of A. basilaris, rare variant

I   Rare variant of A. vertebralis

J   Rare variant of the vertebro-basilar junction with a wide singular A. spinalis ant.

Clinical aspects:
These variants have to be considered in the treatments of aneurysms and during microsurgery of tumors (especially meningiomas), which are penetrated by these arteries

FIG. 156

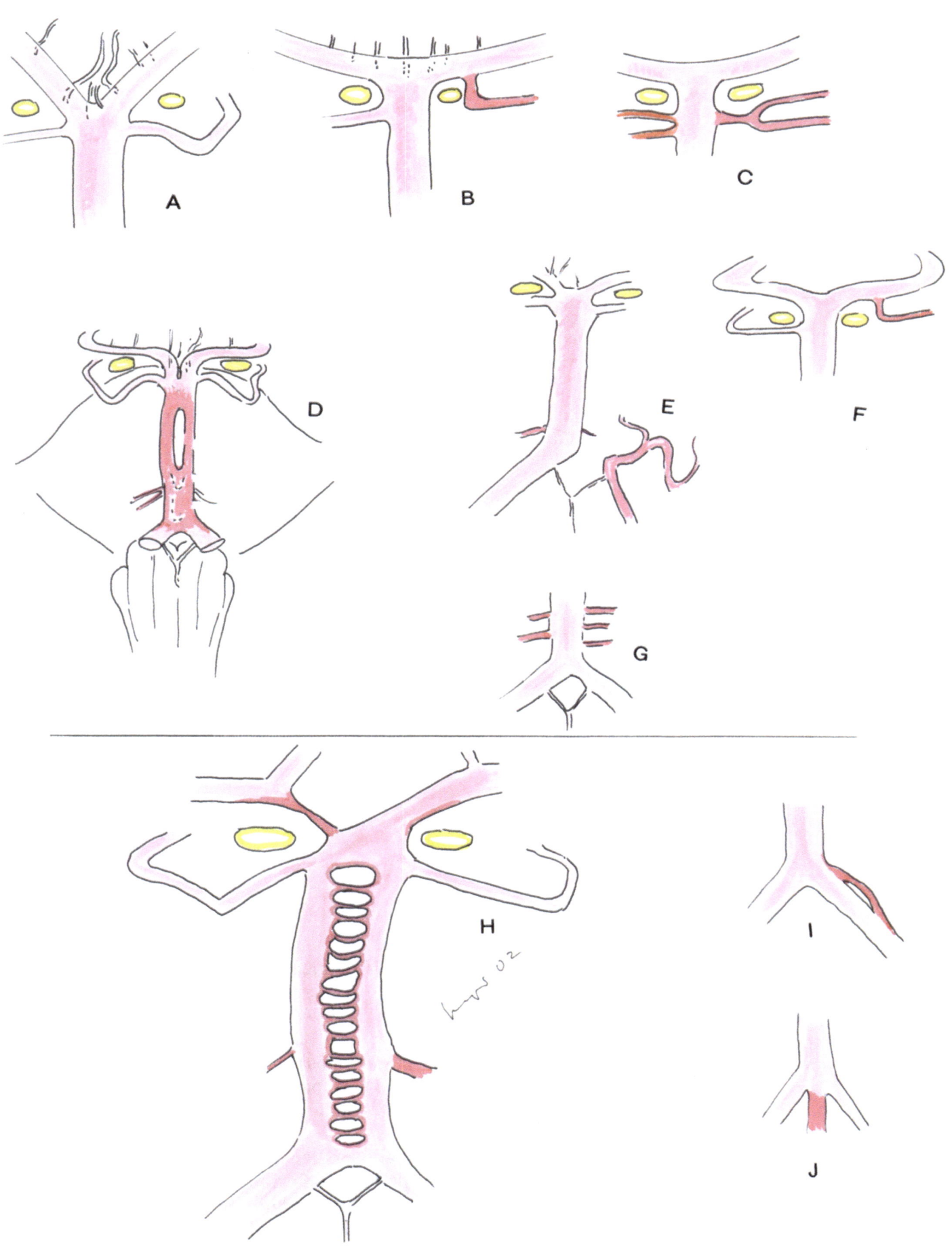

Fig. 157. Continuation of Fig. 156

A and **B**   AICA, distal origin from A. basilaris
C   AICA, origin from A. vertebralis (left-sided)
D   A. basilaris and its branches, no congenital variant (see Fig. 152)

Clinical aspects:
Compression of cranial nerves with consecutive trigeminal neuralgia and others may occur, according to Janetta (1977)

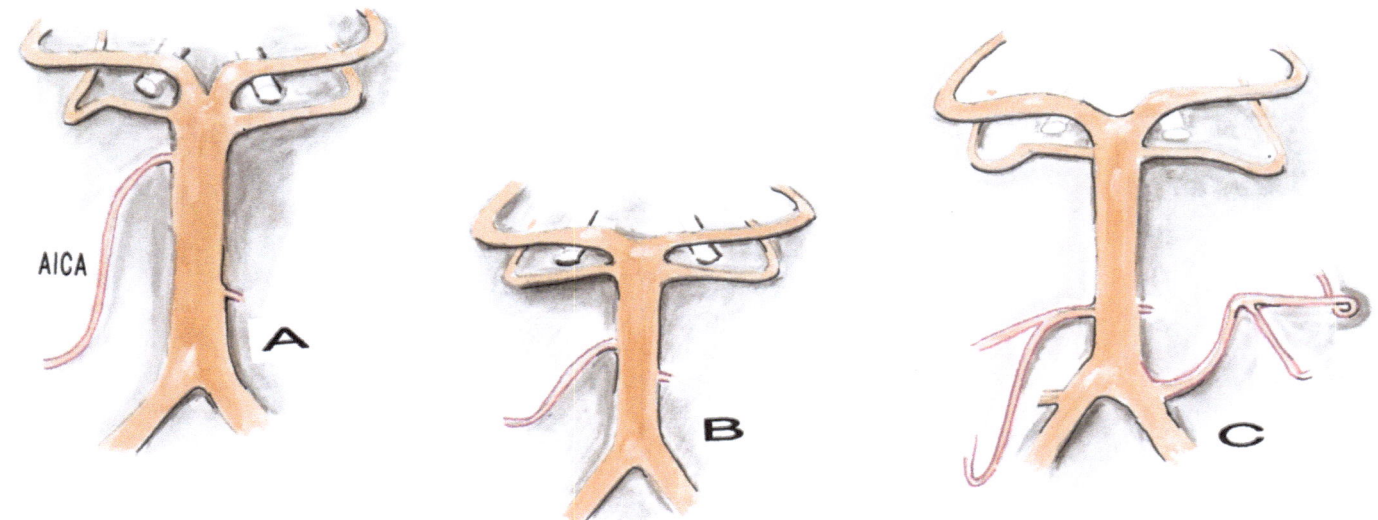

AICA

A

B

C

rare types

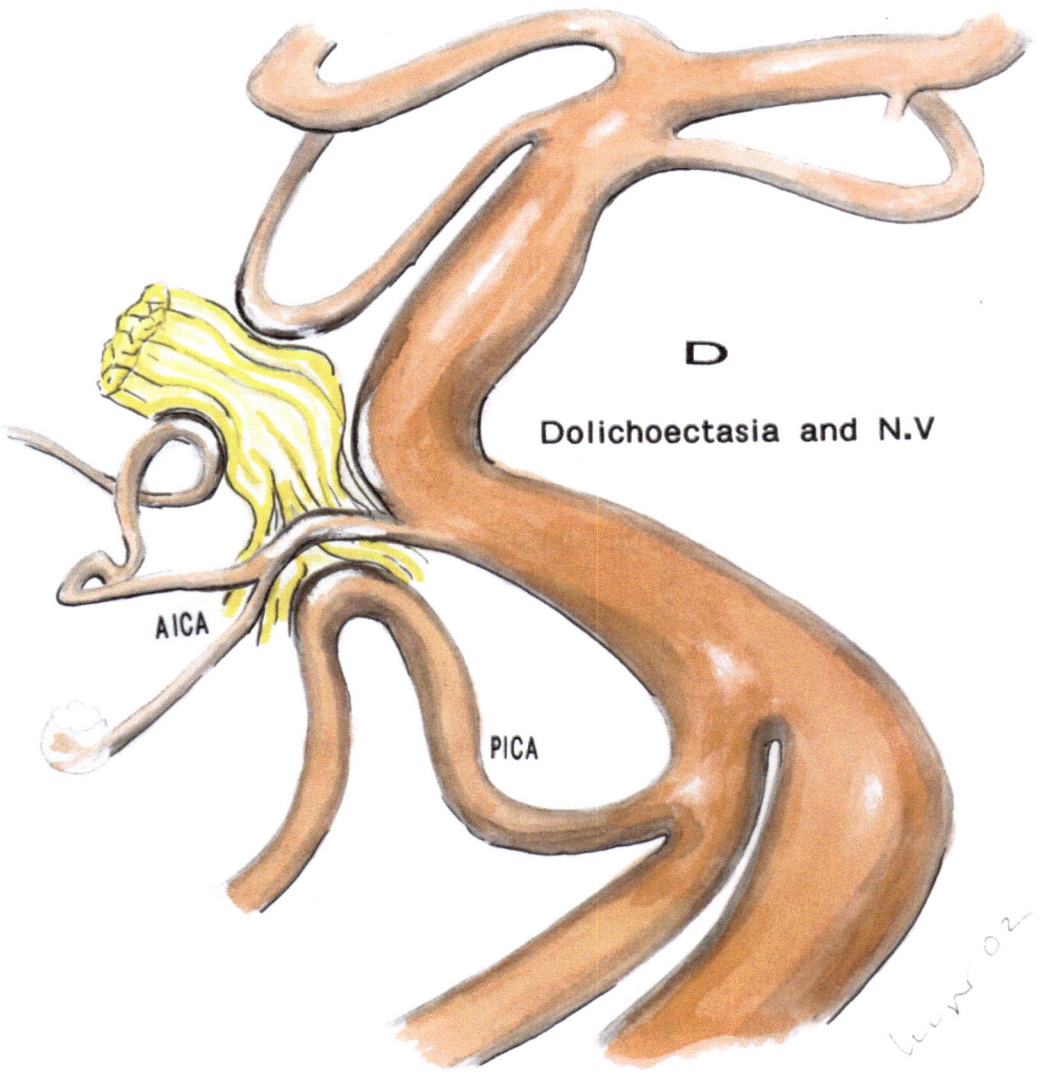

D

Dolichoectasia and N.V

AICA

PICA

Fig. 158. **PICA, proximal trunk,** rare variant

A   Angiography: Normal finding of the typical stenosis of A. vertebralis at its dura penetration point
Right PICA: Normal finding
Left PICA: Extradural origin
B   According to Lang J (1985) p 555, modified
C   Simplified presentation of A, but rightsided

Abbreviations
a   A. vertebralis
b   Lig. deuticul.
c   Tonsilla
d   Labulus biventer
e, f   Duraflaps
g   Craniospinal articulation

Clinical aspects:
In microsurgical procedures, PICA may be mistaken for arteries of the muscles of the neck: Danger for unexpected ligation of PICA!

Danger before starting the trepanation

Preoperative angiogram or angio-MRT should be taken into consideration

FIG. 158

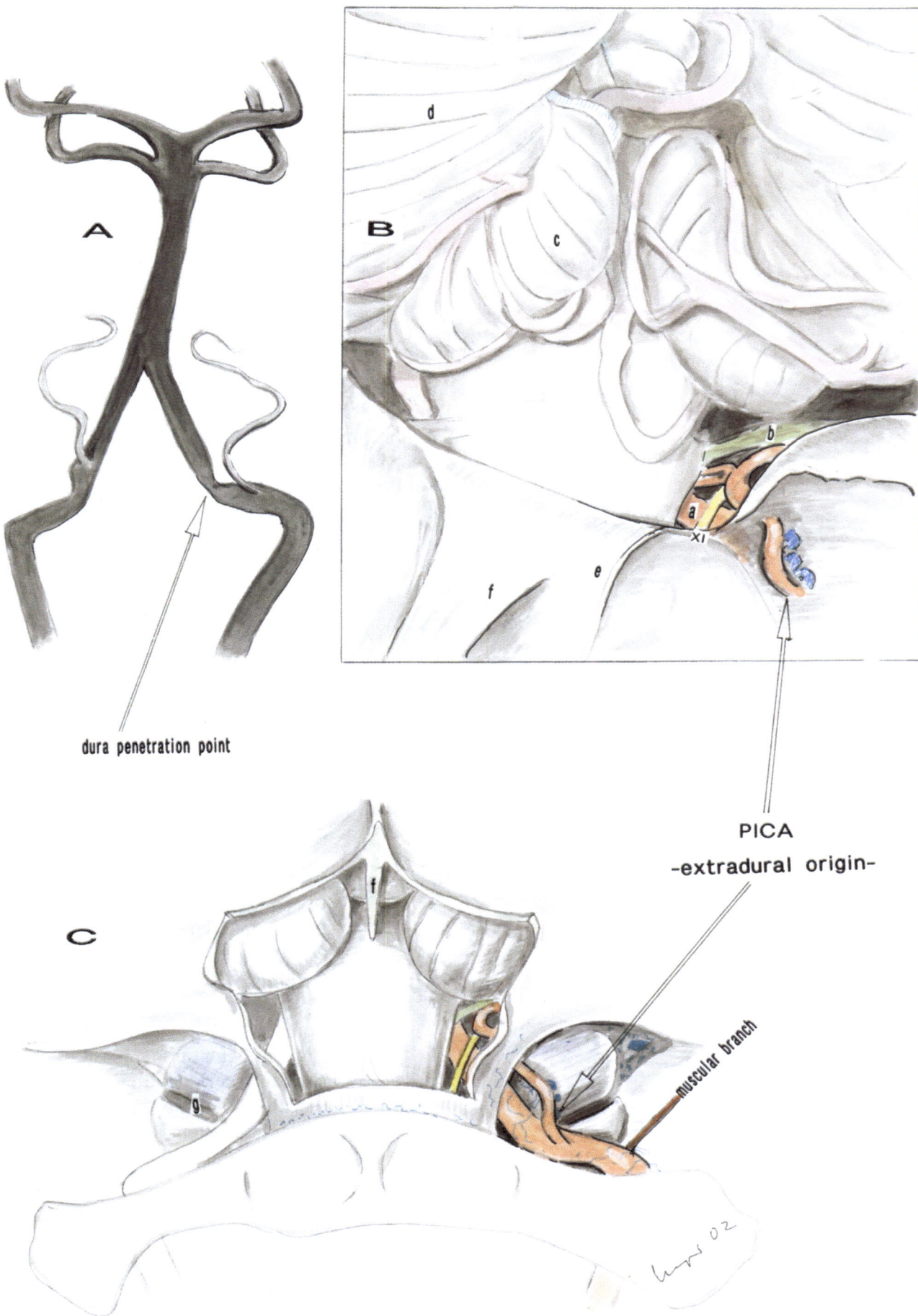

A

dura penetration point

B

PICA
-extradural origin-

C

muscular branch

Fig. 159. **Variants of small branches of A. vertebralis, PICA and AICA**
According to Lang J (1981) p 437, simplified

Variants

a   Aplasia of the caudomedial branch of the AICA, common variant

b   Aplasia of the AICA, common variant

c   Connection of PICA and A. vertebralis, rare variant

c'   Close to the origin of PICA: Rare connection with Medulla oblongata

d   Loop of PICA in a caudal direction, common variant

e   A. spinalis ant. unilateral, common variant

(f   Addendum: Variant of N. hypoglossus, duplication of the nerve in its channel, common variant)

Further Abbreviations

g   Flocculus
h   Bochdalek's plexus
i   Oliva
j   Pyramis of Medulla oblongata
k   Normal relationship of C1 and A. vertebralis
l   Lig. denticulatum
m   typical halter ("Halfter") of A. vertebralis (according to Lang, 1981)
n   Articulatio atlantooccipitalis
o   Massa lateralis atlantis
VI to XII cranial nerves

Clinical aspects:
Small arteries may not to be identified by angiographic methods

FIG. 159

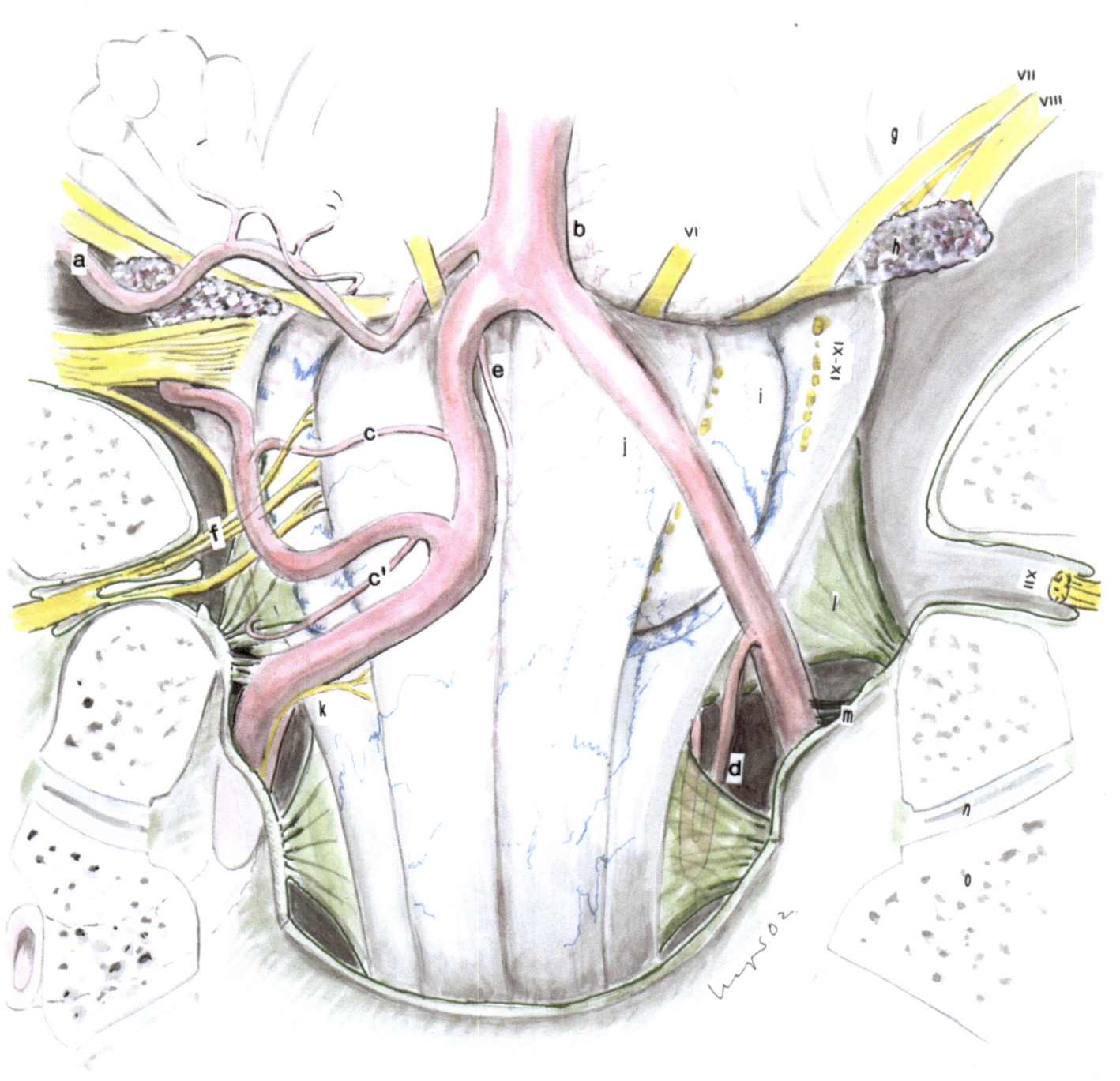

Fig. 160. **PICA, choroid point**

A   This "typical" finding is inconstant (*: choroid point)
Note its relationship with Plexus chor. and Nidus avis of Velum medull. post.

B   PICA is located close to Nidus avis. Choroid point "typical"

C   PICA, common variant

Clinical aspects:
Analysis of angiograms may help to localize the plexus-segment of Velum medullare post., which build the posterior wall of the 4th ventricle. The plexus-free segment of the velum is thin-walled and cannot be identified by MRTs

FIG. 160

PICA

choroid area

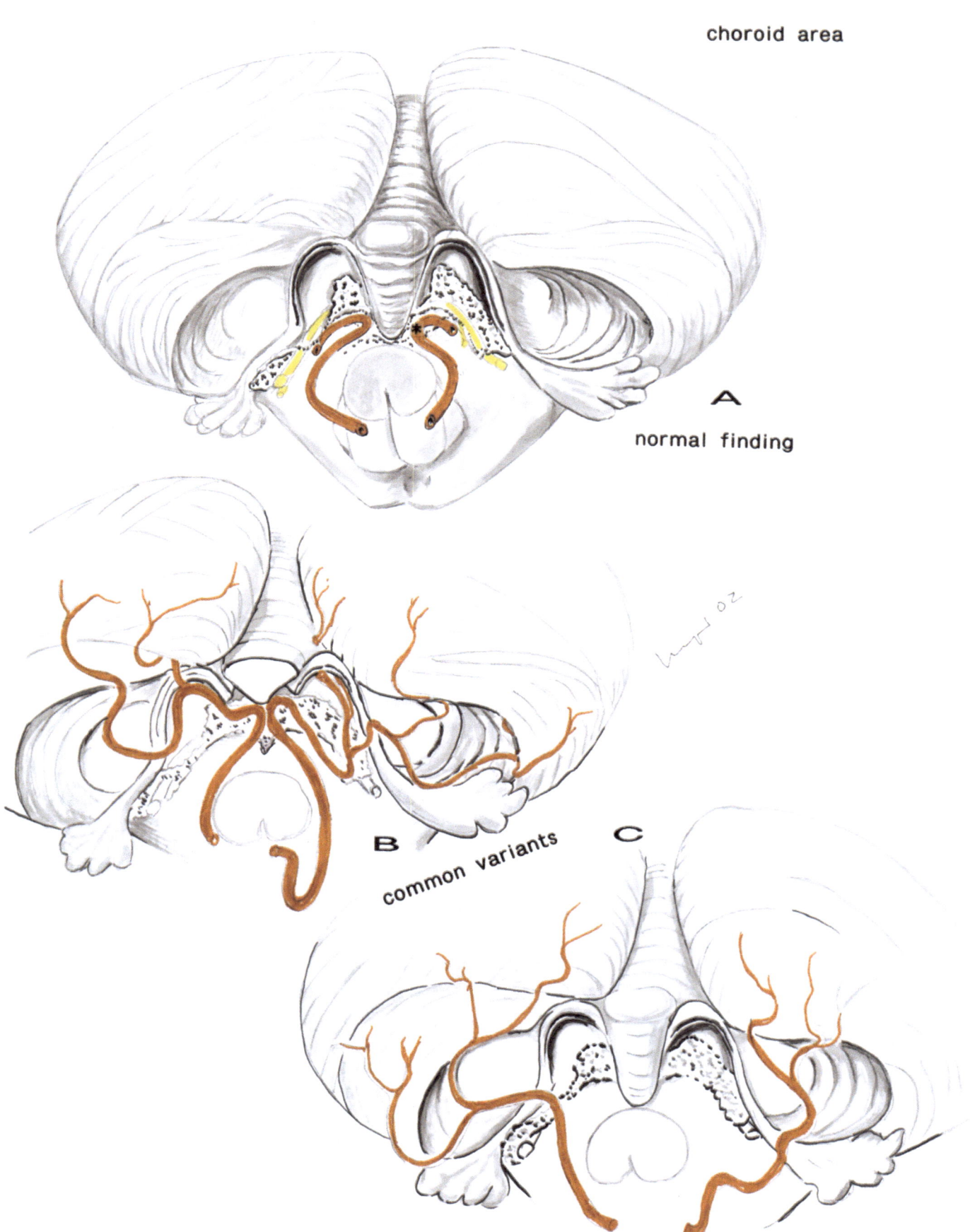

A

normal finding

B

common variants

C

Fig. 161. **PICA and its branches,** further variants

A   A. vermicularis, originating from one PICA, aplasia of contralateral PICA, trifurcation of A. vermicularis
B   For comparison with A

Clinical aspects:
During microsurgery, the high variability of these vessels has to be considered

Abbreviations
a   Aa.vermiculares
b   PICA, lateral main branch(es)
c   choroid branches
d   as c, originating from A. vermicularis
e   PICA, bifurcation
f   Apertura mediana (Magendi)

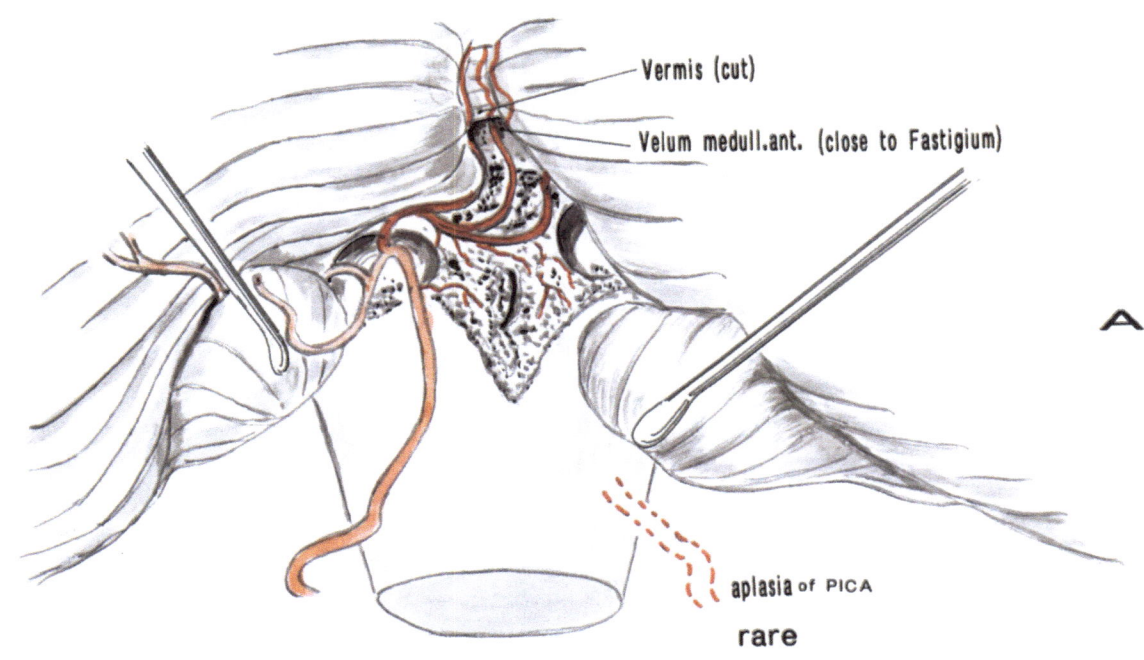

Vermis (cut)

Velum medull.ant. (close to Fastigium)

A

aplasia of PICA

rare

Aa.vermiculares

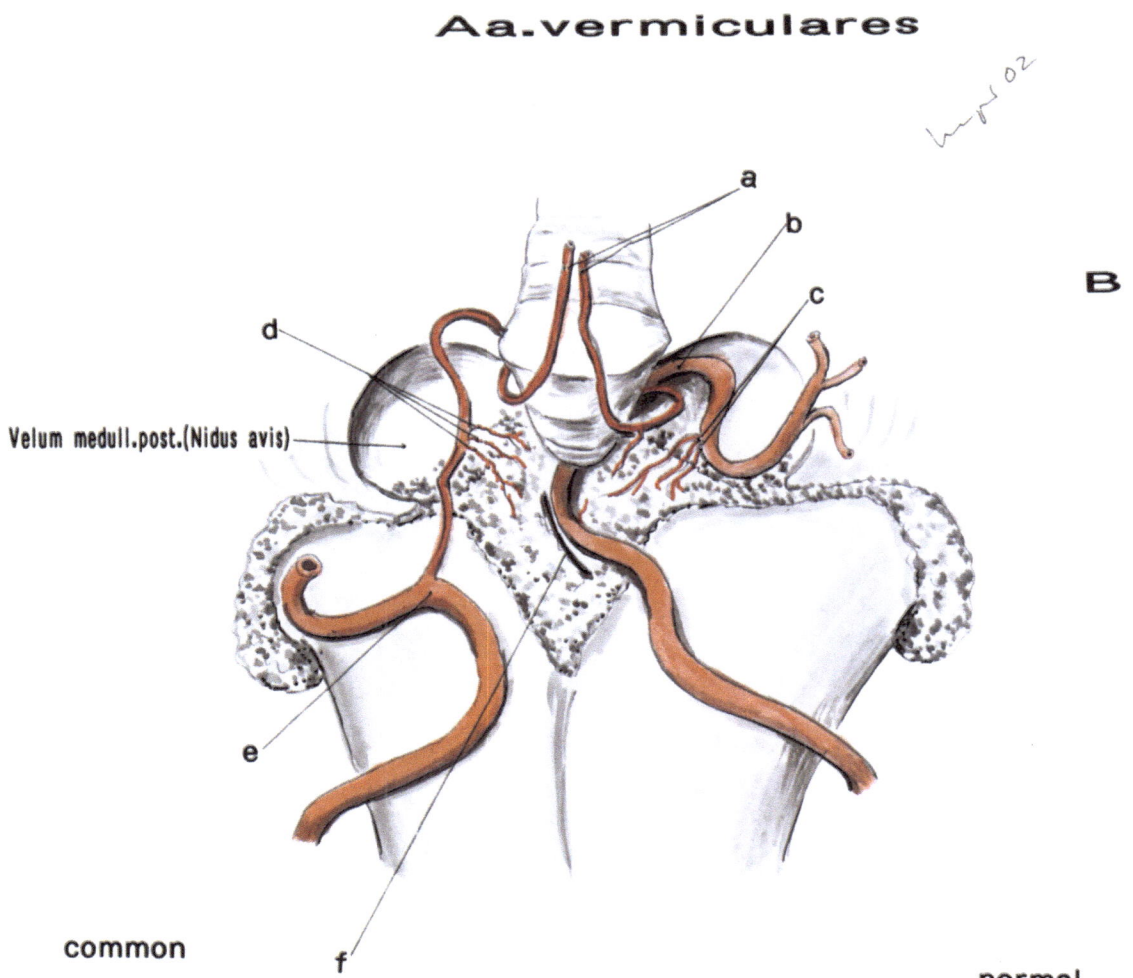

B

a

b

c

d

Velum medull.post.(Nidus avis)

e

f

common

normal

Fig. 162. Continuation of Fig. HS 161

For clinical aspects see Fig. 161

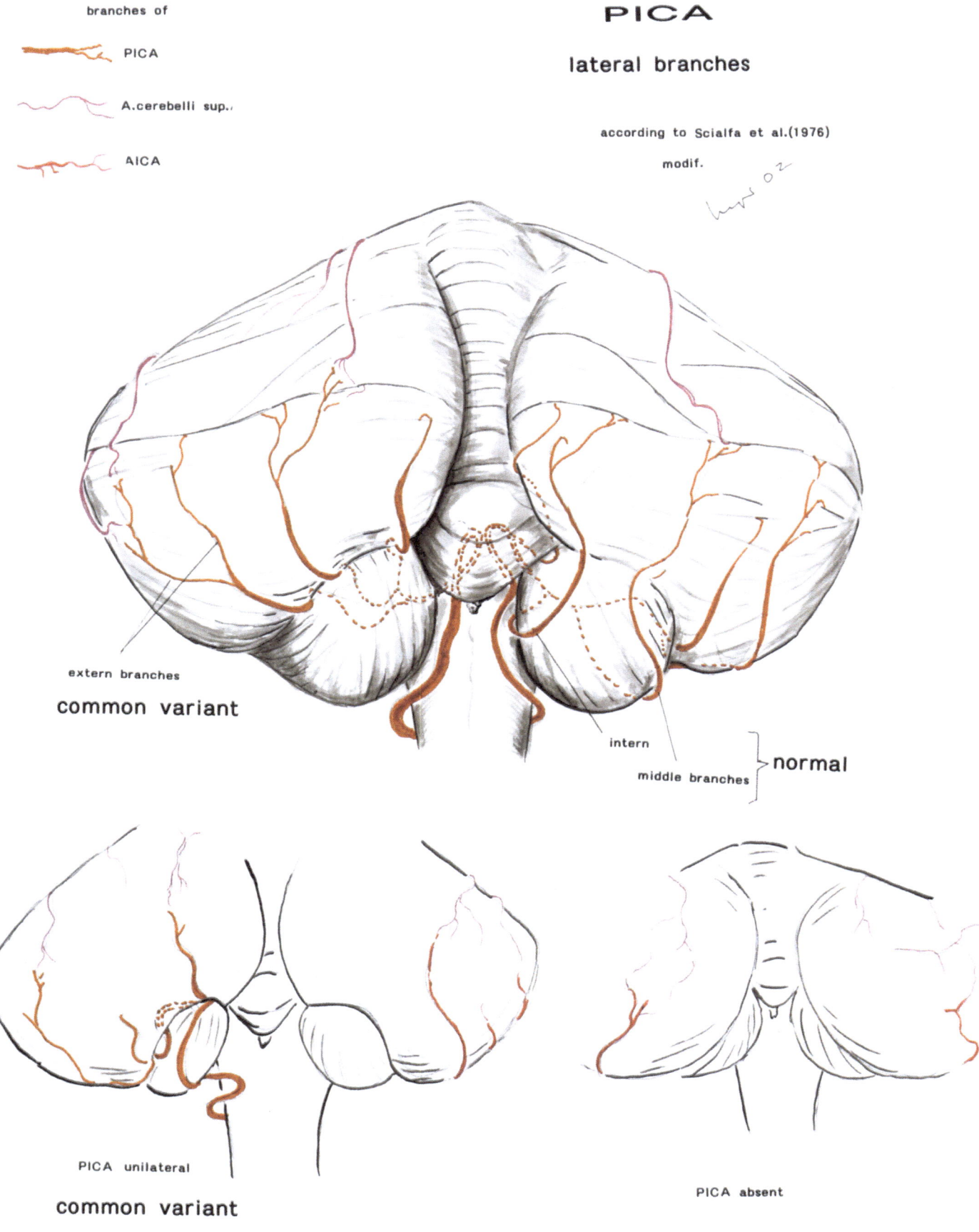

branches of

PICA

A.cerebelli sup.

AICA

PICA

lateral branches

according to Scialfa et al.(1976)

modif.

extern branches

common variant

intern

middle branches

} **normal**

PICA unilateral

common variant

PICA absent

rare

Fig. 163. **AICA, proximal trunk, and A. labyrinthi**

A   labyrinthi (dark colored)
Right: normal
Left: common variant

B and **C**   Duplications of AICA uni- or bilateral are
common variants

D and **E**   More than 2 branches or unilateral absent
AICA are rare variants

For clinical aspects see Fig. 161

FIG. 163

B and C

common variants

B

C

A

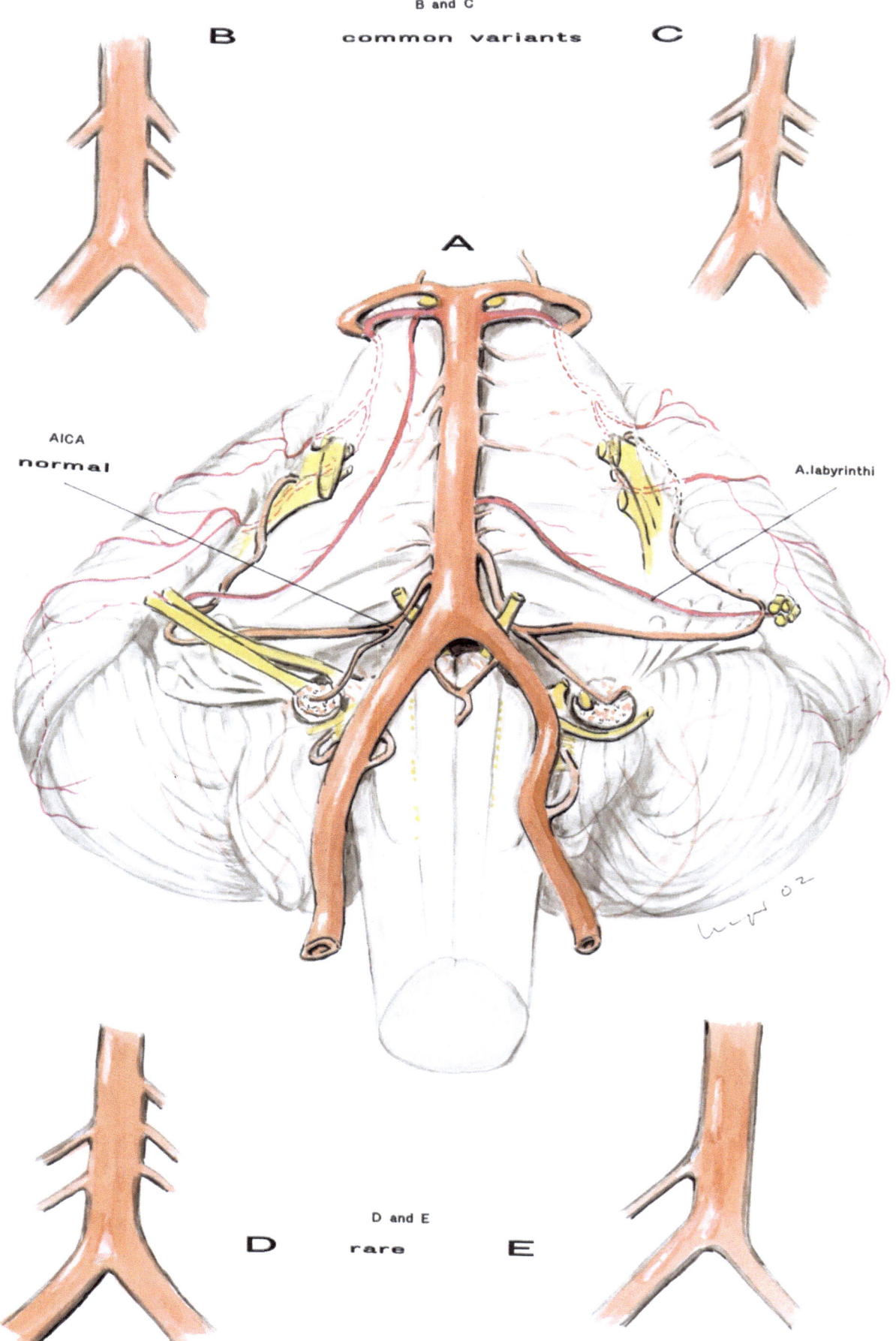

AICA

normal

A.labyrinthi

D and E

D   rare   E

Fig. 164. Continuation of Fig. 163

AICA, proximal ramifications

For clinical aspects see Fig. 161
For special aspects of meatal and trigeminal loop see Fig. 177

FIG. 164

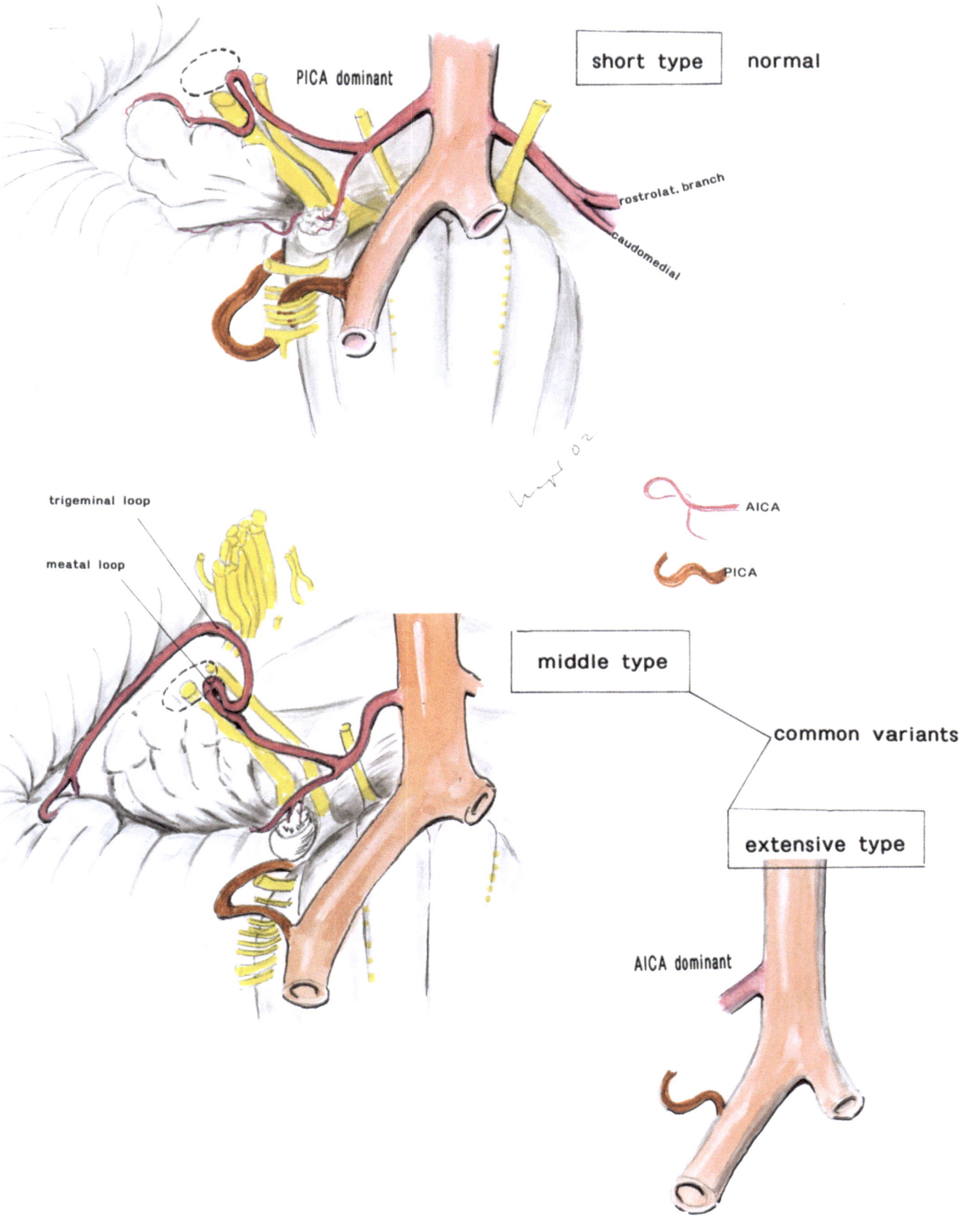

PICA dominant

short type — normal

rostrolat. branch

caudomedial

trigeminal loop

meatal loop

AICA

PICA

middle type

common variants

extensive type

AICA dominant

Figs. 165 and 166. **A. cerebelli sup.; relationship with N. oculomotorius and with the rostral area of Cerebellum**

Clinical aspects:
Variable relationship with aneurysms of the basilar artery and with tumors close to N. oculomotorius and the arteries are well known

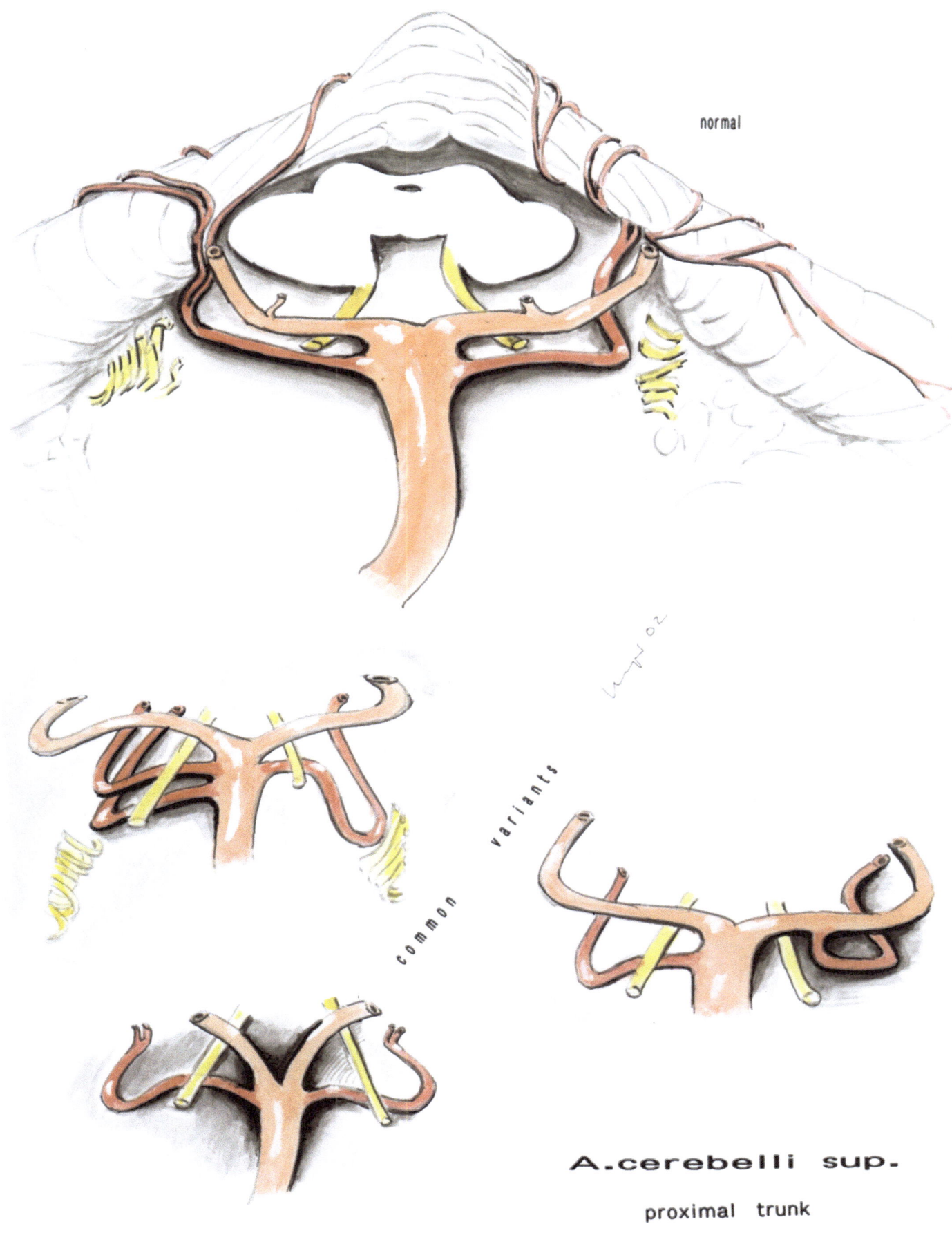

normal

common variants

A. cerebelli sup.

proximal trunk

FIG. 166

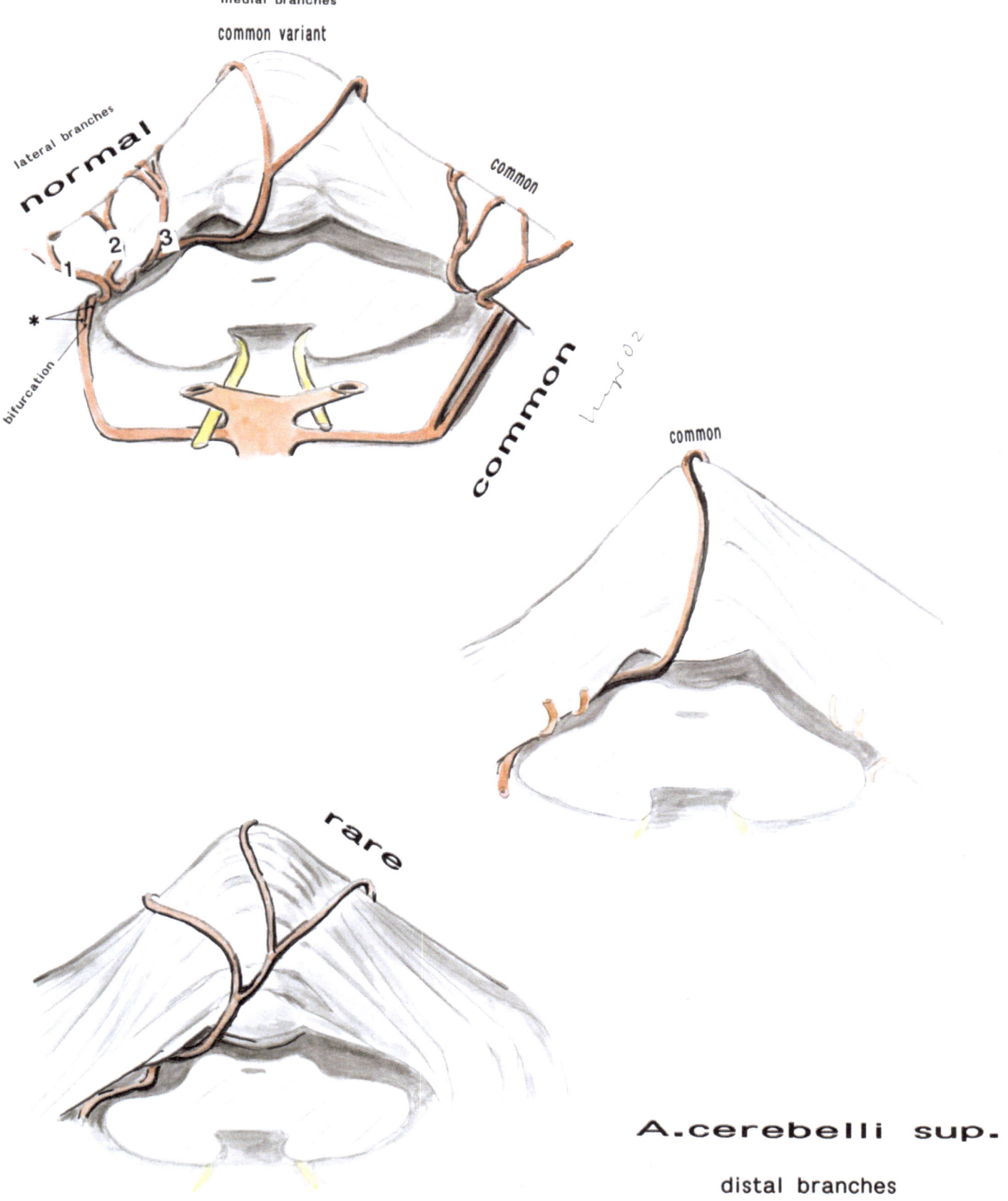

medial branches

common variant

lateral branches

normal

common

common

1 2 3

*

bifurcation

rare

common

A. cerebelli sup.

distal branches

Fig. 167. This historic presentation of Key and Retzius (1875) motivated M.G.Yasargil to analyze arachnoid cisterns*. It was the start of modern microsurgery of the Acoustic neurinoma and other extracerebral tumors of the cerebral base

---

* personal communication in 1976

FIG. 167

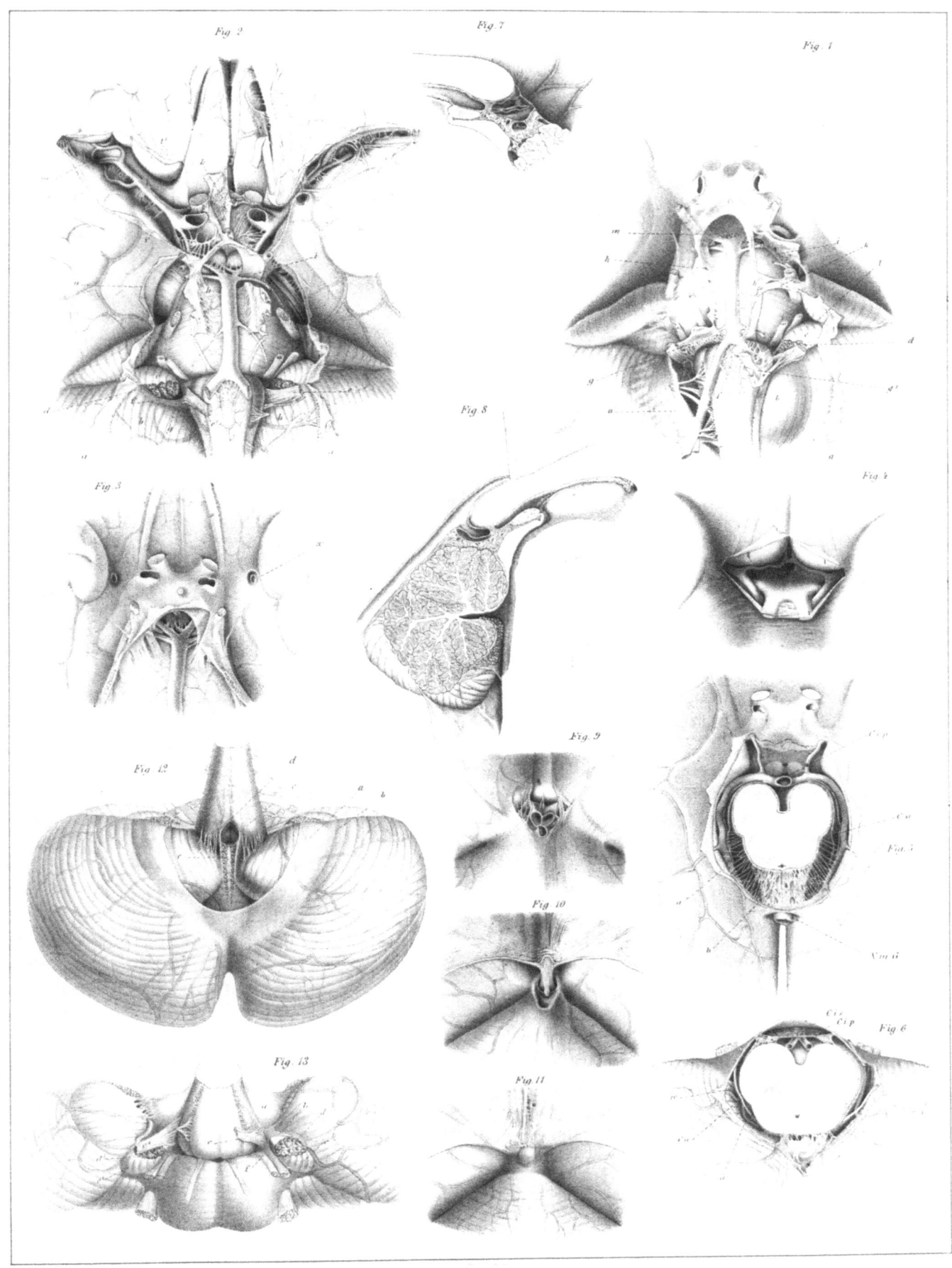

Fig. 168. As Fig. 167, colored and enlarged copies
Cisternal arachnoid layers green colored

A   See Fig. 2 of Key and Retzius
B   See Fig. 1 of Key and Retzius
Arrow in a superior direction: Wall of the cistern of Nn.IX to XI elevated (see right side). This should be avoided in microsurery for acoustic neurinoma.
Arrow in an inferior direction:
Wall of the cistern of Nn.VII and VIII shifted downwards (here together with the wall of the cistern of Nn. IX to XI)
Yasargil's technique for preparation of Cisterna cerebellomedullaris (containing the tumor):
Microsurgical isolation of the wall of Cisterna cerebellomedullaris from the wall of the cistern of Nn.IX to XI, which can be protected by a cottonoid which is placed between both cisternal walls. **This is a very important step of the operation. It may be complicated by adhesions, which may be present, especially close to Flocculus and Bochdalek's plexus segment.**

Another variants are adhesions of the facial nerve and the tumor arachnoidea and defects of the wall of the cerebellopontine cistern.
Close relationships of the AICA with the cisternal wall or direct relationships with the tumor arachnoid layer are rare. If the microsurgeon has understood the structures of arachnoid membranes, then problems by the numerous variants of the course of PICA are rare during the operation of acoustic neurinomas. **But problems by an intratumoral course of AICA are to be expected at surgery for petroclival meningiomas and other basal intracisternal tumors.**
Here the variants of AICA must be analyzed preoperatively.

Abbreviations
a   N. V and Lobulus quadrangularis
b   as a, not overlapped by Lobulus quadrangularis
c   N. XI
d   Pannus of vessels
e   A. spinalis ant.

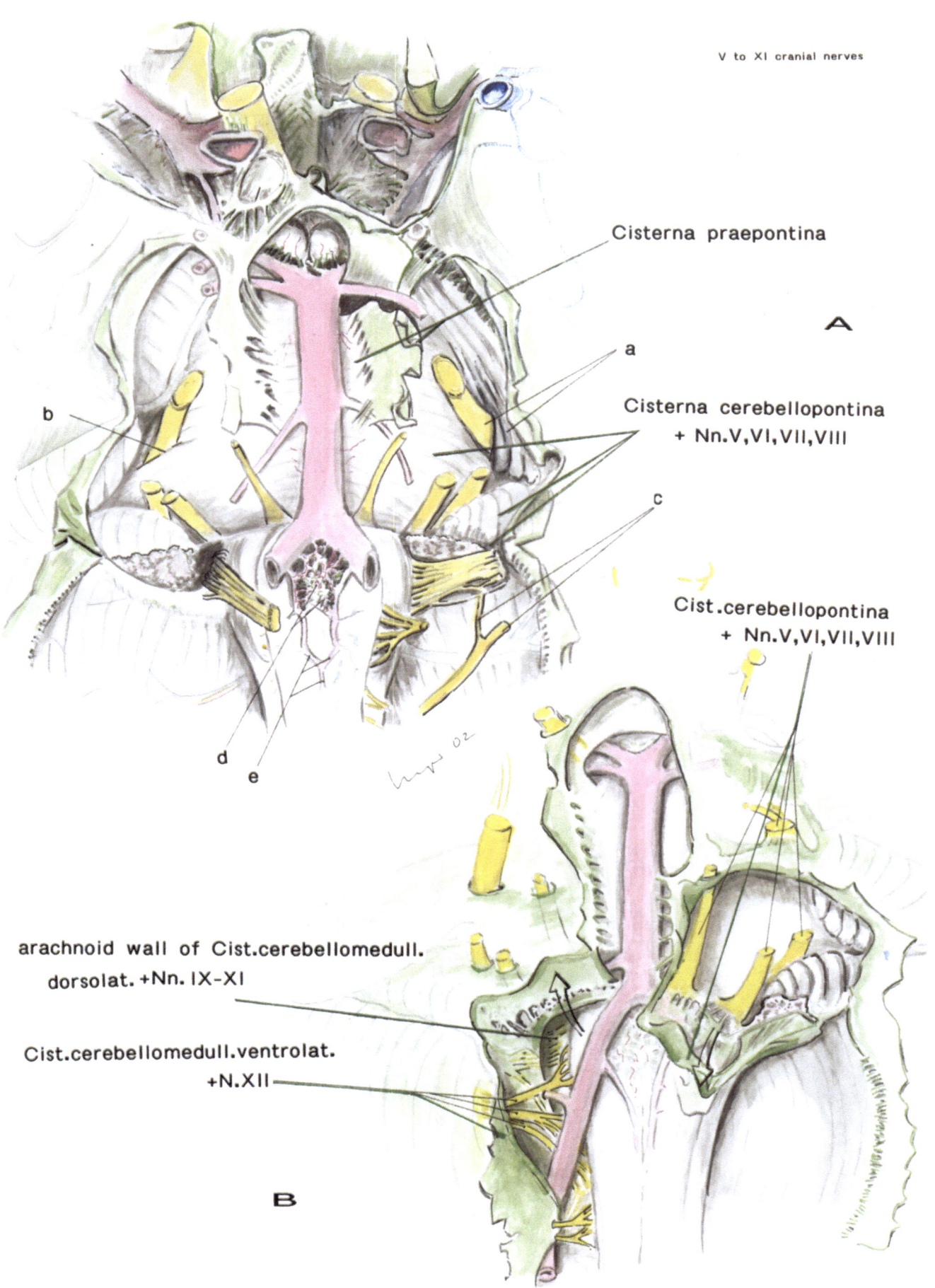

V to XI cranial nerves

Cisterna praepontina

A

a

Cisterna cerebellopontina
+ Nn.V,VI,VII,VIII

c

Cist.cerebellopontina
+ Nn.V,VI,VII,VIII

b

d
e

arachnoid wall of Cist.cerebellomedull.
dorsolat. +Nn.IX–XI

Cist.cerebellomedull.ventrolat.
+N.XII

B

Fig. 169. **Basal cisterns**

A   Schematic presentation. Each cistern contains one or more cranial nerves

B   **Clinical aspects:**
Microsurgery of the acoustic Schwannoma (described by Yasargil,) representing the principles of extracerebral basal microsurgery

Abbreviations

a   arachnoid layer of Cisterna cerebellopontina
a'   arachnoid layer of acoustic neurinoma
b   Cisterna cerebellomedullaris lateralis (cistern of the cranial nerves IX to XI)
c   Cisterna praepontina (containing A. basilaris)
d   Cisterna cerebellomedullaris medialis (cistern of Nn.XII)

FIG. 169

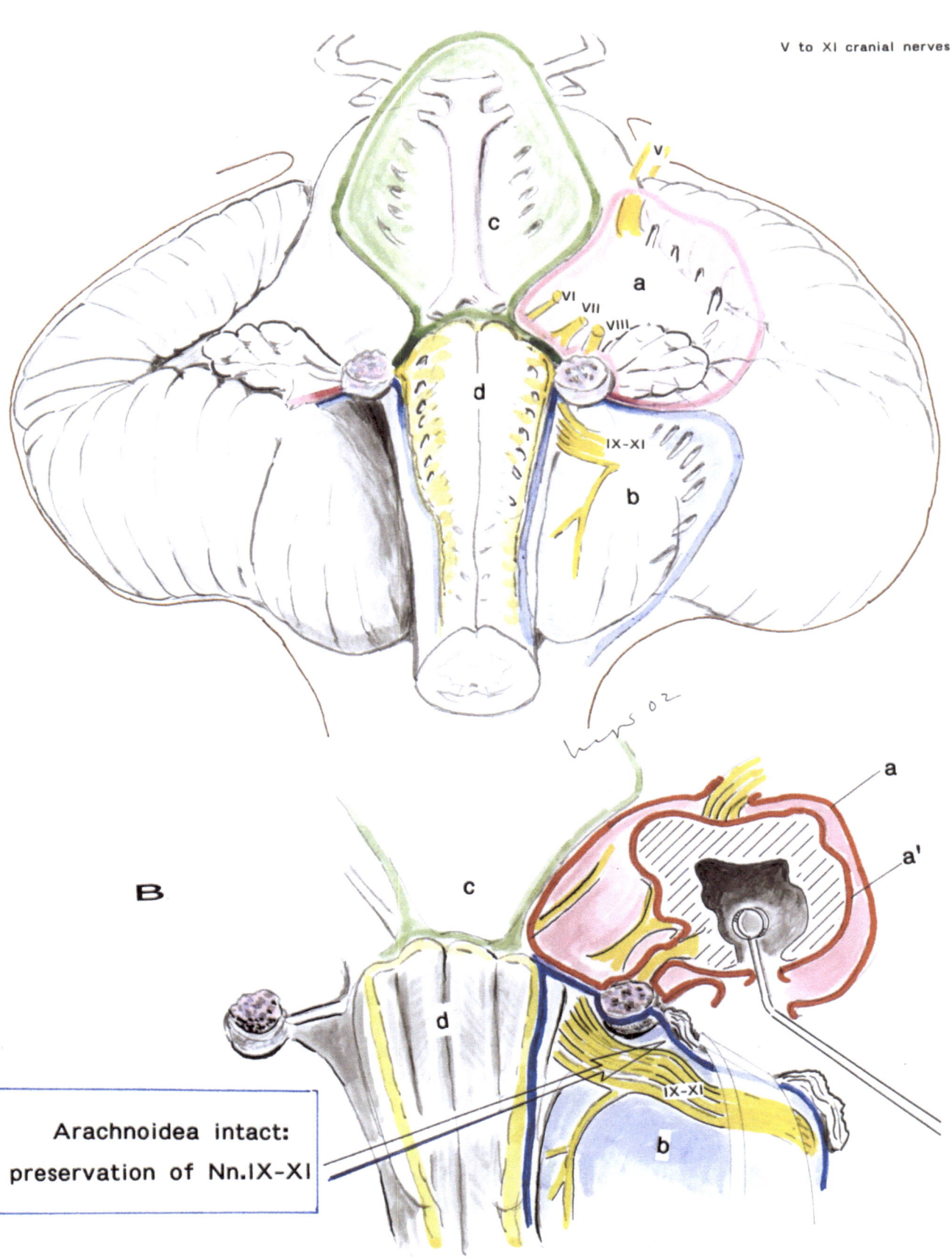

A

V to XI cranial nerves

B

Arachnoidea intact:
preservation of Nn.IX-XI

Fig. 170. **Apertura lateralis (Foramen of Key and Retzius = Foramen Luschkae)**
common variants

A   Typical finding
B   Thin arachnoid layer
C   Thick arachnoid layer, Apertura nearly occluded
D   Apertura occluded by the arachnoid layer. Numerous adhesions with Flocculus

Clinical aspects:
Apertura in A to C seems not to be occluded. But it may be occluded by a thin layer of Velum medullare post. (at its inside: Ependyma). If Apertura mediana (Magendi) is occluded, too, then Hydrocephalus will develop.
Occlusion of Apertura lateralis is a common variant

FIG. 170

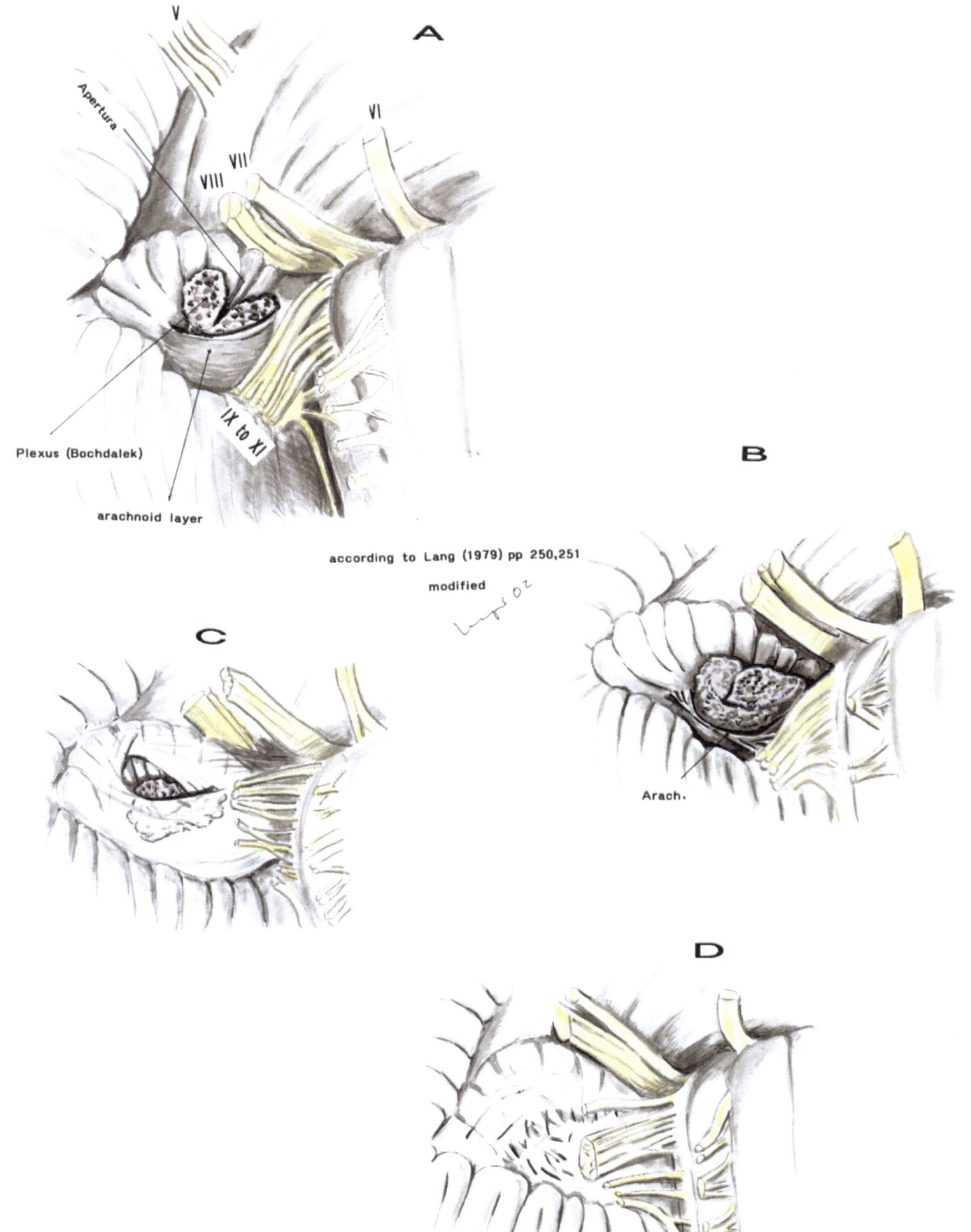

A

V

Apertura

VII

VIII

VI

IX to XI

Plexus (Bochdalek)

arachnoid layer

according to Lang (1979) pp 250,251

modified

B

Arach.

C

D

Fig. 171. Continuation of Fig. 170

Ependyma of Recessus lateralis and surrounding arachnoid layers

A' to **D'**   According to Fig. 170 A to D
A'   Typical finding
B' to **D'**   Common variants

For clinical aspects see Fig. 170

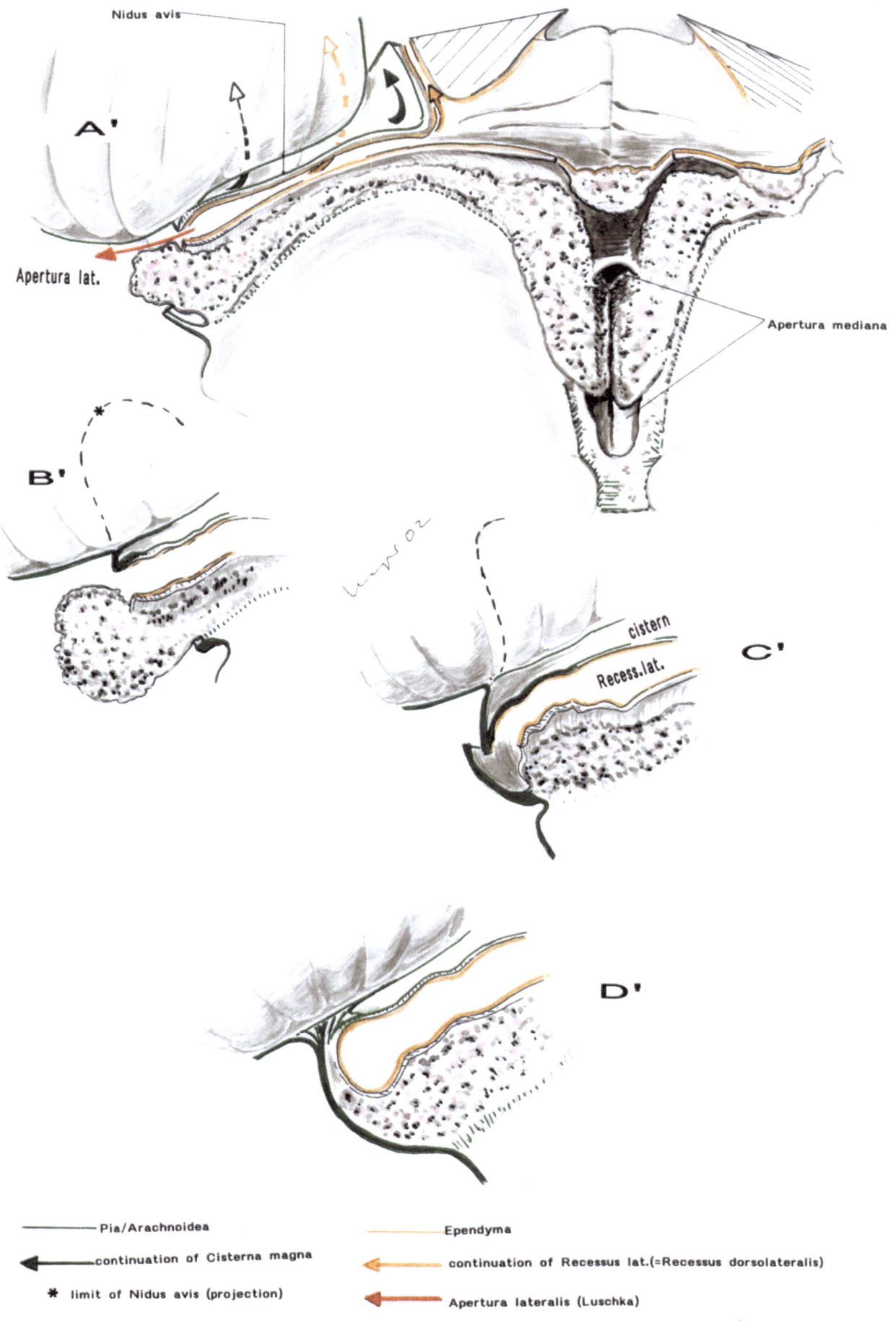

Nidus avis

A'

Apertura lat.

Apertura mediana

B'

*

cistern

Recess.lat.

C'

D'

——— Pia/Arachnoidea   ——— Ependyma

➤ continuation of Cisterna magna   ➤ continuation of Recessus lat.(=Recessus dorsolateralis)

* limit of Nidus avis (projection)   ➤ Apertura lateralis (Luschka)

Fig. 172. **Area of Recessus lateralis**

A   Survey
B   Atretic For. Luschkae
C   For. Luschkae open

FIG. 172

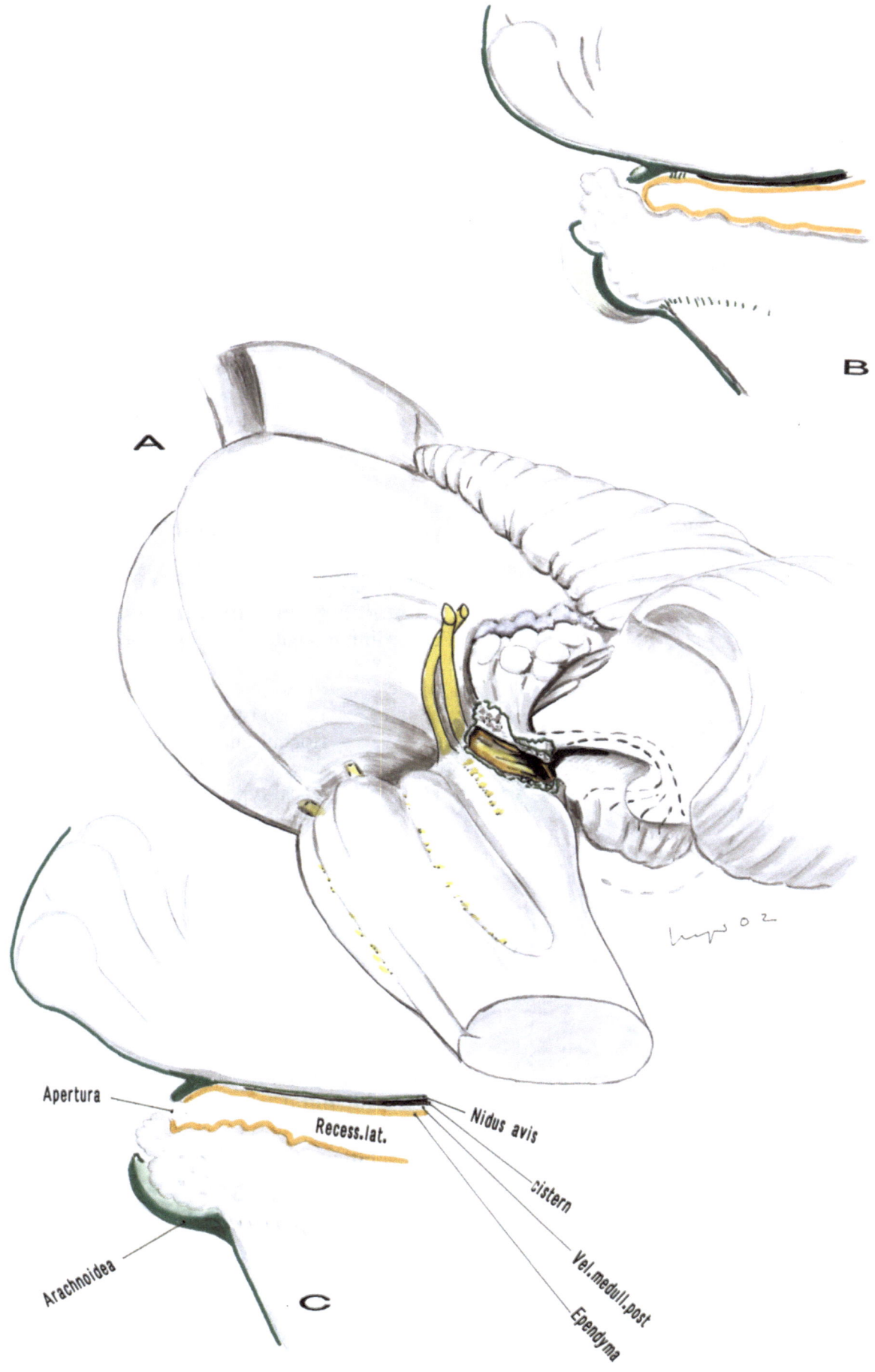

Apertura

Recess.lat.

Nidus avis

cistern

Vel.medull.post

Arachnoidea

Ependyma

C

Fig. 173. Continuation of Fig. 172

Subependymal course of N. statoacusticus (containing Nucleus dorsalis of N. cochlearis), according to anatomical dissections

A	Survey
B and **C**	Dorsal types: N.VIII close to Pedunculus flocculi and Pedunculus cerebellaris medius
D	Basal type close to Medulla oblongata
E	Dorsal type, penetrating Pedunculus flocculi and Pedunculus cerebellaris med. Note the relationships of Ependyma and Arachnoidea with N.VIII in B and C
F	Schematical transection of the area of Apertura lat.

Dotted lines: Projections

Clinical aspects:
Microneurosurgical location of N.statoacusticus, additional to neurophysiological methods

Special literature:
Bochdalek VA (1849), Huang JP, Wolf (1967), Harison JM, Feldman ML (1970), Hitselberger WE, House WF, Edgerton BJ (1984), Kose A-K, Sollmann WP (1998), Klose A-K, Sollmann W-P (1999), Klose A-K (2000)
For further literature see Klose and Sollmann (1999) and Klose (2000)

FIG. 173

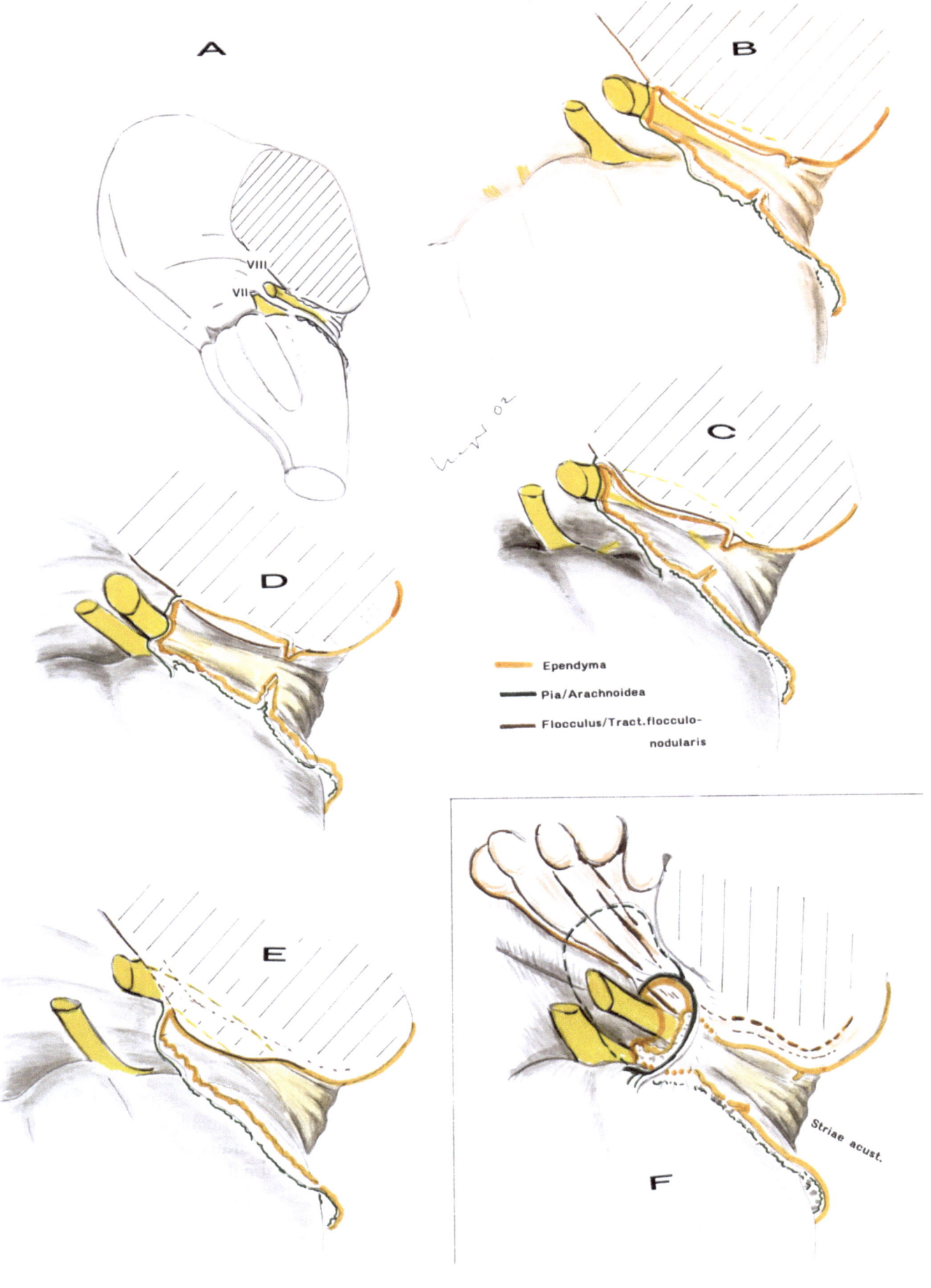

A

B

C

D

VIII

VII

Ependyma

Pia/Arachnoidea

Flocculus/Tract.flocculo-
nodularis

E

F

Striae acust.

Fig. 174. Continuation of Fig. 173

A to **D**   Anatomical dissections, simplified
A' to **D'**   Microanatomical views into the lateral
recessus, transpassing Apertura lateralis,
schematical sketches, according to A to D

Clinical aspects:
See Fig. 173

FIG. 174

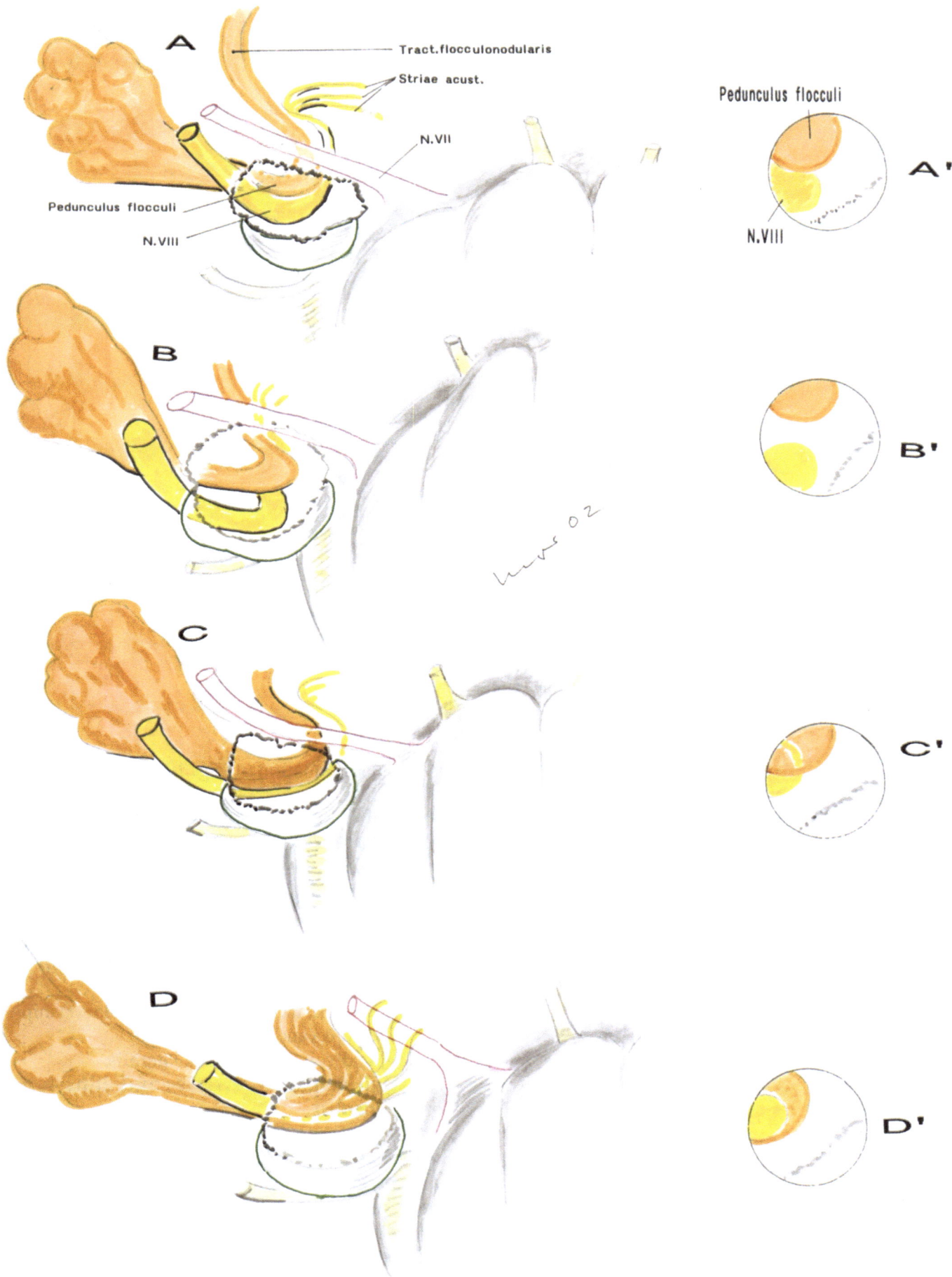

A

Tract.flocculonodularis

Striae acust.

N.VII

Pedunculus flocculi

N.VIII

Pedunculus flocculi

N.VIII

A'

B

B'

C

C'

D

D'

Fig. 175. Continuation of Fig. 174

Vessels of the area of Apertura lateralis and adjacent structures
Further cadaver dissections, sketches according to microanatomical situs

A Caudomedial main branch of AICA and typical branch of a pontine vein with relationship to the dorsal wall of Recessus lateralis

B Typical periolivar veins with relationships to Flocculus (and Recessus lat.?)

C Typical vein of Recessus lateralis.
  High density of arteries and adhesions of Flocculus.

Clinical aspects:
Microsurgical manipulations may endanger vessels close to N.VIII and the Nucleus dorsalis of N. cochlearis.
For literature see Fig. 173

FIG. 175

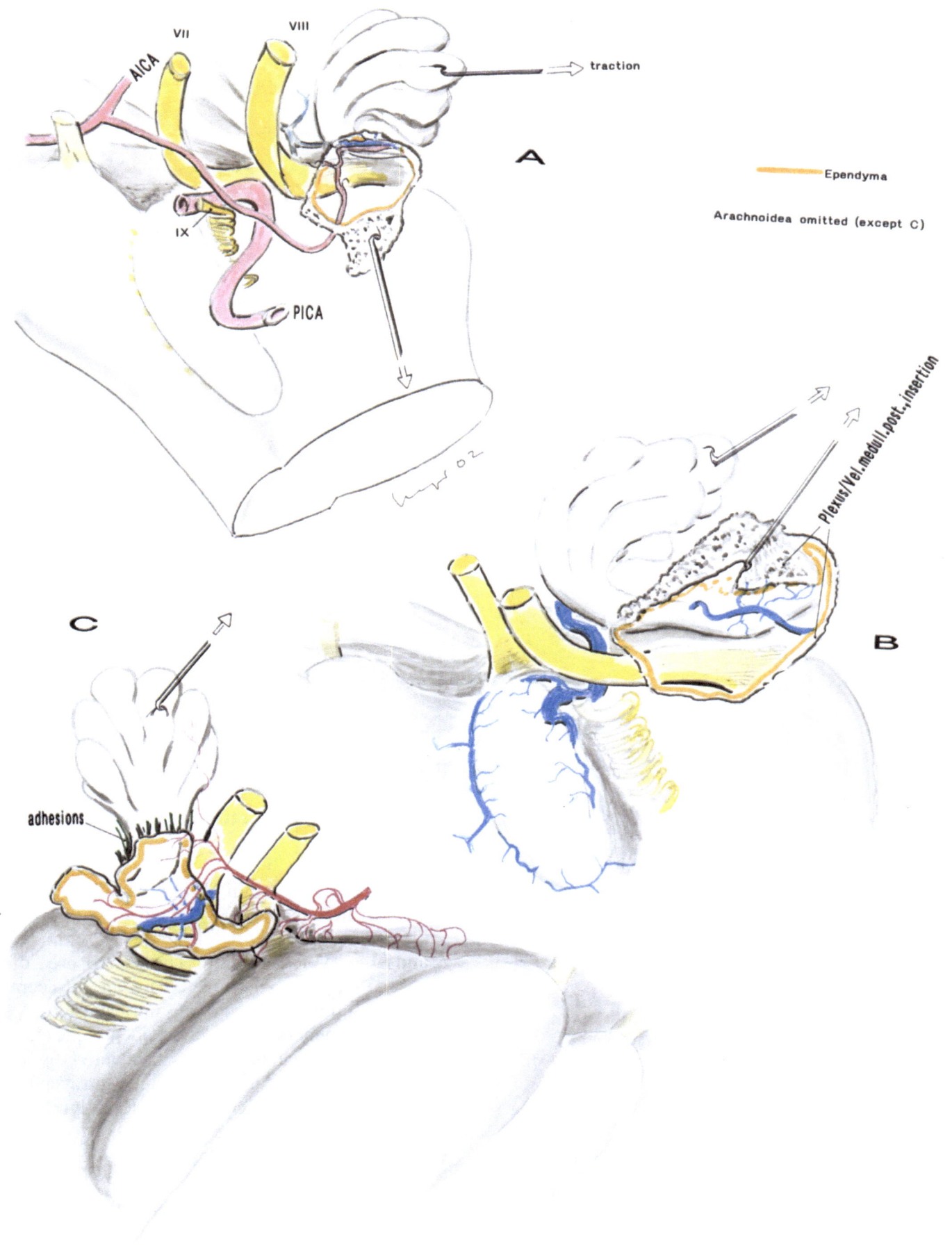

VII

VIII

AICA

→ traction

A

IX

PICA

Ependyma

Arachnoidea omitted (except C)

Plexus/Vel.medull.post.,insertion

B

C

adhesions

Fig. 176. For. Magendi

Apertura mediana of the 4th ventricle (Foramen Magendi)

This apertura is a gap of Velum medullare post. like Apertura lateralis

A and A'   Typical finding: The caudal area of the apertura is covered by an extracerebral (cisternal) and an intraventricular segment of Plexus chorioideus. The cisternal surface of plexus is covered by leptomeningeal layers, the intraventricular surface is covered by Ependyma (= Velum medullare post.)

B   Common variant

For microsurgical aspects of relationship of Apertura mediana and Velum medullare post. with PICA see Figs. 160 to 162

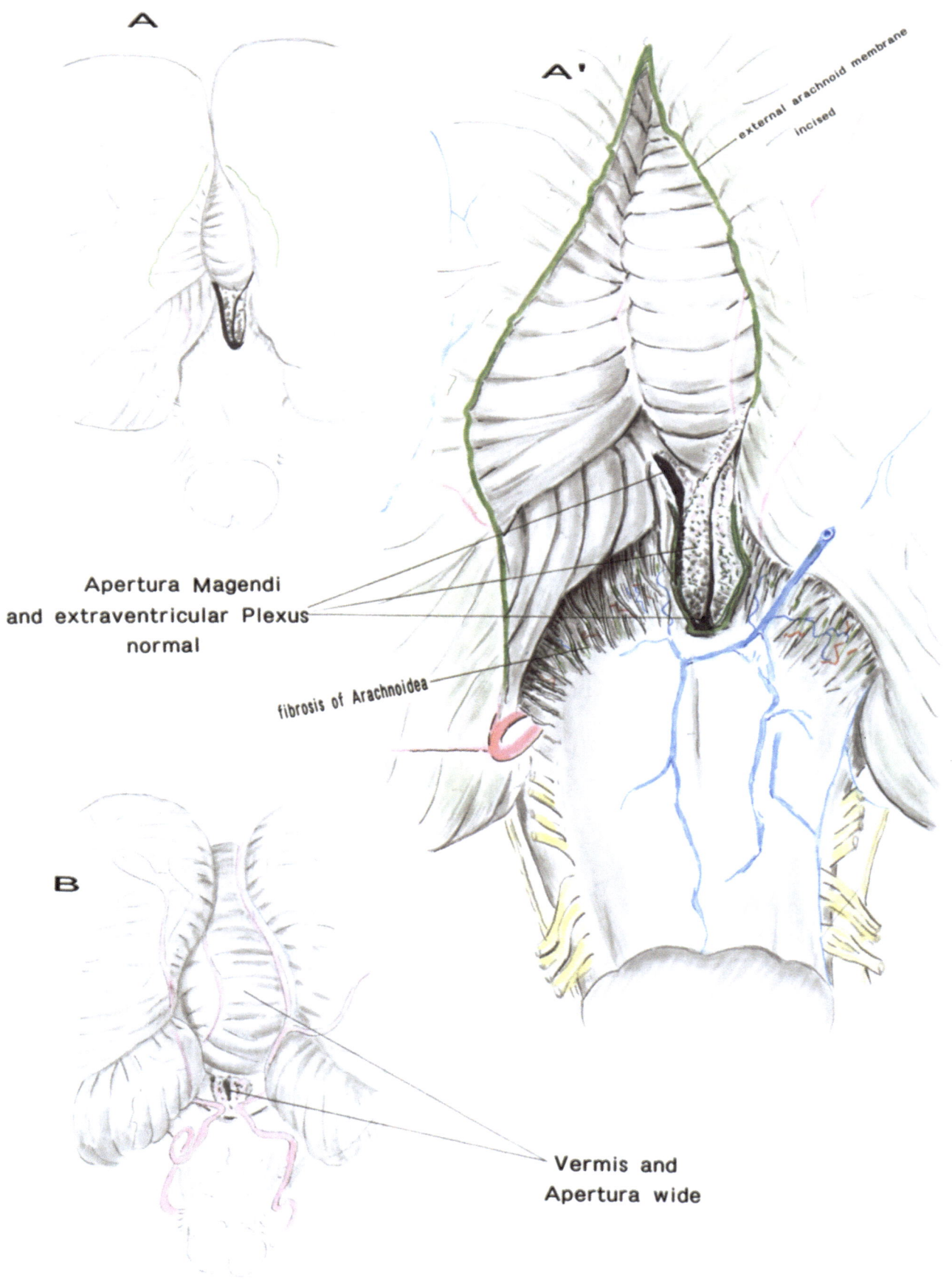

A

A'

external arachnoid membrane
incised

Apertura Magendi
and extraventricular Plexus
normal

fibrosis of Arachnoidea

B

Vermis and
Apertura wide

according to Lang (1979) p 424

modified

Fig. 177. **Meatus acusticus int. and its Fundus, contents**

A   Cranial nerves and meatal loop of AICA, common finding
           * A. labyrinthi (origin variable, see Fig. 163)
           ** A. petrosa
B–F   Further variants *as A
A'   Fundus of Meatus acusticus int.. Its structures may be flattened by tumors

Microsurgical aspects: These variants may be taken into consideration in operations of Neurinomas and meningiomas (and other lesions) of the cerebellopontine angle.

Abbreviations
a   Internal opening of Canalis Fallopii
b   anterior inferior cribriform area, with spirally configurated openings for the cochlear nerves
c and *d*   openings for N. vestibularis sup. and inf.
e   Falciform crest (Crista transversa)
f   Cellulae mastoideae (high variability)

FIG. 177

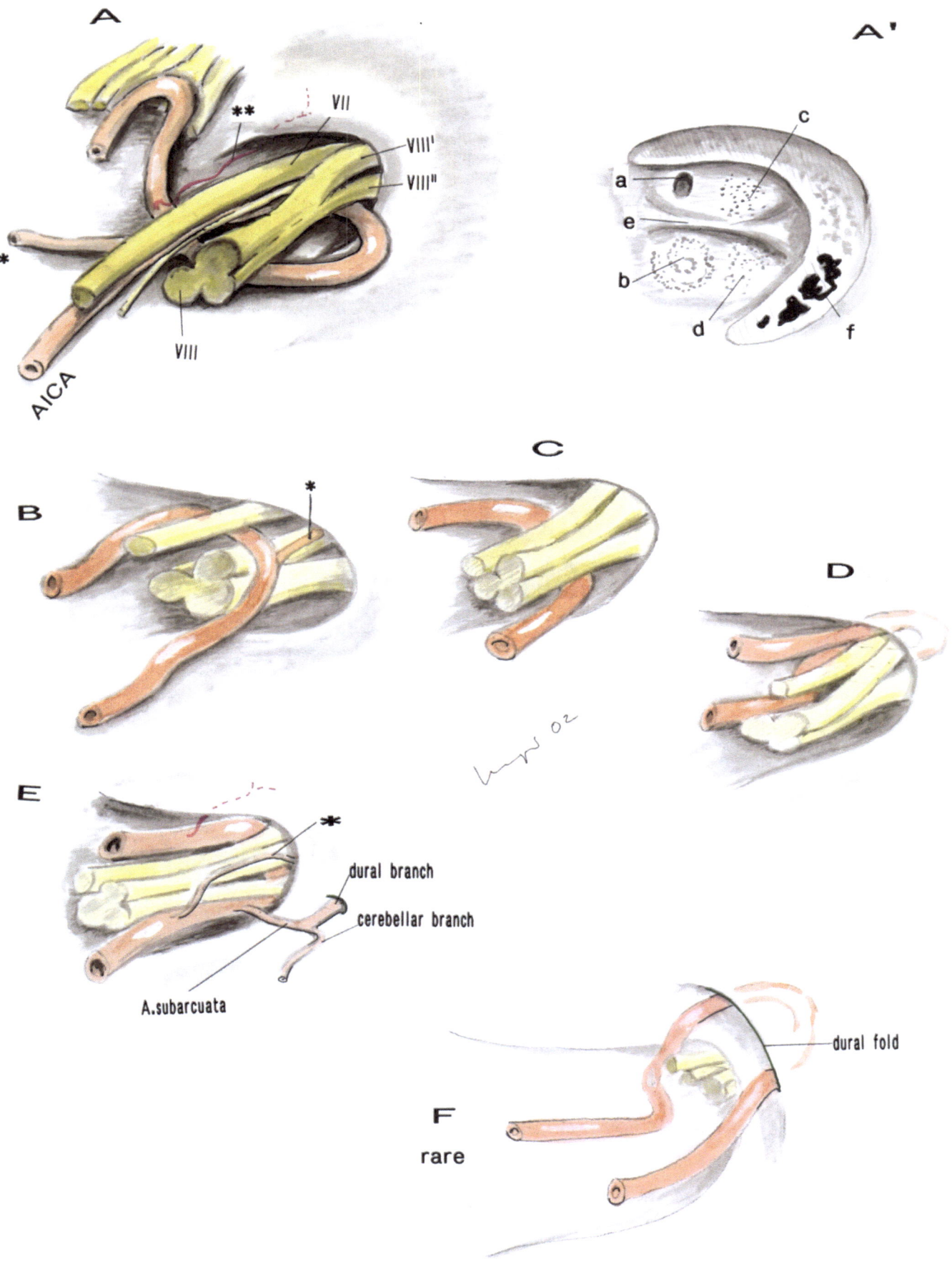

A

A'

VII
VIII'
VIII"

**

*

VIII

AICA

a
c
e
b
d
f

B

C

*

D

E

*

dural branch

cerebellar branch

A.subarcuata

F

rare

dural fold

References

Babin E, Megret M (1973) Variations in the drainage of the basal vein. Neurorad 6, 154-161

Ball T, Schreiber A, Feige B, Wanger M, Lücking CH, Kristeva-Feige R (1999) The role of higher-order motor areas in voluntary movement as revealed by high-resolution EEG and fMRI. NeuroImage 10, 682-694

Bardeleben, Kv, Haeckel H (1908) Atlas der topographischen Anatomie des Menschen, 4. Aufl. Fischer, Jena

Bedford MA (1966) The „cavernous" sinus. Brit J Ophthal 50, 41

Bergland RM, Bronson SR, Thorack RM (1968) Anatomical variations in the pituitary gland and adjacent structures in 225 human autopsy cases. J Neurosurg 28, 93-99

Bergquist E (1973) Tentorial notch and adjacent major vessels in carotid angiography. Acta Radiol [Suppl] No 327

Bergquist E (1975) Fibrous septa in the straight dural sinus. Acta Radiol Stockh 16, 331-336

Bergquist E, Willén R (1974) Cavernous nodules in the dural sinuses. J Neurosurg 40, 330-335

Bien S, Garcia-Cervignon E, Schumacher M, Riche MC, Merland JJ (1988) Erkrankungen der Arteria carotis interna. In: Schumacher M (ed) Diagnostik und Therapie. Schnetztor, Konstanz

Bischoff ThLW (1868) Die Großhirnwindungen des Menschen. Verlag der königl. Akademie München

Bochdalek VA (1849) Neue Beobachtungen im Gebiet der physiologischen Anatomie. II Neu entdecktes Markblatt des Marksegelsystems des kleinen Gehirns. Vierteljahresschrift für die praktische Heilkunde (Prag) 5, 129-132

Bradac GB (1970) The ponto-mesencephalic veins. Neuroradiology 1, 52-57

Browder J, Browder AA, Kaplan HA (1973) Anatomical relationships of the cerebral and dural venous systems in the parasagittal area. Anat Rec 176, 329-332

Browder J, Browder AA, Kaplan HA (1972) The venous sinuses of the cerebral dura mater. Arch Neurol 26, 175-180

Browder J, Kaplan HA (1976) Cerebral dural sinuses and their tributaries. Thomas, Springfield, Ill

Browder J, Kaplan HA, Krieger AJ (1975) Venous channels in the tentorium cerebelli. Surgical significance. Surg Neurol 3, 37-39

Browning H (1953) The confluence of dural venous sinuses. Am J Anat 93, 307-329

Creutzfeldt OD (1983) Cortex cerebri. Leistung, strukturelle und funktionelle Organisation der Hirnrinde. Springer, Berlin Heidelberg New York Tokyo

Dahlstroem L, Frageberg G, Lanner L, Stattin S (1969) Anatomical and angiographic studies of arteries supplying anterior part of temporal lobe. Acta Radiol Stockh 9, 257-263

Daniels DL, Schenck JF, Foster T, Hart H jr, Millent SJ, Meyer GA, Pech P, Haughton VM (1985) Magnetic resonance imaging of the jugular foramen. AJNR 6: 699-703

Danninger P (1980) Mikroanatomie des Sinus sagittalis sup. Inaug Diss Freiburg

Das AC, Hasan M (1970) The occipital sinus. J Neurosurg 33, 307-311

Delmas A, Pertuiset B (1949) Les veines du cortex cérébral: distribution générale, variations, types veineux de distribution. CR Ass Anat 57, 185-193

De Oliveira E, Rhoton AL, Peace D (1985) Microsurgical anatomy of the region foramen magnum. Surg Neural 24, 293-352

Donaghy RMP (1985) Surgery of the dural sinuses. In: Fein JM, Flamm ES (eds) Cerebrovascular surgery, vol 4. Springer, Berlin Heidelberg New York Tokyo, pp 1311-1324

Duvernoy HM (1975) The superficial veins of the human brain. Veins of the brain stem and of the base of the brain. Springer, Berlin Heidelberg New York

Eadie MJ, Jamieson KG, Lennon EA (1964) Persisting carotid-basilar anastomosis. J Neurol Sci 1, 501-511

Ebeling U, Huber P, Reulen HJ (1986) Localization of the precentral gyrus in the CT and its clinical application. J Neurol 233: 73-76

Ebeling U, Reulen HJ, Huber P (1986) Surgery of processes along the pyramidal tract and the internal capsule. In: Samii M (ed) Surgery in and around the brain stem and the third ventricle. Springer, Berlin Heidelberg New York Tokyo, pp 405-409

Ebeling U, Rikli D, Huber P, Reulen HJ (1987) The coronal suture, a useful landmark in neurosurgery? Craniocerebral topography between bony landmarks on the skull and the brain. Acta Neurochir (Wien) 89, 130-134

Ebeling U, Steinmetz H, Huang Y, Kahn T (1989) Topography and identification of the inferior precentral sulcus in MR imaging. AJNR 10, 937-942

Faure J, Binnert D, Michotey P, Salamon G (1971) Étude radio-anatomique des branches collaterales de l'artère carotide i interne (dans son segment intra-caverneux) Neuro-Chirurgie Paris tome 19 no 7, 561-579

Fields WS (1968) The significance of persistent trigeminal artery. Carotid-basilar anastomoses. Radiology 91, 1096-1101

Fisch U (1900) The surgical anatomy of the so-called internal auditory artery. In: Proceedings of the Tenth Nobel Symposium on Disorders of the Skull Base Region. Almqvist & Wiksell, Stockholm

Flanagan JC (1966) Vascular problems of the orbit. Ophthalmology 86, 896-913

Frugoni P, Mingrino S, Giamusso V (1963) Association of cerebral vascular malformation. Coexistence of arteriovenous angioma and persistent carotid-basilar anastomosis (primitive trigeminal artery) Neurochirurgia 6, 74-81

Gado M, Hanaway J, Frank R (1979) Fuctional anatomy of the cerebral cortex by computed tomography. J Comput Assist Tomogr 3, 1-19

Gannon WE (1962) Malformation of brain: persistent trigeminal artery and arteriovenous malformation. Arch Neurol 6, 496-498

Gibbs Leonhard E, Gibbs FA (1934) The cross section areas of the vessels that form the Torcular and the manner in which flow is distributed to the right and to the left lateral sinus. Anat Rec 59, 419-426

Gillilan LA (1973) Anatomy and embryology of the arterial system of the brain tem and cerebellum. In: Vinken PJ, Bruyn GW (eds) Handbook of clinical neurology, vol XI. Vascular diseases of the nervous system. North-Holland, Amsterdam, pp 24-44

Götz GF, Schumacher M (1995) Anomalies of supraaortic vessels. Klinische Neuroradiologie 5, 24-34

Grand W, Hopkins LN (1977) The microsurgical anatomy of the basilar artery bifurcation. Neurosurgery 1, 128-131

Hacker H (1968a) Abflußwege der sylvischen Venengruppe. Radiologe 8, 383-387

Hacker H (1968b) Venenabflüsse des Gehirns. Deutscher Röntgenkongreß 1967 Part A. Thieme Stuttgart

Hardy DG, Rhoton AL (1978) Microsurgical relationships of the superior cerebellar artery and the trigeminal nerve. J Neurosurg 49, 669-678

Hardy DG, Peace DA, Rhoton AL (1980) Microsurgical anatomy of the superior cerebellar artery. Neurosurgery 6, 10-28

Hardy DW, Rhoton AL Jr (1978) Microsurgical relationships of the superior cerebellar artery and the trigeminal nerve. J Neurosurg 49, 669-678

Harrison JM, Feldmann ML (1970) Anatomical aspects of the cochlear nucleus and superior olivary complex. In:von Neff WD (ed) Contributions to sensory physiology vol 4. Academic Press, NY

Hasan M, Das AC (1969) A note on the Faly cerebelli. Acta Anat Basel 74, 624

Henle J (1871) Handbuch der systematischen Anatomie des Menschen, Bd 3, 2. Abt. Braunschweig, Vieweg

Hermann E, Seeger W (1965) Die diagnostische Bedeutung des Verlaufs der Arteria basilaris und deren großen Endäste im Vertebralis-Angiogramm. Dtsch Z Nervenheilk 187, 531-538

Hitselberger WE, House WF, Edgerton BJ (1984) Cochlear nucleus implant. Otolaryngol. Head Neck Surg 92, 52-54

Hölzel G (1977) Die lateralen subependymalen Ventrikelvenen. Inaug. Diss. Freiburg i.Br.

Hoydt: See Unsöld R, Seeger W (1989)

Huang YP, Ito J, Pinner J (1975) Angiographic anatomy and circulation time of the veins of the posterior fossa. In: Kitamura K, Newton Th, (eds) Symposium on Recent Advances in Neuroradiology. Igaku Shoin, Tokyo

Huang YP, Wolf BS (1965) The veins of the posterior fossa, superior or Galenic group. Am J Roentgenol 95, 808-821

Huang JP, Wolf BS (1967) Precentral cerebellar vein in angiography. Acta Radiol Stockh 5, 250-262

Huang JP, Wolf BS (1967) The vein of the lateral recess of the fourth ventricle and its tributaries. Am J Roentg Rad Ther Nucl Med 101, 1-21

Huang YP, Wolf BS (1974) The basal cerebral vein and its tributaries. In: Newton Th, Potts DG (eds) Radiology of the skull and brain, Vol 2, Book 3, pp 2111-2154

Huang JP, Wolf BS (1974) Veins of the posterior fossa. In: Newton Th, Potts DG (eds) Radiology of the skull and brein. Mosby, St. Louis, pp 2155-2219

Huang YP, Wolf BS, Antin SP, Okudere T (1968) The veins of the posterior fossa, anterior or petrosal draining group. Am J Roentgenol 104, 36-56

Huang YP, Wolf BS, Okudera T (1966) Angiographic anatomy of the inferior vermian vein of the cerebellum. Acta Radiol Stockh 9, 327-344

Huber P (1979) Zerebrale Angiographie für Klinik und Praxis. In: Krayenbühl H, Yasargil MG (Hrsg) Stuttgart, Thieme

Hütter B (1999) Neuropsychologische und verhaltensmedizinische Folgen von Subarachnoidalblutungen. Habil, Aachen

Hütter B (2000) Neuropsychological sequelae of subarachnoid hemorrhage and its treatment. Springer, Wien New York

Jannetta JP (1977) Observations on the etiology of trigeminal neuralgia, hemifacial spasm, acoustic nerve dysfunction and glossopharyngeal neuralgia. Neurochirurgia 20, 145-154

Janetta PJ, Rand RW (1966) Microanatomy of the trigeminal nerve. Anat Rec 154, 362

Kaplan HA (1959) The transcerebral venous system: an anatomical study. Arch Neurol Psychiat Chicago 1, 148-152

Kaplan HA (1984) Results of obliteration of specific cerebral veins and dural venous sinuses: Animal and human studies. Chapter 9. In: Kapp JP, Schmidek HH (eds) The cerebral venous system and its disorders. Grune & Stratton Inc. Orlando, San Diego, San Francisco, pp 275-282

Kaplan HA, Browder J (1973) Atresia of the rostral superior sagittal sinus substitute parasagittal venous channels. Neurosurg 38, 602-607

Kaplan HA, rowder A, Browder J (1973) Narrow and atretic transverse dural sinuses: Clinical significance. Ann Otol Rhinol Laryngol 83, 351-354

Kaplan HA, Browder J (1976) Cerebral Dural Sinuses and their Tributaries. Springfield Ill, ChC Thomas

Kaplan HA, Browder J (1976) Neurosurgical consideration of the features of the cerebral dural sinuses and their tributaries. Clin Neurosurg 23, 155-169

Kaplan HA, Browder AA, Browder J (1972) Atresia of the rostral superior sagittal sinus: Associated cerebral venous patterns. Neuroradiology 4, 209-211

Kaplan HA, Browder A, Browder J (1973) Narrow and atretic transverse dural sinuses: Clinical significance. Ann Otol (St. Louis) 82, 351-354

Kaplan HA, Browder A, Browder J (1973) Nasal venous drainage and the Foramen caecum. Laryngoscope 83, 327-329

Kaplan HA, Browder J, Howard EM, Browder AA (1974) Vascular spaces of the middorsal dura mater. Arch Pathol 97, 173-177

Kaplan HA, Browder J, Knightly JJ, Rush BJ, Browder A (1972) Variations of the cerebral dural sinuses at the torcular Herophili – importance in radical neck dissection. Am J Surg 124, 456-461

Kaplan HA, Browder J, Krieger AJ (1975) Venous channels within the intracranial dural partitions. Radiology 115, 641-645

Kaplan HA, Ford DH (1966) The brain vascular system. Elsevier, Amsterdam New York

Kempe LG (1968) Operative neurosurgery. Vol I. Springer, Berlin Heidelberg New York

Kempe LG (1970) Operative neurosurgery, Vol 2: Posterior fossa, spinal chord, and peripheral nerve disease. Springer, Berlin Heidelberg New York

Key A, Retzius G (1875) Studien in der Anatomie des Nervensystems und des Bindegewebes, Band 1. Stockholm, Norstad

Klose AK (2000) Die topographische Anatomie des Nucleus cochlearis – eine mikroanatomische Studie zur Implantation von Hirnstammelektroden. Thesis, Hannover

Klose AK, Sollmann WP (1998) Landmarks for implantation of the cochlear nucleus – a microanatomical study. 49th Annual Meeting of the German Society of Neurosurgery, Hannover. Zbl Neurochir [Suppl] 61

Klose AK, Sollmann WP (1999) Anatomical variations of landmarks for implnatation at the cochlear nucleus. In: J Laryng Otol [Suppl] 27. Proceed of the 2nd internat. Auditory Brainstem Implant Sympos, Freiburg/Germany

Krause F (1905) Zur Freilegung der hinteren Felsenbeinfläche und des Kleinhirns. Beiträge Klin Chir 37, 728

Krause F (1908/1911) Chirurgie des Gehirns und Rückenmarks, Bd 1 und 2. Urban und Schwarzenberg, Berlin

Krause F (1926) Operative Freilegung der Vierhügel nebst Beobachtung über Hirndruck und Dekompression. Zbl Chir 53, 2812-2819

Krayenbühl HM, Yasargil MG (1957) Die vaskulären Erkrankungen im Gebiet der A. vertebralis und basilaris. Thieme, Stuttgart

Kribs M, Kleihues P (1971) The recurrent artery of Heubner. In: Zülch KJ (ed) Cerebral circulation and stroke. Springer, Berlin Heidelberg New York, p 41 ff

Kristeva R, Cheyne D, Deecke L (1991) Neuromagnetic fields accompanying unilateral and bilateral voluntary movements: topography and analysis of cortical sources. Clin Neurophysiol 81, 284-298

Lamp, Morris L (1961) The carotid-basilar artery: A report and discussion of five cases. Clin Radiol 12, 179-186

Lang J (1972) Cisterna fissurae transversae, Ventriculus Vergae und Defektbildungen des Septum pellucidum. Gegenbaurs Morph Jahrb 118, 539-572

Lang J (1972) Kopf Teil B Gehirn und Augenschädel. Springer, Berlin Heidelberg New York

Lang J (1981) Neuroanatomie der Nn. Oticus, trigeminus, facialis, glossopharyngeus, vagus, accessorius und hypoglossus. Arch Otorhinolaryngol 231, 1-69

Lang J (1981) Klinische Anatomie des Kopfes. Springer, Berlin Heidelberg New York

Lang J (1985) Kopf Teil A. Springer, Berlin New York Tokyo

Lang J (1985) Anatomy of the brainstem and the lower cranial nerves, vessels and surrounding structures. Am J Otol [Suppl] 1, 19

Lasjanias P, Berenstein A (1987) Surgical neuroangiography 1 Functional anatomy of craniofacial arteries. Springer, Berlin Heidelberg New York Tokyo

Liliequist B (1959) The subarachnoid cisterns. Anatomic and roentgenologic study. Acta Radiol Stockh [Suppl] 185

Martin RG, Grant JL, Peace D, Theiss C, Rhoton AL Jr (1980) Microsurgical relationship of the anterior inferior cerebellar artery and the facial-vestibulocochlear nerve complex. Neurosurgery 6, 493-507

Matsuno H, Rhoton AL, Peace D (1988) Microsurgical anatomy of the posterior fossa cisterns. Neurosurgery 23, 58-80

Matricali B, van Dulken H (1981) Aneurysm of fenestrated basilar artery. Surg Neurol 15, 189-191

Matsushima T, Rhoton AL, de Oliveira E et al (1989) Microsurgical anatomy of the veins of the posterior fossa. J Neurosurg 59, 63-105

Matsushima T, Suzuki SO, Fukui M et al (1989) Microsurgical anatomy of the tentorial sinuses. J Neurosurg 71, 923-928

McLaughlin MR, Jannetta JP, Clyde BL, Subach BR, Comey CH, Resnick DK (1999) Microvascular decompression of cranial nerves: lessons learned after 440 operations. J Neurosurg 90, 1-8

Marino R (1976) The anterior cerebral artery: I. Anatomo-radiological study of its cortical territories. Surg Neurol 5, 81-87

McCord GM, Goree JA, Jiminez JP (1972) Venous drainage to the inferior sagittal sinus. Radiol 105, 583-589

Mitterwallner F v (1955) Variationsstatistische Untersuchungen an den basalen Hirngefäßen. Acta Anat (Basel) 24, 371

Morgagni GB (1799) See Seeger W (1964)

Nieuwenhuys R (1988) The human central nervous system. Springer, Berlin Heidelberg New York Tokyo

Ono M, Ono M, Rhoton AL Jr, Barry M (1984) Microsurgical anatomy of the region of the tentorial incisura. J Neurosurg 60, 465-399

Pachtmann H, Hilal SK, Wood EH (1974) The posterior choroidal arteries. Neuroradiology 112, 343-352

Padget DH (1948) The development of the cranial arteries in the human embryo. Contr Embryol Carnegie Inst Washington 32, 205-261

Padget DH (1956) The cranial venous system in man with reference to development, adult configuration and relation to the arteries. AM J Anat 98, 307-355

Padget DH (1957) The development of the cranial venous system in man from the viewpoint of comparative anatomy. Contrib Embryol Carneg Inst 36, 79-140

Parkinson D, Shields CB (1974) Persistent trigeminal artery: its relationship to the normal branches of the cavernous carotid artery. J Neurosurg 40, 244-248

Perlmutter D, Rhoton AL (1976) Microsurgical anatomy of the anterior cerebral-anterior communicating-recurrent artey complex. J Neurosurg 45, 259-272

Plets C (1969) The arterial blood supply and architecture of the posterior wall of the third ventricle. Acta Neurochir (Wien) 21

Rauber-Kopsch (1906) Lehrbuch der Anatomie des Menschen, Bd 2, Knochen, Bänder. Thieme Leipzig

Renn WH, Rhoton AL (1975) Microsurgical anatomy of the sellar region. J Neurosurg 43, 288-298

Rhoton AL jr (1974) Microsurgery of the internal acoustic meatus. Surg Neurol 2, 311-318

Rhoton AL, Buza R (1975) Microsurgical anatomy of the jugular foramen. J Neurosurg 42, 541-550

Rieger P, Huber G (1983) Fenestration and duplicate origin of the left vertebral artery. Report of three cases. Neuroradiology 25, 45-50

Rowland LP, Mettler FA (1940) Relation between the coronal suture and cerebrum. J Comp Neurol 89, 21-40

Sacki N, Rhoton AL jr (1977) Microsurgical anatomy of the upper basilar artey and the posterior circle of Willis. J Neurosurg 46, 563-578

Salamon G (1971) Atlas de la vascularisation artérielle du cerveau chez l'homme. Atlas of the arteries of the human brain. Paris: Sandoz.

Salamon G, Huang YP (1976) Radiologic Anatomy of the Brain. Springer Berlin Heidelberg New York

Samii M, Jannetta PJ (1981) The cranial nerves. Springer, Berlin Heidelberg New York

Seeger W (1964) Neurochirurgia 137, 1-10

Seeger W (1978) Atlas of topographical anatomy of the brain and surrounding structures. Springer, Wien New York

Seeger W (1980) Microsurgery of the brain 1. Springer, Wien New York

Seeger W (1980) Microsurgery of the brain 2. Springer, Wien New York

Seeger W (1982) Microsurgery of the spinal cord and surrounding structures. Springer, Wien New York

Seeger W (1983) Microsurgery of the cranial base. Springer, Wien New York

Seeger W (1984) Microsurgery of cerebral veins. Springer, Wien New York

Seeger W (1985) Differential approaches in microsurgery of the brain. Springer, Wien New York

Seeger W (1986) Planning strategies of intracranial microsurgery. Springer, Wien New York

Seeger W (1987) Anatomical dissections for use in neurosurgery, vol 1. Springer, Wien New York

Seeger W (1988) Anatomical dissections for use in neurosurgery, vol 2. Springer, Wien New York

Seeger W (1990) Strategies of microsurgery in problematic brain areas – with special reference to NMR. Springer, Wien New York

Seeger W (1993) The microsurgical approaches to the target areas of the brain. Springer, Wien New York

Seeger W (1995) Microsurgery of intracranial tumors, vol 1. Springer,

Wien New York

Seeger W (1995) Microsurgery of intracranial tumors, vol 2. Springer, Wien New York

Seeger W (2000) Microanatomical aspects for Neurosurgeons and neuroradiologists. Springer, Wien New York

Seeger W, Zentner J (2002) Neuronavigation and neuroanatomy. Springer, Wien New York

Spalteholz W (1906) Handatlas der Anatomie 3, p 640, Hirzel Leipzig

Spee F von (1896) Kopf. In: Bardelebens Handbuch der Anatomie des Menschen Band I/2. Fischer, Jena, S 1908

Szdzuy D, Lehmann R, Nickel B (1972) Common trunk of the Anterior Cerebral Arteries Neurorad 4, 51-56

Schaeffer JP (1924) Some points in the regional anatomy of the optic pathway, with especial reference to tumors of the hypophysis cerebri and resulting ocular changes. Anat Rec 28, 243-279

Schlesinger B (1939) The venous drainage of the brain with special reference to the galenic system. Brain 62, 274-291

Schumacher M, Wakhloo AK (1994) An orbital arteriovenous malformation in a patient with origin of the ophthalmic artery from the basilar artery. AJNR 15, 550-553

Stephens RB, Stilwell DL (1969) Arteries and veins of the human brain. Thomas, Springfield Ill

Teal JS, Rumbaugh CL, Bergeron RT, Scanlan RL, Segall HD (1972) Persistent carotid-superior cerebellar artery anastomosis: a variant of persistent trigeminal artery. Radiology 103, 335-341

Tsai FY, Wadley D, Angle JF, Alfieri K, Byars S (1990) Superselective ophthalmic angiography for diagnostik and therapeutic use. AJNR AM J Neuroradiol 11, 1203

Tsai FY, Mahon J, Woodruff JV, Roach JF (1975) Congenital absence of bilateral vertebral arteries with occipital-basilar anastomosis. AJR Am J Roentgenol 124, 281-286

Unsöld R (1982) Zur computertomographischen Differentialdiagnose der Erkrankungen des Sehnerven. Graefes Arch Clin Exp Ophthalmol 219, 124-138

Unsöld R (1983) Zur computertomographischen Diagnose von Läsionen der vorderen Sehbahn. Fortschr Ophthalmol 218, 124-138

Unsöld R, Degroot J, Newton TH (1980) Images of the optic nerve. Anatomic CT-correlation. AJNR 1, 317-323

Unsöld R, Newton TH, Berninger W (1980) The value and clinical application of multiplanar computer reformations in orbital diagnosis. In: Margulis AR, Gooding CA (eds) Diagnostic Radiology. Proceedings of the Annual Postgraduate Course in Diagnostik Radiology, San Francisco, March 3-7, 1980. University of California, Department of Radiology, San Francisco

Unsöld R, Newton TH, Hoydt WF (1980) CT-Examination technique of the optic nerve. J Comp Assist Tomogr 4, 560-563

Unsöld R, Seeger W (1989) Compressive optic nerve lesions at the optic nerve canal. Springer, Berlin Heidelberg New York, p 67

Valavanis A, Schubiger O (1983) High-resolution CT of the normal and abnormal fallopian canal. Am J Neuroradiol 4, 748

Vesalius A (1543) See Seeger (1978) p 287 ff

Wolf BS, Huang YP (1963) The insula and deep middle cerebral venous drainage system: Normal anatomy and angiography. Am J Roentgenol 90, 472-489

Wolf BS, Huang YP, Newman CM (1963) Lateral anastomotic mesencephalic vein and other variations in drainage of basal cerebral vein. Am J Roentgenol 89, 411-422

Wolf BS, Huang YP, Newman CM (1963) The superficial sylvian venous drainage system. Am J Roentgenol 89, 398-410

Wolf BS, Huang YP, Newman CM (1963) The superficial sylvian venous drainage system. Am J Roentgenol 89, 398-422

Yamamoto I, Kageyama N (1980) Microsurgical anatomy of the pineal region. J Neurosurg 53, 205-221

Yasargil MG (1978) Mikrochirurgie der Kleinhirnbrückenwinkel-Tumoren. In: Plester D, Wende S, Nakayama N (Hrsg) Kleinhirnbrückenwinkel-Tumore. Diagnostik und Therapie. Springer, Berlin Heidelberg New York, pp 215-257

Yasargil MG (1984-1996) Microneurosurgery. Thieme, Stuttgart New York

I. Microsurgical anatomy of the basal cisterns and vessels of the brain (1984)

II. Clinical considerations, surgery of the intracranial aneurysms and results (1984)

III.A AVM of the brain, history, embryology, pathological considerations, diagnostic studies, microsurgical anatomy (1987)

III.B AVM of the brain, clinical considerations, general and special operative techniques, surgical results, non-operated cases, cavernous and venous angiomas, neuroanesthesia (1988)

IV.A CNS tumors: Surgical anatomy, neuropathology. Neuroradiology, neurophysiology, clinical considerations, operability, treatment options (1994)

Yasargil MG, Kasdaglis K, Jain KK, Weber HP (1976) Anatomical observations of the subarachnoid cisterns of the brain during surgery. J Neurosurg 44, 298-302

Subject Index

SpringerMedicine

Henri M. Duvernoy

The Human Brain Stem and Cerebellum

Surface, Structure, Vascularization,
and Three-Dimensional Sectional Anatomy with MRI

In collaboration with J. F. Bonneville, E. A. Cabanis,
F. Cattin, J. Guyot, and M. T. Iba-Zizen. With drawings by J. L. Vannson.
1995. VII, 430 pages. 168 figures.
Hardcover **EUR 212,–**
(Recommended retail price). Net-price subject to local VAT.
ISBN 3-211-82503-7

This atlas of the brain stem and cerebellum is the sequel to the author's "The Human Brain". Its first part describes the surface of the brain stem and cerebellum as well as their location in the posterior cranial fossa. Furthermore it describes the structures of the brain stem and cerebellum which is followed by a brief survey of their functions, enabling the reader to obtain on overall view of the role both of nuclei and fasciculi.
Finally the vascular network is analyzed in detail (superficial pial vessels and intranervous territories of deep vessels). The second part of the book provides the reader with an understanding of the sectional anatomy on the basis of three-dimensional views and a comparison with MRI views.

".. For those in training in neuroradiology, neurosurgery and neurology this book is excellent ...
I will refer to it constantly".

R. Paxton/Neuroradiology

"... a thorough and excellent account of the anatomy of the brain stem and cerebellum ...
an excellent reference and would be of great value to both trainees and senior surgeons".

British Journal of Neurosurgery

SpringerWienNewYork

Sachsenplatz 4–6, P. O. Box 89, A-1201 Wien, Fax +43-1-330 24 26, e-mail: books@springer.at, Internet: **www.springer.at**
New York, NY 10010, 175 Fifth Avenue • D-14197 Berlin, Heidelberger Platz 3 • Tokyo 113, 3–13, Hongo 3-chome, Bunkyo-ku

SpringerMedicine

Henri M. Duvernoy

The Human Brain

Surface, Blood Supply,
and Three-Dimensional Sectional Anatomy

In collaboration with P. Bourgouin, E. A. Cabanis, F. Cattin, J. Guyot,
M. T. Iba-Zizen, P. Maeder, B. Parratte, L. Tatu, and F. Vuillier.
With drawings by J. L. Vannson
Second, completely revised and enlarged edition.
1999. VII, 491 pages. 272 figures, partly in colour.
Hardcover **EUR 213,–**
(Recommended retail price). Net-price subject to local VAT
ISBN 3-211-83158-4

The recent progress of medical imaging due to CT, MRI, and the three-dimensional reconstruction of cerebral structures calls for a better understanding of the anatomy of the brain. Therefore, this book comprises serial sections – 2 mm thick – of the cerebral hemispheres and diencephalon in the coronal, sagittal, and axial planes. So as to point out the level of the sections more accurately, each section is shown from different angles emphasizing the surrounding hemisphere surfaces. This three-dimensional approach has proven to be extremely useful to apprehend the difficult anatomy of the gyri and sulci of the brain. Certain complex cerebral structures such as the occipital lobe, the deep grey matter (basal ganglia and thalamus), and the vascularization are demonstrated in greater detail.

The second edition of this successful atlas has been completely revised and updated. 44 serial sections have been added showing the brain in much greater detail. Mostly two MRI views of improved quality are presented with almost every section. A chapter on the vascular anatomy of the brain with beautiful color drawings has been added.

"This wonderful atlas by Duvernoy and colleagues is a must in every institute dealing with neurological patients and neuroimaging. The atlas is very meticulously made. A correlation is given between MRI and morphology ... The atlas is very good in giving the precise localisation ..."

Clinical Neurology and Neurosurgery

Reviews of the first edition:

"... What differentiates this textbook of anatomy from standard neuroanatomical textbooks is its presentation of the brain slices in such a way as to allow for better 3-D visualization of the anatomy ... The quality of the MR images and photographs of the anatomical sections is excellent ..."

Journal of Neurosurgery

"... This atlas will be extremely precious for neurologist, neurosurgeons, and neuroradiologists."

Acta Neurologica Belgica

SpringerWienNewYork

Sachsenplatz 4–6, P. O. Box 89, A-1201 Wien, Fax +43-1-330 24 26, e-mail: books@springer.at, Internet: www.springer.at
New York, NY 10010, 175 Fifth Avenue • D-14197 Berlin, Heidelberger Platz 3 • Tokyo 113, 3–13, Hongo 3-chome, Bunkyo-ku

SpringerMedicine

Wolfgang Seeger

Microanatomical Aspects for Neurosurgeons and Neuroradiologists

In Collaboration with J. Zentner and M. Schumacher.
2000. VII, 423 pages. 201 figures, partly in colour.
Hardcover **EUR 267,–**
(Recommended retail price) Net-price subject to local VAT.
ISBN 3-211-83376-5

Modern diagnostic imaging and operative approaches have witnessed significant improvements in our times. Computerassisted methods are in use in all microsurgical fields. Neuronavigation, novel stereotactic methods, endoscopic procedures, magnetic resonance imaging, ultrasound and the progress in pre- and intraoperative epilepsy diagnostics must be mentioned in particular in this connection.

However, the insights of neuroanatomy and neurophysiology have not become obsolete thereby, on the contrary: such knowledge is imperative and a prerequisite for all neurosurgeons, nowadays more than ever before. Otherwise, excellent modern approaches are liable to fall into discredit if microanatomical aspects are neglected.

The goal of this book is two-fold: first, to guide the resident towards a fruitful application of anatomical basics in visualizing and operative techniques. Second, to draw attention to as many anatomical norm variants as possible to forestall complications during surgery. Standard methods, such as the pterional approach, often confront the surgeon with a range of anatomical variants.

"..This book is the result of a large experience and presents numerous illustrations which speak for themselves ... Wolfgang Seeger's book, thanks to its simplicity, pedagogical efficiency and to the enormous amount of knowledge it gathers, can be one of the 'companion guides' of the practitioner neurosurgeon. It will also be very useful to the neuroradiologist who will know more about the technical preoccupations of the neurosurgeon beyond the mere diagnosis of the lesion and about the elements the neurosurgeon needs to visualise in order to perform surgical gestures in the best conditions possible."

Surgical and Radiologic Anatomy

"..This is a valuable book for neurosurgeons and neuroradiologists."

British Journal of Neurosurgery

"... truly a monumental volume. It is beautifully illustrated and a pleasure to handle ..."

Journal of Neurology, Neurosurgery, and Psychiatry

Sachsenplatz 4–6, P. O. Box 89, A-1201 Wien, Fax +43-1-330 24 26, e-mail: books@springer.at, Internet: **www.springer.at**
NewYork, NY 10010, 175 Fifth Avenue • D-14197 Berlin, Heidelberger Platz 3 •Tokyo 113, 3–13, Hongo 3-chome, Bunkyo-ku

SpringerMedicine

Wolfgang Seeger,
Josef Zentner

Neuronavigation
and Neuroanatomy

2002. VII, 419 pages. With 200 coloured figures.
Hardcover **EUR 248,–**
(Recommended retail price)
Net-price subject to local VAT.
ISBN 3-211-83741-8

Neuronavigation enables the surgeon to define each cranial and cerebral structure before and during surgery but the problem of brainshifting remains. This atlas shows drawings of anatomical landmarks for neuronavigation for preoperative planning.

The authors show the relationships between bony landmarks which are unchanged during the operation and landmarks which are no more available after opening of the skull but still recognizable during the operation, e.g. by ultrasonic sector scan. It further includes the description of many important anatomical variants, which are important for microsurgeons when using minimal invasive modern techniques (endoscopy, sterotaxy) to avoid errors and complications. The book describes unknown projections for MRI and CT which may be adapted for special surgical problems.

The anatomical drawings are the result of a twenty-five-years study of the topographical anatomy of the brain and the surrounding structures combined with the experience of modern microsurgery.

Springer Wien New York

Sachsenplatz 4–6, P. O. Box 89, A-1201 Wien, Fax +43-1-330 24 26, e-mail: books@springer.at, Internet: **www.springer.at**
New York, NY 10010, 175 Fifth Avenue • D-14197 Berlin, Heidelberger Platz 3 • Tokyo 113, 3–13, Hongo 3-chome, Bunkyo-ku

Springer-Verlag
and the Environment

WE AT SPRINGER-VERLAG FIRMLY BELIEVE THAT AN international science publisher has a special obligation to the environment, and our corporate policies consistently reflect this conviction.

WE ALSO EXPECT OUR BUSINESS PARTNERS – PRINTERS, paper mills, packaging manufacturers, etc. – to commit themselves to using environmentally friendly materials and production processes.

THE PAPER IN THIS BOOK IS MADE FROM NO-CHLORINE pulp and is acid free, in conformance with international standards for paper permanency.